Telecommunication
Transmission
Systems

Other McGraw-Hill Communications Books of Interest

AZEVEDO • *ISPF*

BALL • *Cost-Efficient Network Management*

BERSON • *APPC: A Guide to LU6.2*

BLACK • *Network Management Standards*

BLACK • *TCP/IP and Related Protocols*

BLACK • *The V Series Recommendations*

BLACK • *The X Series Recommendations*

BROWN, SIMPSON • *The OSI Dictionary of Acronyms*

CHORAFAS • *The Complete LAN Reference*

COOPER • *Computer and Communications Security*

DAYTON • *Integrating Digital Services*

EDMUNDS • *SAA/LU6.2: Distributed Networks and Applications*

FOLTS • *The McGraw-Hill Compilation of Open Systems Standards*

FORTIER • *Handbook of LAN Technology*

HA • *Digital Satellite Communications*

HEBRAWI • *OSI: Upper Layer Standards and Practices*

HELD, SARCH • *Data Communications*

HELDMAN • *Global Telecommunications*

HUGHES • *Data Communications*

INGLIS • *Electronic Communications Handbook*

KESSLER • *ISDN*

KESSLER, TRAIN • *Metropolitan Area Networks*

KIMBERLY • *Electronic Data Interchange*

KNIGHTSON • *Standards for Open Systems Interconnection*

LEE • *Mobile Cellular Telecommunications Systems*

NEMZOW • *Ethernet Management Guide*

PELTON • *Voice Processing*

RADICATI • *Electronic Mail*

RANADE • *Advanced SNA Networking*

RANADE • *Introduction to SNA Networking*

RANADE • *VSAM: Performance, Design, and Fine Tuning*

RANADE, RANADE • *VSAM: Concepts, Programming, and Design*

RHEE • *Error Correction Coding Theory*

ROHDE, BUCHER • *Communication Receivers*

SABIN • *Single-Sideband Systems and Circuits*

SCHLAR • *Inside X.25*

To order or receive additional information on these or any other McGraw-Hill titles, in the United States please call 1-800-822-8158. In other countries, contact your local McGraw-Hill representative.   MH92

Telecommunication Transmission Systems

Microwave, Fiber Optic, Mobile Cellular Radio, Data, and Digital Multiplexing

Robert G. Winch

McGraw-Hill, Inc.

New York   San Francisco   Washington, D.C.   Auckland   Bogotá
Caracas   Lisbon   London   Madrid   Mexico City   Milan
Montreal   New Delhi   San Juan   Singapore
Sydney   Tokyo   Toronto

Library of Congress Cataloging-in-Publication Data

Winch, Robert G.
    Telecommunication transmission systems : microwave, fiber optic, mobile cellular radio, data, and digital multiplexing / Robert G. Winch.
       p.   cm.
    Includes index.
    ISBN 0-07-070964-5
    1. Telecommunication systems. I. Title.
TK5101.W48 1993
621.382—dc20   92-26723
                                                    CIP

    4 5 6 7 8 9 0   DOC/DOC 9 8 7 6

ISBN 0-07-070964-5

The sponsoring editor for this book was Daniel A. Gonneau, the editing supervisor was Stephen M. Smith, and the production supervisor was Suzanne W. Babeuf. It was set in Century Schoolbook by McGraw-Hill's Professional Book Group composition unit.

Printed and bound by R. R. Donnelley & Sons Company.

This book is printed on acid-free paper.

Contents

Preface   xi

Chapter 1. Introduction   1

  1.1 **Transmission Media**   3
  1.2 **Digitization**   8
  1.3 **Digital Microwave Radio System Configuration**   10
  1.4 **The Optical Fiber System Configuration**   12
  1.5 **Mobile Radio Systems**   13
  1.6 **Data Communications and the Network**   14
  1.7 **International Standards**   16

Chapter 2. Digital Multiplexing (Baseband Composition)   17

  2.1 **Pulse Code Modulation**   18
      2.1.1 **Sampling**   18
      2.1.2 **Quantization**   20
      2.1.3 **Encoding and Decoding**   27
      2.1.4 **Recent Coding Developments**   28
      2.1.5 **PCM Transmission Formats**   30
      2.1.6 **Formats for 30-Channel PCM Systems** 32
      2.1.7 **Frame Alignment**   32
      2.1.8 **Multiframe Alignment**   37
      2.1.9 **Formats for 24-Channel PCM Systems** 38
  2.2 **Line Codes**   39
  2.3 **Asynchronous Higher-Order Digital Multiplexing (CCITT)**   47
      2.3.1 **The Digital Multiplexing Hierarchy** 47
      2.3.2 **Second-Order Multiplexing (2 to 8 Mb/s)**   48
      2.3.3 **Positive Pulse Stuffing (or Justification)**   50
      2.3.4 **The 8-Mb/s Frame Structure**   52
      2.3.5 **Third-Order Multiplexing (8 to 34 Mb/s)**   54
      2.3.6 **Fourth-Order Multiplexing (34 to 140 Mb/s)**   55
      2.3.7 **Fifth-Order Multiplexing (140 to 565 Mb/s)**   56
  2.4 **Asynchronous Higher-Order Digital Multiplexing (North America)**   58
      2.4.1 **The Digital Multiplexing Hierarchy** 58
      2.4.2 **DS1-C Multiplexing (1.544 to 3.152 Mb/s)**   58
      2.4.3 **DS2 Multiplexing (1.544 to 6.312 Mb/s)**   60

v

2.4.4 DS3 Multiplexing (6.312 to 44.736 Mb/s)   60
2.4.5 DS4 Multiplexing (44.736 to 274.176 Mb/s)   63
2.5 Synchronous Digital Multiplexing   63
2.5.1 Network Node Interface   65
2.5.2 Synchronous Transport Signal Frame   66
2.5.3 Synchronous Transport Module Frame   73
2.5.4 Comparison of Asynchronous and Synchronous Interfaces   76
2.5.5 SDH Multiplexing Structure Summary   77
2.6 Multiplexing Digital Television Signals   80
2.6.1 Digitization of TV Signals   81
2.6.2 Analog Color TV   82
2.6.3 Video Compression Techniques   83
2.6.4 A Typical Digital TV Transmission System   84
2.6.5 High-Definition Television   86

Chapter 3. Signal Processing for Digital Radio Communications   89
3.1 Modulation Schemes   89
3.1.1 Bandwidth Efficiency   89
3.1.2 Pulse Transmission through Filters   90
3.1.3 Pulse Amplitude Modulation   92
3.1.4 Frequency Shift Keyed Modulation, Minimum Shift Keying (MSK),
        and Gaussian MSK   93
3.1.5 Phase Shift Keyed Modulation   95
3.1.6 Quadrature Amplitude Modulation   97
3.1.7 Comparison of Modulation Techniques   99
3.1.8 Demodulation   105
3.2 Error Control (Detection and Correction)   110
3.2.1 Forward Error Correction   111
3.2.2 Automatic Request for Repeat   120
3.2.3 Trellis-Coded Modulation   121
3.3 Spread Spectrum Techniques   124
3.3.1 Pseudorandom Noise Generation   127
3.3.2 Spread Spectrum Systems   129
3.3.3 Direct Sequence Spread Spectrum   130
3.4 Access Techniques for Mobile Communications   132
3.4.1 Frequency Division Multiple Access   132
3.4.2 Time Division Multiple Access   132
3.4.3 Code Division Multiple Access   133

Chapter 4. The Microwave Link   135
4.1 Antennas   138
4.1.1 Antenna Gain   138
4.1.2 Beamwidth   141
4.1.3 Polarization   143
4.1.4 Antenna Noise   144
4.1.5 High-Performance Antennas   146
4.1.6 Antenna Towers   146
4.2 Free Space Propagation   149
4.3 Atmospheric Effects   151
4.3.1 Absorption   151

4.3.2   Refraction   151
4.3.3   Ducting   157
4.4   Terrain Effects   157
4.4.1   Reflections   157
4.4.2   Fresnel Zones   159
4.4.3   Diffraction   163
4.5   Fading   166
4.5.1   Flat Fading   166
4.5.2   Frequency Selective Fading   167
4.5.3   Factors Affecting Multipath Fading   168
4.6   Availability   169
4.6.1   Performance Objectives   171
4.7   Diversity   178
4.7.1   Space Diversity   179
4.7.2   Frequency Diversity   181
4.8   Link Analysis   182
4.8.1   Hop Calculations   182
4.8.2   Passive Repeaters   186
4.8.3   Noise   191

Chapter 5. Digital Microwave Radio Systems and Measurements   201
5.1   System Protection   201
5.1.1   Diversity Protection Switching   201
5.1.2   Hot-Standby Protection   202
5.1.3   Combining Techniques   204
5.1.4   IF Adaptive Equalizers   206
5.1.5   Baseband Adaptive Transversal Equalizers   211
5.2   Digital Microwave Radio Systems   220
5.2.1   140-Mb/s DMR with 16-QAM   220
5.2.2   Digital Microwave Radio Transceiver Components   222
5.2.3   140-Mb/s DMR with Higher Modulation Levels (64-QAM or
        256-QAM)   253
5.2.4   Low-Capacity DMR   256
5.3   Performance and Measurements   261
5.3.1   RF Section Tests   262
5.3.2   IF Section Tests   265
5.3.3   Baseband Tests   266
5.4   Comparison between Analog and Digital Microwave Radio   282
5.4.1   Composition of the Baseband   283
5.4.2   FM Analog Microwave Radio   284
5.4.3   Measurements   289
5.4.4   Summary   293

Chapter 6. Introduction to Fiber Optics   295
6.1   Introduction   295
6.2   Characteristics of Optical Fibers   295
6.2.1   Numerical Aperture   296
6.2.2   Attenuation   297
6.2.3   Dispersion   300

6.2.4 Polarization   303
6.2.5 Fiber Bending   304
6.3 Design of the Link   306
6.3.1 Power Budget   308
6.4 Optical Fiber Cables   310
6.4.1 Cable Construction   310
6.4.2 Splicing, or Jointing   311
6.4.3 Installation Problems   321
6.5 Fiber Optic Equipment Components   323
6.5.1 Light Sources   323
6.5.2 Light Detectors   330
6.5.3 Polarization Controllers   333
6.5.4 Amplifiers   335
6.5.5 Modulators   338
6.5.6 Couplers   341
6.5.7 Isolators   343
6.5.8 Filters   344
6.5.9 Photonic Switches   347
6.5.10 Solid-State Circuit Integration   348

Chapter 7. Optical Fiber Transmission Systems   355
7.1 Intensity Modulated Systems   358
7.2 Coherent Optical Transmission Systems   359
7.2.1 Nonlinear PLL   365
7.2.2 Balanced PLL   366
7.3 Optical Multiplexing   368
7.3.1 Wave Division Multiplexing   368
7.3.2 Frequency Division Multiplexing   370
7.4 Repeaters   372
7.4.1 Regenerative Repeaters   372
7.4.2 Optical Repeaters   373
7.5 Systems Designs   374
7.5.1 Junction Routes (Interexchange Traffic)   374
7.5.2 Long-Haul Links   378
7.5.3 Local Area Networks and Subscriber Loops   381
7.5.4 BER Improvement   388
7.6 Optical Fiber Equipment Measurements   389
7.6.1 Cable Measurements   390
7.6.2 Line Terminal Measurements   391

Chapter 8. Mobile Radio Communications   399
8.1 Introduction   399
8.2 Cellular Structures and Planning   400
8.3 Frequency Allocations   407
8.4 Propagation Problems   409
8.4.1 Field Strength Predictions   410
8.4.2 Effects of Irregular Terrain   414

8.5   Antennas   416
      8.5.1   Base Station Antennas   417
      8.5.2   Mobile Unit Antennas   419
8.6   Types of Mobile Systems   420
8.7   Analog Cellular Radio   423
8.8   Digital Cellular Radio   426
      8.8.1   Digital Narrowband TDMA   426
      8.8.2   Future Digital Cellular Radio   435
8.9   Portable Radio Telephones and PCNs   436
8.10 Data over Cellular Radio   439
8.11 Cellular Rural Area Networks for Developing Countries   440

Chapter 9. Data Transmission and the Future Network   443
9.1   Standards   444
9.2   Data Transmission in an Analog Environment   449
      9.2.1   Bandwidth Problems   449
      9.2.2   Modems   449
9.3   Packet Switching   451
      9.3.1   Packet Networks   453
      9.3.2   The X.25 Protocol   454
      9.3.3   Network Interface Protocols   459
      9.3.4   Optical Packet Switching   460
9.4   Local Area Networks   461
      9.4.1   CSMA/CD (Ethernet)   463
      9.4.2   Token Rings   464
      9.4.3   10Base-T (Ethernet)   467
      9.4.4   Fiber Distributed Data Interface   467
9.5   WANs and MANs   473
9.6   ISDN   476
      9.6.1   Narrowband ISDN   476
      9.6.2   Broadband ISDN   487
9.7   Data Communications Testing   500
Appendix: Standards   502

Bibliography   507
Index   527

Preface

The telecommunications market in 1990 was approximately US$500 billion and is anticipated to be US$1 trillion by the year 2000. The telecommunications industry is experiencing rapid growth, especially in the global implementation of digital telecommunications equipment. Also, telecommunications is gradually merging with data communications (computer information transfer). This expanding market should be a significant source of job creation over the next decade and beyond.

In today's world, technical advancement is occurring so rapidly that it is very difficult for most engineers and technicians to stay up-to-date with the enormous amount of literature being created in each discipline. Most of us can only hope to keep up with developments occurring in a relatively narrow field. Telecommunications is a vast technical subject and it is the intention of this book to cover a wide area, while focusing on some specific aspects of the subject in great detail.

This text is designed for the *graduate engineer* or *senior technician*. The amount of mathematics has been purposely reduced to a minimum and emphasis is placed on the practical application of the theory. It is intended that the material discussed be a shortcut to experience, to give the practicing engineer/technician a better understanding of how existing telecommunications systems have evolved and an insight into their future development.

To ensure international compatibility, telecommunications equipment must be designed to conform to international standards. The construction, operation, and maintenance of the latest equipment is described with an acute awareness of these international standards. In this respect I am indebted to the International Telecommunication Union (ITU) for granting me authorization to reproduce information of which the ITU is the copyright holder. Due to the limitations of space, only parts of the ITU Recommendations are given or in some instances only a reference is made to the original source. For further information, the complete ITU publications can be obtained from the ITU General Secretariat, Sales Section, Place des Nations, CH-1211 Geneva 20, Switzerland.

The main themes of this book are microwave radio, fiber optics, mobile radio telephone, and data communications. Each topic starts from fundamental principles and progresses to present-day technology, so the material is also of interest to managerial staff who may not have had time to keep up with all the latest technical advances.

Chapter 1 is an introduction which sets the scene for the rest of the book by describing the configuration for each of the main terrestrial telecommunication systems: microwave radio, optical fiber, and cellular radio.

Chapter 2 describes the digitization of voice signals and how voice, data, or video channels can be combined by the time-division multiplexing technique. The new and very innovative synchronous digital hierarchy (SDH) is also described. A brief description of television digitization is included as a prelude to future HDTV.

Chapter 3 discusses modulation techniques which are evolving to enhance bandwidth efficiency and error control to improve performance. The spread-spectrum technique is described, highlighting its potential benefits for future mobile communications systems.

Chapter 4 describes how the theory behind microwave communications leads to the physical appearance of today's systems. Attention is paid to the effects of terrain and atmospheric conditions on microwave propagation. The system performance is characterized and methods of enhancing performance are discussed.

Chapter 5 is a detailed study of digital microwave radio systems design and the measurements that are made to evaluate operational performance.

Chapters 6 and 7 give an overview of the evolution of fiber optics from the early step-index fiber communication systems up to future soliton-transmission coherent-detection systems. Chapter 6 emphasizes the characteristics of the fibers and components used in the systems, and Chap. 7 deals with systems design for various lengths of optical communication links, from short-distance local area networks to transoceanic distances.

Chapter 8 details the development of cellular radio communications, with a comparison of the two major systems presently in existence, namely, the IS-54 system in North America and the GSM system in Europe.

Chapter 9 is a brief introduction to data communications, stating some of the international standards used to establish equipment compatibility and describing the basics of packet switching, LANs, and ISDN.

I am grateful to authors, publishers, and companies who have granted permission to reproduce figures and photographs from previous publications. Finally, I cannot find adequate words to express my gratitude to my wife, Elizabeth, for her patience, dedication, and expert proofreading of the manuscript, without which this project would not have been completed.

Robert G. Winch

Telecommunication
Transmission
Systems

1

Introduction

It has been estimated that the telecommunications market in 1990 was in the region of US$500 billion and projections indicate this could move to US$1 trillion by the year 2000. One tends to consider the subject of telecommunications in relation to the developed, or industrialized, world without realizing that this comprises the minority of the world's population. The developing world is a potentially enormous market for the future.

It is recognized that in all regions of the world there is a close correlation between a country's per capita gross national product (GNP) and its telephone density. This is illustrated in Fig. 1.1 as a graph for several countries of GNP against the number of telephone lines per 100 persons. While it is recognized that the telephone is a catalyst in promoting economic growth, it cannot be claimed that the greater economic growth of a particular country is a direct consequence of increasing the telephone density.

The number of telephone lines per 100 persons has been used in many statistical surveys to indicate the state of development of a country. In some respects this is unfortunate because in many developing countries as much as 70 to 90 percent of the population lives in the rural areas. A figure of less than one telephone line per 100 people has been described in many economics and telecommunications reports as indicating a very low state of development. To achieve a penetration of telephone services to everyone in the world, a goal of ensuring a telephone is within 5-km reach of everyone by the year 2000 has been proposed. Although it is universally accepted as a desirable objective, it will probably be much later than the year 2000 before it comes to fruition.

It is well known that it is very expensive to provide telecommunication links to low-population-density areas and the return on the in-

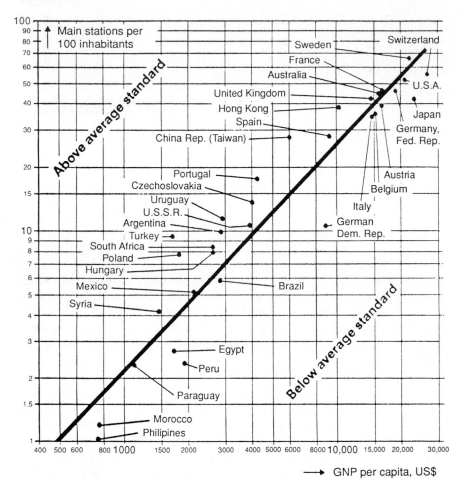

Figure 1.1 Graph of telephone density against GNP (1989). [*Reproduced with permission from ITU, "Development of Telecommunications Equipment Industries," Sombeek, P.*, World Telecommunications Forum, *Africa Telecom 90 (Harare, Zimbabwe).*]

vestment is minimal. With urban area telecommunications networks in developing countries requiring a large influx of funds for upgrading or expansion, it is hard for such countries with very limited financial resources to allocate funds to the rural areas.

Meanwhile, technology in the developed world is in a state of positive feedback, accelerating at a pace which is so fast that it is hard to keep up with the literature on the subject. International standards are essential to ensure that new equipment has compatibility, on a global level, with existing and future equipment. The international standards organizations are having a hard time keeping current with technological advancements. The objective of the following chapters is to provide some insight into the present and future technological trends and also to give technical details of many aspects of present-day, high-usage, digital telecommunications transmission equipment.

1.1 Transmission Media

The telecommunications objective is to produce high-quality voice, video, and data communication between any pair of desired locations, whether the distance between locations is 1 or 10,000 km. The distance between the two locations determines the type of transmission equipment used for setting up the interconnection. Communication within a building, described as a local area network (LAN), is done using metallic wires or optical fibers. Any routing of information within the building is done by a switch on the premises known as a private branch exchange (PBX).

When the distance is extended to a neighboring building or to span a distance within a village, town, or city, the local telephone network is usually used. This entails making a connection to the nearest switching exchange by a pair of copper wires, routing the initiating party to the desired receiving party, and completing the connection on the receptor's pair of copper wires, which are also connected to the nearest exchange. The term *switching exchange* is also known as *central office,* or CO, and they are used interchangeably in the rest of this text. The connection between the CO and the customer is called the *local loop,* while the term *subscriber* is also used for customer; they are also used interchangeably. In the event that the connection is within the same neighborhood, the two parties are connected to the same exchange, but if the connection is across town, routing from one exchange to another is necessary. It is at this stage that the choice of technology becomes important in the overall cost of the network. In the early days of telecommunications, all interexchange traffic was done using numerous pairs of copper wires, one pair for each interconnection. This was very cumbersome since interexchange cables were required which incorporated hundreds or thousands of copper pairs. Techniques known as multiplexing were subsequently devised for passing multiple simultaneous telephone calls (referred to as *traffic*) down one pair of copper wires. More recently, optical fibers have been introduced to fulfill this role. Future networks will connect the COs to the customers, in what is known as the local loop, using optical fibers, but the manner in which this should be done is still being debated in many parts of the world. If the amount of traffic between two locations is sufficiently large, it may be cost effective to set up a fixed (dedicated) connection. This is often done between company buildings within a city using a microwave radio link. The mobile telephone also comes into the local loop category.

The next stage of interconnection is intercity, or long distance. The contenders to fulfill this role are microwave radio, optical fiber, and satellite. A competition has emerged between these three technologies. Microwave and satellite communications are far more mature technologies, whereas fiber optics is still in its infancy. Taking this into account, the rapid progress that has already been made by fiber optics over the past 10 years indicates that it is in a good position to "win the race" and be the dominant technology in the future. Many people see the impact of fiber optics on telecommunications as being similar to the invention of the transistor and its effects on computer technology.

The other telecommunications medium which is a competitor for use in transmission systems is the geostationary satellite link. Satellite links have several advantages. Mainly, the broadcast nature of satellites is very attractive for television transmission. The information transmitted from a satellite can be received over a very large area, enabling it to serve a whole continent simultaneously. Also, the cost of satellite communications is *independent of the distance* between the source and the destination (i.e., 1 and 5000 km cost the same). However, the satellite system only becomes cost competitive with microwave and optical fiber systems when the distance is large (e.g., greater than 500 km). There are some situations which are ideally suited to satellite communications. A country like Indonesia, for example, consists of hundreds of small islands. It is cheaper to use a dedicated satellite for telecommunications than to interlink all of the islands by microwave or optical fiber systems. A similar situation exists in very mountainous regions where there are hundreds of villages within the valleys of the mountain range.

The technology used by satellite communications overlaps terrestrial microwave radio technology to a large extent. The radio nature and operating frequencies are the same. The main differences lie in the scale of the components. Since the satellite link is 36,000 km long, high-power transmitters and very low-noise receivers are necessary. Also, the size and weight of the satellite electronics must be kept to an absolute minimum to minimize launch costs. Considerable interest has recently been devoted to the very small aperture terminal (VSAT) satellite technology. As the definition of VSAT implies, these systems have earth station terminals which use small antennas of only 1 to 4 m in diameter. This is a significant reduction in the 30-m-diameter antennas used in the original earth station designs of the 1970s. The use of such small-diameter antennas enables business organizations to use satellite communications cost effectively, since they can place the complete earth stations on the roof of their buildings or within a small area of their properties. Again, long distance and broadcast-type transmission produce the highest cost effectiveness. Perhaps one of the main disadvantages of satellite communications is the propagation delay. It takes approximately a quarter of a second for the signal to travel from the earth up to the satellite and back down again. This is not a problem for two-way speech communication, provided echoes are removed from the system by sophisticated electronic circuits. If satellites are used for intercontinental communications, three geostationary satellites are needed for complete global coverage. To speak to someone at a place on the earth diametrically opposite, or outside the "vision" of one satellite, a double satellite hop is required, which produces a propagation delay of about 0.5 s. Some user discipline is required in this situation, because interrupting the speaker as occurs in normal conversation results in a very disjointed conversation. This delay is totally unsatisfactory for many people. For data communications and data over voice channels, such as Telefax, etc., there is no problem with this delay time. The double hop delay can be improved a little by satellite-to-satellite transmission. The cost of this type of system is considerably less, since one of the earth stations is eliminated from the connection. The future of satellite communications is a very debatable subject. It is the author's belief that satellite communications will eventually be used mainly for broadcast purposes, with global optical fiber connectivity providing the vast majority of video telephone traffic and data services.

Another application of satellite communications could be to provide a global mobile telephone system. To elaborate, many telecommunications organizations are already catering to the demands of the urban population of many countries by offering a telephone handset which uses ultra-high-frequency (UHF) radio technology to make the interconnection between the CO and the customer. If this demand increases as rapidly as forecasted and also on a global scale, the operation of a global mobile telephone network may be appropriate. Such a system has recently been proposed by Motorola company. The project, called *Iridium*, would entail (or 66) 77 satellites operating in low earth polar orbit to provide a global cellular network structure. The estimated cost of this project is in excess of 2.5 billion 1990 U.S. dollars. Whether this project literally gets off the ground remains to be seen. At present, the bulk of all long and medium distance telephone traffic is transmitted over terrestrial-style microwave radio and optical fiber links. For this reason this book is restricted to microwave radio and optical fiber technologies. These are primarily digital electronics technologies. Chapter 2 describes the process of digital multiplexing, which is a means of combining voice, video, and data channels into one composite signal ready for transmission over the microwave radio or optical fiber link. This composite signal is usually referred to as the *baseband* (or BB). Chapters 3 to 7 describe how the two technologies have acquired their present-day capabilities. There are applications for which microwave radio systems may never be replaced by fiber optics. There are situations where they complement each other within the same network and some situations where both are applicable, in which case a prudent choice has to be made between one or the other based on economics. The choice is not always easy. Mobile communications for interconnecting the customer to the CO is discussed in Chap. 8. Telecommunications involves more than voice telephone interconnectivity. Data information transmission is becoming an increasingly important telecommunications requirement and Chap. 9 is devoted to that subject.

Economics is the driving force which determines the fate of a new technology. No matter what the benefits may be, if the cost is too high, the new technology will have only limited application. Relatively low cost combined with improved performance will undoubtedly ensure fiber optics a global acceptance. A large portion of the world has made a substantial investment in microwave communications systems. The introduction of fiber optics does not mean that the existing microwave radio equipment has to be scrapped. As higher capacities (more voice, video, or data channels) become necessary, the optical fiber systems can be installed and will work side by side with the microwave equipment. Many third world countries that are only in the early stages of expanding their communications networks are in an excellent position to be able to take immediate advantage of the new fiber optic equipment and consequently "leapfrog" the copper wire and microwave-based technologies.

Before entering into technical details, some obvious statements about microwave radio and fiber optic systems can be made. Microwave links use radio wave propagation from point to point, whereas fiber optic links have a continuous cable spanning the distance from point to point. This very obvious difference between the two types of system automatically has the outcome of defining some applications for which both techniques are applicable and, conversely, indicates some applications for which each is excluded. For example, in a mountainous terrain, microwaves can "hop" from peak to peak across a mountain range quite adequately, unimpeded by intervening rocks, forests, rivers, etc. Similarly, microwave radio systems can interlink chains of islands whose distances are relatively close without concern for underwater cabling techniques or the depth of the water. When link security is a problem in unsettled parts of the world, microwave radio systems are usually the preferred choice. Cable, whether optical fiber, coaxial, or twisted copper pair, cannot be as well protected as a microwave station. Cable suspended between poles is particularly vulnerable to sabotage or severe weather conditions. Underground placement of optical fiber cable is considered to be the better arrangement. The cost unfortunately is quite high (usually at least twice as much as overhead installation of new cables). Unintentional damage of cables by agriculture and construction activities is by no means rare. In fact, such occurrences in some places can be so frequent that the statistics are too embarrassing to publish. Over relatively flat terrain, an optical fiber cable may be a preferable choice at first sight because of lower cost. If this terrain is mainly hard rock, though, the installation cost could rise rapidly if a lot of blasting is required, making cable a less attractive choice.

In cities, fiber cable can very easily replace or be added to twisted pair or coaxial cable in existing ducts. If the ducts are full and the twisted pair or coaxial equipment is too new to retire, a microwave radio system would then be the preferred option. Several cities have excessively high water tables, causing severe electrical problems. Here is an excellent application for optical fiber cable because it is not metallic. The installation of fiber optic cable can be made with little or no additional cost so that water does not cause any problems. This nonmetallic nature of the optical fiber is also useful in other applications. For example, in power-generating stations electromagnetic induction can play havoc with communications equipment that uses metallic cable. Optical fibers are impervious to electromagnetic interference. Furthermore, since they are completely dielectric materials, there is no possibility of a short circuit. This is very desirable in chemical factories or areas where explosions could be caused by sparks from two wires being short circuited.

There are many other instances where the choice between microwave radio and optical fiber systems is not so clearly defined. When considering a high-capacity backbone route over hundreds, or even thousands of kilometers, most telecommunications companies and/or authorities have made the decision not to install any more twisted pair or coaxial cable. The next decision is whether to use microwave radio or fiber optics. The decision may be in favor of microwaves for very rugged terrain and fiber optics for very flat terrain. When the region under consideration contains both very flat areas and very mountainous areas, a combination may be suitable. The problem of which to choose is compounded by dynamic economic conditions. Prediction of future advances in technology is very desirable. Although this cannot be done with any certainty, keeping a close watch on research and development results can define the trends and make future projections possible.

For example, if a "spur" route from a backbone is required to supply a small village, present economics dictates that a microwave radio system would be cheaper than an optical fiber system. The number of channels required increases with the population to be serviced. There is a specific population size at which the price of microwave radio equipment equals that of the optical fiber equipment (including cable). As time progresses the price of optical fiber equipment is coming down. To what extent it will come down in the future is subject to debate. Also, the population of the village may increase or decrease depending on many factors. This apparently simple situation is already starting to develop into a complex problem. It also appears that any decision will involve at best an educated guess. To use the word *guess* is unscientific so *prediction based on statistical analysis* will be used instead to describe the process by which these difficult decisions are usually made. One consoling fact is that both optical fiber and microwave radio systems can be upgraded in capacity. For the microwave radio system, additional transmitters and receivers are required.

Similarly, the fiber optic terminal equipment can be changed to increase the capacity. This is true only if the high-bandwidth (capacity) cable is installed to allow for future expansion. Microwave radio systems have the added flexibility of being able to redirect the path of a link to accommodate changing communication requirements. The equipment can be readily moved from one location to another, the free space propagation media being conveniently amenable and omnipotent. The installed cable, unfortunately, cannot be similarly moved without incurring significant additional cable costs. The atmosphere in which microwaves are propagated is not without its problems, as Chap. 4 addresses in detail.

The high-quality cost effectiveness of the communication link is a prerequisite for a successful system. Cost and quality are, as usual, interrelated. The way in which they are related is very complex and leads the discussion to a technical level.

1.2   Digitization

The major quality improvement obtained in digital transmission systems is due to the receiver signal recovery technique (i.e., regeneration). In analog transmission systems, each repeater retransmits the received signal and also retransmits the noise. The noise accumulates at each repeater, so after a certain transmission length the signal-to-noise ratio (S/N) will be so poor that communication is not possible. In digital transmission systems, each repeater "regenerates" the original received signal and retransmits the signal free of noise. Theoretically, therefore, digital transmission has no transmission length limit. However, in reality, there is a phenomenon known as *jitter*. As described later, this is a pulse position noise observed as small variations of the pulse zero crossing points of the digital bit stream from their precise positions (see Fig. 5.60). Jitter accumulates because of its introduction by several electronic circuits within a digital transmission system. Excessive jitter causes unacceptable bit errors to occur and therefore limits the maximum link length capability of the digital system. In analog systems the S/N determines the quality of the link or channel. In digital systems, it is the bit error ratio (BER) that determines the quality of the link or channel. For example, a BER of 1×10^{-6} means that 1 bit per million transmitted bits has been received incorrectly. To summarize, the advantages of the digital systems over analog are:

- All subscriber services such as telephony, telegraphy, high-speed data, TV, facsimile, etc., can be sent via the same transmission medium. Consequently, the concept of the *integrated services digital network* (ISDN) can be realized.

- Utilization of the higher unused radio frequency (RF) bands. RF bands in the region of 10 GHz or above are unsuitable for analog systems because of high attenuation by rain, fog, etc. Whereas the S/N in analog systems decreases linearly with fading of the RF carrier, the BER in digital systems is unaffected by fading until the received RF level abruptly approaches the threshold value. These characteristics are discussed in Chap. 4 (see Fig. 4.36).

- High immunity against noise makes digital transmission *almost* independent of link length.

- Use of integrated circuits makes digital systems economical and alignment free.

- Easy maintenance, based on go/no-go type of measurement.

- Synergistic integration of digital transmission systems such as optical fiber, digital satellite, and digital microwave radio systems with digital exchanges.

As networks become more digitized, the combination of the time-division multiplexing (TDM), time-division switching, digital radio,

and optical fiber systems is considerably more economical and technically flexible than the corresponding analog networks.

1.3 Digital Microwave Radio System Configuration

Figure 1.2 shows a simplified microwave link incorporating just one
regenerative repeater and two end terminal stations. The terminal
stations house switching exchanges which connect the customers to
the long distance paths. In this illustration, a large number of customers (e.g., around 2000) are multiplexed together ready for transmission over the microwave link. The signal is converted to the microwave frequency (around 6 GHz) and transmitted over a path of
typically 30 to 60 km from station A to the receiving antenna at the
repeater station. The repeater either (1) simply amplifies the signal
and sends it off on its journey using a different microwave frequency
to minimize interference, or (2) it completely regenerates the individual pulses of the bit stream before reconverting the signal back to a
microwave beam for onward transmission. Station B receives the microwave signal, processes it, and unravels the individual channels
ready for distribution to the appropriate customers at this end of the
link.

Figure 1.3 is a simplified block diagram showing the major differences between analog microwave radio (AMR) and digital microwave
radio (DMR) transmitters. At the intermediate frequency (IF) and
above, the two systems are very similar. The IF-to-RF conversion
shown here is done by the heterodyne technique. The modulated signal is mixed with an RF local oscillator to form the RF signal which is
then amplified and filtered ready for transmitting from the antenna.

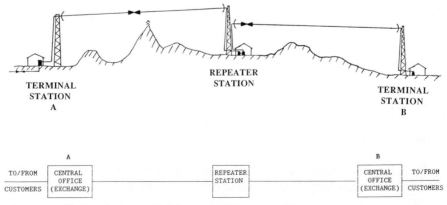

Figure 1.2 Basic microwave link incorporating a repeater.

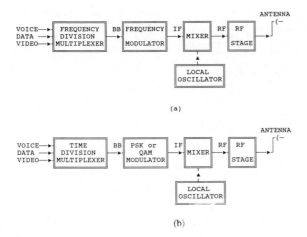

Figure 1.3 Comparison between analog (a) and digital (b) microwave radio transmitters.

For both analog and digital systems, there are one or two variations on this theme. For example, the digital information can directly modulate the RF signal without going through an IF stage. This is called *direct RF modulation.* Another technique is to use a frequency multiplier to convert the IF signal to the RF signal. The advantages, disadvantages, and fine details of these systems are highlighted in Chap. 5.

The major differences between the analog and the digital microwave radios lie in:

1. The composition of the baseband

2. The modulation techniques

3. The service channel transmission (not shown in Fig. 1.3)

The baseband is the combined, multiple voice, data and/or video channels which are to be transmitted over a telecommunications transmission system.

As indicated in the simplified diagram of Fig. 1.4, the receivers differ mainly in the demodulation technique and demultiplexing of the baseband down to the voice, data, or video channels. For the AMR receiver (Rx), the incoming RF signal is downconverted, frequency demodulated, and then frequency division demultiplexed to separate the individual voice, data, or video channels.

For the DMR receiver, the incoming RF signal is similarly downconverted to an IF prior to demodulation. Coherent demodulation is preferred. However, for coherent demodulation, exact transmitted carrier frequency and phase of the modulated signal must be obtained at

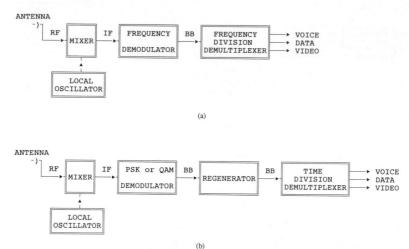

(a)

Figure 1.4   Comparison between analog (*a*) and digital (*b*) microwave radio receivers.

the receiver. A way around this problem is to use differential encoding and decoding as described in Chap. 5. The demodulated signal is subsequently restored to its original transmitted bit stream of pulses by the regenerator. Finally, the time-division demultiplexer separates the individual voice, data, or video channels for distribution to their appropriate locations.

Long distance DMR link systems use regenerative repeaters. *Regenerative* is a term used when the signal goes through a complete demodulation-regeneration-modulation process (Fig. 1.5). Regenerators can interface with digital multiplex equipment for data insert and drop applications. In *regenerative* repeaters the noise and distortion is largely removed in the regeneration process and so there is *no noise accumulation,* only jitter accumulation.

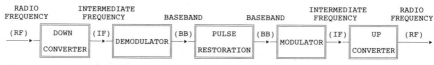

Figure 1.5   A regenerative repeater.

1.4   The Optical Fiber System Configuration

The optical fiber link (Fig. 1.6) has some similarities to the microwave link. Both systems transmit the same output from the digital multiplexer (i.e., the baseband). The bit stream in the case of the optical fiber system can be used directly to turn a laser on and off to send

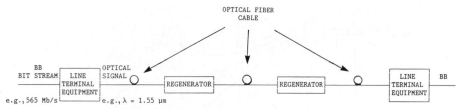

Figure 1.6 The optical fiber link.

light pulses down the fiber cable. The more modern optical fiber systems use a heterodyning or homodyning technique to improve the overall performance of the system. Regenerators are used at intervals to boost the signal, as in the microwave radio system. The distance between regenerators is gradually increasing for optical fiber systems as technology improves. Conversely, the line-of-sight microwave radio system regenerator spacing is limited by a physical, not a technological, constraint (i.e., the curvature of the earth). The increasing regenerator spacing of optical fiber systems is an important factor in enabling overall costs to be reduced. The spacing of optical regenerators is limited by the attenuation and dispersive (pulse broadening) characteristics of the fiber. As usual, the receiver incorporates a detector, an amplifier, and a means of restoring the original baseband bit stream ready for demultiplexing to voice, data, or TV signals.

The debate rages as to which systems will become dominant in the future. It is widely believed that fiber optics is not only here to stay, but it will transform our lifestyles in the decades to come.

1.5 Mobile Radio Systems

So far, the systems mentioned above are fixed terminal or point to point in nature. Also, whether the choice of system equipment used for an interconnection is microwave radio, optical fiber, or satellite is irrelevant to the customer. The customer has recently become acutely aware of telecommunications equipment in the form of the portable radio telephone or the car telephone. Although these are the major mobile systems, other important mobile systems include aircraft telephones and telephones to ships and trains. All of these systems are examples of radio technology being used directly *in the local loop*. Historically, mobile telephones were mainly installed in vehicles. They were initially rather heavy and cumbersome devices. With the acceleration of electronic circuit miniaturization over recent years, the portable telephone market has mushroomed. In the industrialized world the day may not be too far away when city dwellers consider a portable telephone to be as much a part of their daily attire as a wrist

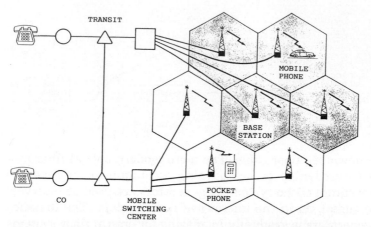

Figure 1.7 The cellular radio schematic. (*Adapted from* Ericsson Review no. 3, *1987, Soderholm, G., et al., Fig. 1.*)

watch. The term *city dweller* is used because portable telephones link up to the main national telephone network via very expensive equipment installed in the main city or urban areas. Eventually, coverage of an entire country will no doubt be accomplished, although the economics of such an enterprise may be prohibitive even for many developed countries.

Radio waves have a large attenuation as they travel through the atmosphere, so in order to keep the required customer transmitter power to an acceptably low level, the distance between the customer and the nearest *base station* to link into the telephone network must be kept as small as possible. This has led to the necessity for numerous base stations arranged in a type of honeycomb, otherwise known as the cellular structure (see Fig. 1.7). As one can appreciate, each customer uses his or her own communication frequency for the duration of each call. Because of the limited frequency spectrum available, the number of subscribers using the systems appears to be quite small. However, the cellular system lends itself to frequency reuse, whereby subscribers in different areas use identical frequencies of communication without suffering any interference with each other. The intricacies of this and the other above-mentioned systems will be described in detail in Chap. 8.

1.6   Data Communications and the Network

The transmission of digital information is considered to be a bit stream, which is a sequence of millions of 1s and 0s that represent a combination of many voice and data channels or a video channel. The information to be communicated starts as analog information which is digitized for transmission then reconverted to analog signals at the destination. There are inevitable analog-to-digital (A/D) and digital-to-analog (D/A) conversions in this type of network. While telephone traffic was, and still is, the primary aim of the telecommunications network, a growing presence of data traffic has developed. Broadly speaking, data traffic means computer-to-computer communication.

During the transition from an analog to a digital network, there must be some mechanism to accommodate a combination of analog and digital traffic. New transmission equipment installed in the network in the developed world, and increasingly in the developing world, is digital in nature. In this case, as stated above, there is an unavoidable A/D conversion process for information that starts as analog signals. However, for digital computers the transmission path could be digital throughout. If the transmission involves, for example, a subscriber's PC to be connected to a computer mainframe in a downtown office, the subscriber loop must have data transmission capability.

There is a fundamental problem which arises when trying to merge data transmission into the telephone network designed for analog telephone traffic—bandwidth. The telephone network was initially designed to have 300- to 3400-Hz bandwidth for a voice channel. This is not compatible with megabit-per-second data rates (bit rates) which are ideally required by computers. Fortunately, the rates achievable at present are fast enough to allow a reasonable amount of information transfer, but the required data rates are increasing rapidly. The obvious solution that springs to mind is to use a wide bandwidth transmission medium such as optical fiber cable for data transmission. Supplying optical fiber to every customer is a very expensive enterprise but will probably become a reality for many countries near the year 2000. The bandwidth capability of the optical fiber would allow a variety of services to be provided including videophone, video for TV, stereo music channels, etc. Such a broadband integrated services digital network (B-ISDN) would incorporate broadcast facilities, which are already causing heated debate in several countries about the legal details. For example, how is a cable TV company affected if the telephone company has a better distribution system?

Before fiber-to-the-home (FTTH) becomes a reality and bandwidth is no longer a concern, an interim ISDN has been proposed which extends a 64-kb/s path out to the customers' premises. This is a considerable improvement over the early data bit rates of 9.6 or 14.4 kb/s using A/D and D/A converters, called *modems*, over an analog telephone network 3.1-kHz bandwidth circuit.

It is difficult to discuss data transmission without referring to the network as a whole. In particular, this encompasses the subject of switching, which plays an even greater major role in data communications than ordinary voice communications. For example, speech and video signals are real time in nature. Instantaneous processing and transmission are required so that there is no irritating delay as is sometimes experienced on satellite links. There is an important difference here when considering data which does not have this real-time constraint. Instead of using the conventional speech circuit switching, data can be split up into packets and transmitted when convenient time slots become available in a time division multiple access (TDMA) type of medium. This *packet switching* has become a significant force in data communications today. This and other data communications subjects will be discussed in Chap. 9.

1.7   International Standards

Throughout the rest of this text there are frequent references to the recommendations made by the International Telegraph and Telephone Consultative Committee (CCITT) and the International Radio Consultative Committee (CCIR). These organizations are part of the International Telecommunication Union (ITU), which is a member organization within the United Nations. These two committees have global representation. Since these forums are apolitical in nature, there has been considerable cooperation between its member nations over the past few decades, regardless of political bias.

Standards are essential. A simple example to illustrate the necessity for standards is the requirement for the receiver of a communications system in one country to operate on the same frequency as the message originator's system in another country. Obviously, if they are operating on different frequencies, there will be no communication. This is a trivial situation, but without consensus from all member states, not only frequency problems but also a multitude of equipment incompatibilities and quality disparities would exist.

There are bodies other than the ITU which are usually national rather than international in nature. On a communications level, as the world shrinks in size, there is a growing need to harmonize the national standards of individual nations with the international standards of the ITU.

2

Digital Multiplexing (Baseband Composition)

The digital baseband signal transmitted by digital microwave radios or optical fiber line terminal equipment is formed by TDM instead of analog frequency division multiplexing (FDM) as it was previously. This digital multiplexer equipment is considerably simpler and cheaper than its analog counterpart. The TDM technique involves periodically sampling numerous channels and interleaving them on a time basis. The resulting transmitted bit stream of 1s and 0s is unraveled at the receiving end so that the appropriate bits are allocated to the correct channels to reconstruct the original signal. The digital multiplexer packages channels in groups of 24 for North America and Japan and 30 elsewhere (CCITT). The pulse code modulation (PCM) scheme is universally used for this purpose. For higher-capacity links, several 24- or 30-channel bit streams are digitally multiplexed to a higher-order bit rate [e.g., 140 Mb/s provides 1920 channels (CCITT)]. This higher-order multiplexing can be done on either a *synchronous* or *asynchronous* basis. Since the inception of PCM in the 1970s, the systems manufactured have been almost completely of the asynchronous type. Recent (1988) international standards have been derived to ensure that the next generation of digital multiplexers will be of the synchronous type. There are some benefits to the synchronous variety which will become clear in the following. However, because of the enormous quantity of asynchronous multiplexers in existence, it will be several years before the transformation gains momentum. Both types will be outlined in the higher-order multiplexer section.

2.1   Pulse Code Modulation

2.1.1   Sampling

One of the first steps in the conversion of an analog signal into a digital one is the process of sampling. A sample is the magnitude of a modulating signal at a chosen instant, usually represented by the voltage of the modulating signal. Sampling is the process of measuring amplitude values at equal intervals of time (i.e., periodically, as shown in Fig. 2.1). The sampling rate for periodic sampling is the number of samples per unit time. For example, for telephony the sampling rate is 8000 samples per second, or 8 kHz. The sampling period is therefore ⅛ kHz (i.e., 125 μs). This sampling period is fixed by the *Shannon's sampling criterion,* which states that the sampling frequency must be at least double the highest frequency to be transmitted. Within one sampling period, samples of several telephone channels can be sequentially accommodated. This process is called TDM. The principle is used in all PCM systems. In the 30-channel PCM sys-

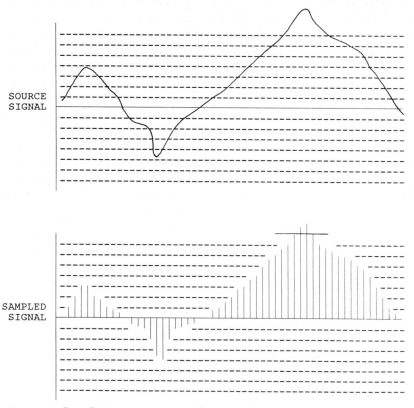

Figure 2.1   Sampling.

tem, samples of 30 telephone channels are available for transmission in each sampling period.

A sampled signal contains complete and unambiguous information about the source signal as long as the sampling frequency is at least twice the highest frequency f_s of the source signal B. It can be shown mathematically that if the source signal has a spectrum as in Fig. 2.2a, the sampled signal will have a spectrum as in Fig. 2.2b. This consists of a number of subspectra, the first of which (number 1) lies in the frequency range 0 to B and is identical to the spectrum of the source signal. Subspectrum number 3 is identical to number 1 but moved f_s Hz in frequency (f_s = the sampling frequency). Number 2 is a reflected image of number 3 in f_s, and so on to infinity at a cycle of f_s Hz. Since the spectrum of the sampled signal coincides with that of the source signal in the $f_s + B$ band, we can conclude that the sampled signal contains all of the information about the source signal, provided that f_s is greater than $2B$. If this condition is not satisfied, the situation

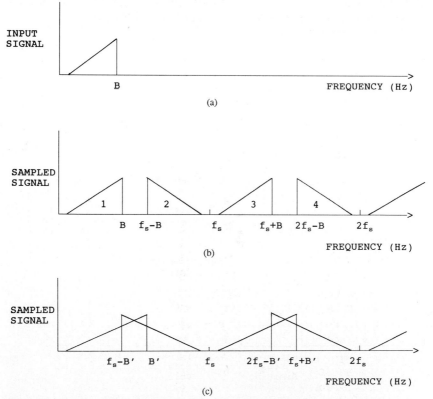

Figure 2.2   Spectra for sampled signals. (a) Input signal spectrum; (b) sampled signal spectrum for $f_s > 2B$; (c) sampled signal spectrum for $f_s < 2B$.

in Fig. 2.2c will arise. The subspectra of the sampled signal will overlap each other, and this will result in the loss of information about the original source signal. This phenomenon is called *aliasing distortion*. Depending on the designer, the multiplexing itself can be performed after the sampling process or after the encoding process.

2.1.2 Quantization

Quantizing.   Quantizing is a process by which analog samples are classified into a number of adjacent quantizing intervals. Each interval is represented by a single value called the quantized value, which is equal to the reconstructed sample obtained in decoding. Quantizing usually involves encoding as well. The quantization interval is one of the intervals, or value ranges, into which the working range is divided. A quantizing interval is limited by two decision values (see Fig. 2.3).

Working range.   This is the permitted range of values of an analog signal divided into quantizing intervals. In quantizing telephone signals, the range of signals to be transmitted is divided into 256 intervals (for the A-Law compander characteristic).

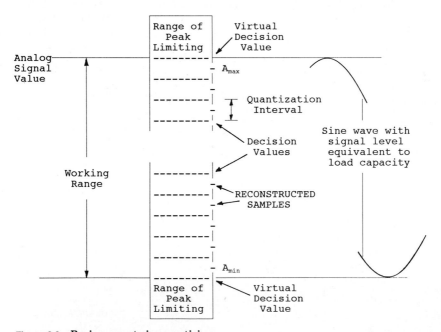

Figure 2.3 Basic concepts in quantizing.

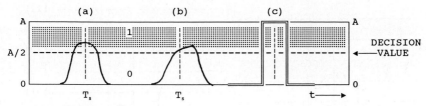

Figure 2.4   Decision value.

Decision value.   This is a reference value defining the boundary between quantizing intervals or, in regenerating a digital signal, the threshold between two signal value ranges (see Fig. 2.4):

1. Pulse without interference but decreased in amplitude by cable attenuation (Fig. 2.4a)
2. Pulse with superimposed interference whose magnitude is only just greater than $A/2$ at the sampling instant (Fig. 2.4b)
3. Regenerated pulse (Fig. 2.4c)

Figure 2.4 shows the principle of pulse regeneration of a binary signal. Complete regeneration is possible by sampling at instant T_s (center of the signal element). If the signal value at this instant is above the decision value $A/2$, a "mark" (binary 1) is detected. If it is below, a "space" (binary 0) is detected. A bit error occurs when a noise disturbance exceeds $A/2$ at the point where no pulse is supposed to be present.

Virtual decision value.   The virtual decision value is one of the two "hypothetical" decision values located at the ends of the working range, obtained by extrapolation of the real decision values (Fig. 2.3). The virtual decision values identify the range of signal values which are transmitted without "peak limiting" occurring.

Reconstructed sample.   This is a quantized sample generated at the output of a decoder when a specific "character signal" is applied at its input. A signal value is usually reconstructed which corresponds to the mean value (center value) of the associated quantizing interval.

Uniform quantizing.   This is quantizing in which the quantizing intervals are of equal size. Figure 2.5 shows how the uniform quantizing scheme converts the pulse amplitude modulated (PAM) signals into

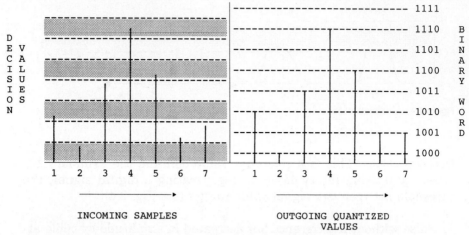

Figure 2.5   Uniform quantization with symmetrical binary values.

the 4-bit codewords containing the PAM data. For both the CCITT and North American recommended PCM systems, the codeword has 8 bits. The first bit (the most significant bit) gives the polarity of the PAM signal, and the next 7 bits denote the magnitude of the PAM signals. To transmit code words of 8 bits, $2^8 = 256$ quantizing intervals are necessary.

Quantization noise.   The distortion produced by quantization is significant. Since at the receiving end the analog signal is reconstructed from discrete amplitude samples, it is not an exact replica of the original signal. Because of the finite number of steps in the quantizing process, the input signal can only be approximated as shown in Fig. 2.6. For example, if there are 128 quantization steps and the maximum speech amplitude is, say, 1.28 V, each step has a width of 10 mV. All pulse amplitudes lying between 280 mV and 290 mV will therefore be encoded as 285 mV, the mean value between those steps. This difference between the input signal and the quantized output signal is called the *quantizing noise*. It is evident that the quantizing noise can be reduced by increasing the number of quantizing steps, but this would amount to increasing the number of bits in the code word designating the amplitude information. An increase in the number of bits per word can be accommodated only with reduced pulse width. This

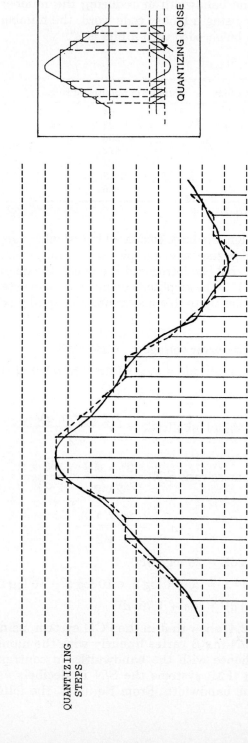

Figure 2.6 Distortion due to the quantization process. Solid line is original signal, dashed line is reconstructed signal.

means increasing the bandwidth or reducing the number of multiplexed channels. By using a binary code word, the number of amplitude levels that can be encoded with n bits is 2^n:

n = number of bits in the code	Number of quantizing steps
6	64
7	128
8	256
9	512
10	1024
11	2048
12	4096

Assuming that the signal is large compared to a single quantizing step, the errors introduced in successive samples will be uncorrelated. So, the maximum error that could be introduced is one-half the quantizing step (i.e., $\pm dV/2$, where dV is the amplitude of one quantizing step).

Since all error values up to this maximum are equally possible and randomly occur,

$$\text{rms error voltage } e' = dV/2\sqrt{3} \qquad (2.1)$$

$$\text{Peak-to-peak amplitude} = \text{step width} \times \text{number of steps}$$

$$= 2^n \times dV$$

$$\text{rms signal amplitude} = \frac{1/2 \times 2^n \times dV}{\sqrt{2}} \qquad (2.2)$$

$$\text{Signal-to-quantized-noise ratio S/N} = \frac{2^n \times dV}{2\sqrt{2}} \times \frac{\sqrt{3} \times 2}{dV}$$

$$= \frac{2^n \times \sqrt{3}}{\sqrt{2}} \qquad (2.3)$$

$$\text{S/N in dB} = 20 \log (\text{S/N})$$

$$= 20n \times \log 2 + 10 \log 3 - 10 \log 2$$

$$(\text{S/N}) \text{ dB} = (6n + 1.76) \text{ dB} \qquad (2.4)$$

The results of Eq. (2.4) show that in the PCM system, using uniform quantization, the S/N in dB varies linearly with the number of bits in the word, and hence with the bandwidth. In contrast, for frequency modulation (FM) systems the S/N in decibels varies with the logarithm of the bandwidth. From Eq. (2.4), the following S/N can be achieved depending upon the number of bits in the code word:

Number of bits per code	S/N (dB)
7	43.76
8	49.76
10	61.76
12	73.76

A system transmitting speech signals must be able to accommodate signals of about a 60-dB dynamic range (i.e., voltage range of 1000:1). To achieve this with uniform quantization, the number of bits per code word should be at least 12 (4096 quantization steps).

With uniform quantization, the quantization distortion for signals with small amplitudes is greater than that for signals with larger amplitudes. Also, in telephony, the probability of the presence of smaller amplitudes is much greater than that of larger amplitudes. Low signal levels must therefore be amplified more than stronger signals to achieve a reasonably constant S/N. In analog techniques this is done by *com*pressing the dynamic range at the transmitting end and ex-*panding* it at the receiving end (companding). In PCM, a similar result is achieved by making the quantization steps wider for higher-level signals and narrower for lower-level signals. This is called *instantaneous companding* and requires a nonuniform quantizing characteristic for the compander.

Nonuniform quantizing.   This is quantizing in which the quantizing intervals are not of equal size. Small quantizing intervals are usually allocated to small signal values (samples) and large quantization intervals to large samples so that the signal-to-quantization distortion ratio is nearly independent of the signal level. This is the method usually used for quantizing telephone channels.

Companding.   This is a process in which compression is followed by expansion. Compression is a process in which the effective gain applied to a signal is varied as a function of the signal magnitude. The effective gain is greater for small signals than for large signals. Expansion is the reverse process of compression. When the input and output signals represent instantaneous values, this is called instantaneous companding. Instantaneous companding was previously used with quantization to achieve the overall effect of nonuniform quantization. Because the ear's response to sound is proportional to the logarithm of

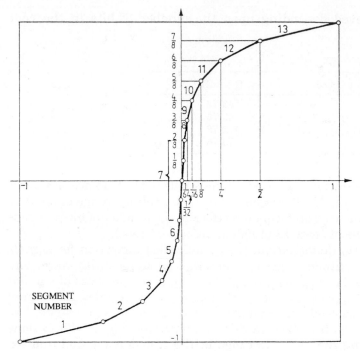

Figure 2.7   The 13-segment compander characteristic according to the *A*-Law.

the sound amplitude, the compression curve recommended by the CCITT and used by all equipment manufacturers has an approximate logarithmic characteristic. The required logarithmic curve is approximated by several linear segments. The 13-segment compander characteristic according to CCITT Recommendation G.711 is shown in Fig. 2.7. This uses the so called *A*-law with the value of $A = 87.6$. It gives an improvement of 26 dB in the S/N for 8-bit coding. This compander characteristic is approximated by the equations:

$$Y = \frac{1 + \ln Ax}{1 + \ln A} \qquad \text{for } 1/A < x < 1$$

$$Y = \frac{Ax}{1 + \ln A} \qquad \text{for } 0 < x < 1/A$$

where x = normalized input level
Y = normalized quantized steps
ln = natural logarithm

In North America and Japan the μ-Law characteristic is used for companding. This is represented by the function:

$$y = \text{sgn}(x) \frac{\ln(1 + \mu x)}{\ln(1 + \mu)}$$

where x = input amplitude
  $\text{sgn}(x)$ = polarity of x
    μ = amount of compression

The result is similar to the *A*-Law characteristic.

Figure 2.8 illustrates how the *compressed* signal at the transmitting end when combined with the *expanded* signal at the receiving end produces a linear result. The signal heard by the person at the receiving end is therefore a faithful reconstruction of the transmitted signal, except for a small quantization noise.

2.1.3 Encoding and decoding

Encoding (or coding) is the conversion of an analog sample, within a certain range of values, into an agreed combination of digits. In PCM it is the generation of "character signals" (code words) allocated to quantized intervals. These represent the quantized samples.

Encoder or coder. In PCM, *quantizing* and *encoding* are very closely related. In actual implementation, the encoder or coder is usually one complete device providing both quantizing and encoding.

Operating principle of the iterative encoder. The method used for evaluation and encoding is as follows. At the beginning of each encoding period, the first bit is assumed to be a 1 if the sample is positive and 0 if it is negative. The sample is then compared with an analog reference voltage corresponding to half of the maximum level. If the sample amplitude is greater than this reference voltage, this reference is maintained during the rest of the comparison (i.e., $B2 = 1$). If the sample amplitude is smaller than this reference voltage, this reference voltage is removed (i.e., $B2 = 0$). In this manner, the reference voltage corresponding to each of the code bits is tried in turn until all of the code bits have been determined. At the end of the comparison, the code

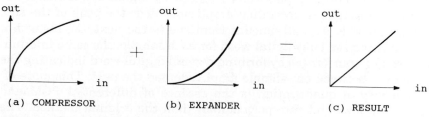

(a) COMPRESSOR   (b) EXPANDER   (c) RESULT

Figure 2.8 Compression + expansion = linear result.

word corresponding to the sample amplitude is directly available at the output of the circuit controlling the reference voltages. The output of this circuit is then stored so that the encoder can process the next sample. The iterative A/D conversion decision sequence is shown in the "decision tree" of Fig. 2.9.

Decoding.   Decoding is the inverse process of encoding. In PCM it is a process in which a "reconstructed sample" is produced, corresponding to a character signal or code word. Since this involves reading the word directly and does not involve an iterative process, decoding is approximately 50 times faster than encoding. Decoding must be done at the receiver in order to retrieve the encoded signals in the form of PAM samples.

2.1.4   Recent coding developments

The standard bandwidth for voice channels in a digital PCM transmission system has been set at the customary 300- to 3400-Hz value (as in CCITT Rec. G.711). Recent developments in speech coding techniques have led to CCITT Rec. G.721, which allows a reduction of the transmitted bit rate to 32 kb/s while preserving audio quality. This implies that, for a given transmission bit rate, the number of voice channels transmitted can be *doubled.* Alternatively, if the original 64-kb/s transmission rate is maintained, the quality of the voice signals can be improved. In other words, if the sampling rate is now 16 kHz instead of 8 kHz, the bandwidth of the encoded signal can be extended to approximately 7 kHz. The resulting quality enhancement is particularly noticeable when transmitting music. A technique has been developed for reducing the sampling rate, called *adaptive differential PCM* (ADPCM). This technique has now advanced to the point where highly intelligible, but noticeably degraded, speech can be transmitted at a rate as low as 800 b/s.

The ADPCM works on the following principle. When successive samples are quantized, because of the gradual nature of the variation of the source signal (see Fig. 2.1), a lot of redundant information is transmitted. For example, when a sine wave signal is quantized, the successive sample quantization amplitudes near the peak of the sine wave vary only a small amount from one to the next. So, instead of transmitting the long digital word for each sample, the same information can be transmitted by forming a small digital word indicating the difference between one sample amplitude and the next. This more efficient way of quantization is the essence of differential PCM and leads to a lower bit rate per channel than the original method. The inclusion of an adaptive predictor circuit for tracking the trend of the

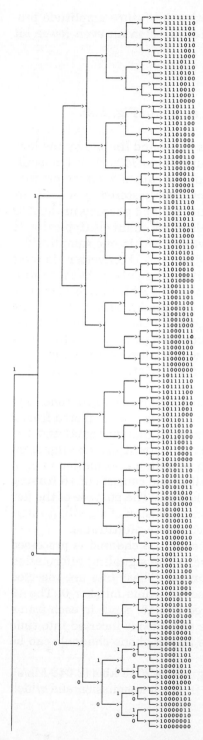

Figure 2.9 The decision tree for iterative A/D conversion (only positive values shown).

sampled signal and statistically predicting its future amplitude provides further transmission efficiency, resulting in an even lower bit rate per channel.

2.1.5 PCM transmission formats

A typical PCM transmission concept is shown in Fig. 2.10.

Transmission path.   The speech signal is first band limited by the low-pass filter so that only the frequency band 300 to 3400 Hz is transmitted. The speech signal is then sampled at the rate of 8 kHz to produce the PAM signal. The PAM signal is temporarily stored by a *hold* circuit so that the PAM signal can be quantized and encoded in the A/D converter. Samples from a number of telephone channels (typically 24 or 30) can be processed by the A/D converter within one sampling period of 125 μs. These samples are applied to the A/D converter via their respective gates selected by the transmit timing pulses. At the output of the A/D converter, the speech samples exit as 8-bit PCM code words. These code words from the speech path are combined with the *frame alignment* word, service bits, and the signaling bits in the multiplexer to form *frames* and *multiframes*. They are then passed on to the high-density bipolar 3 (HDB3) line encoder, which converts the binary signals into bipolar (pseudo-ternary) signals for transmission over the wire-line, the DMR, or the optical fiber cable.

In the European CCITT systems, each frame contains 32 time slots of approximately 3.9-μs duration. The time slots are numbered from 0 to 31. Time slot 0 is reserved for the frame alignment signal and service bits. Time slot number 16 is reserved for multiframe alignment signals and service bits and also for the signaling information of each of the 30 telephone channels. Each multiframe consists of 16 frames, so the time duration of one multiframe is 2 ms. The purpose of the formation of multiframes is to allow the transmission of signaling information for all 30 channels during one complete multiframe.

The signaling information for each telephone channel is processed in the signaling converter, which converts the signaling information into a maximum of 4-bit codes per channel. These bits are inserted into time slot 16 of each PCM frame except frame number 0. The 16 frames in each multiframe are numbered 0 to 15. Since in each frame signaling information from 2 telephone channels is inserted into time slot 16, signaling information from the 30 telephone channels can be transmitted within one multiframe.

The transmission rate of the PCM signals is 2048 kb/s (2.048 Mb/s). This is controlled by the timing clocks in the transmission end which

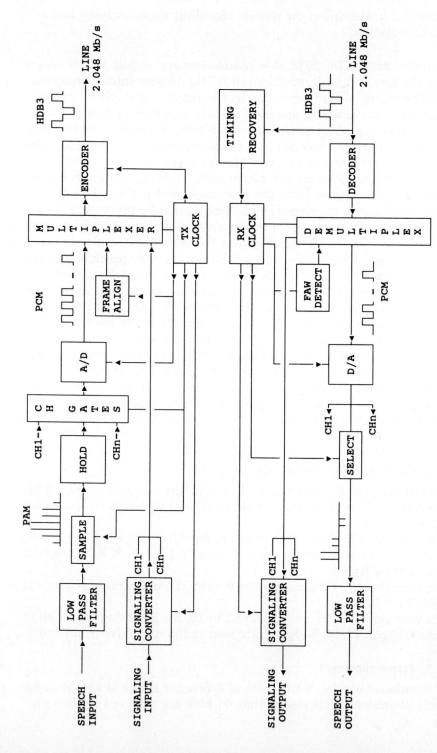

Figure 2.10 The basic PCM asynchronous transmission concept.

control the processing of the speech, signaling, synchronizing, and service information.

Reception path. The 2048-kb/s pseudo-ternary signal which comes from the line is first decoded by the HDB3 decoder into a binary signal. This signal is then separated by the input demultiplexer or separator into the respective speech channels, together with supervisory information (signaling, etc.). The speech codes are sent to the D/A converter, the signaling bits are sent to the signaling converter, and the frame alignment bits and service bits for alarms, etc., are sent to the frame alignment detector and alarm unit. The timing signals for the receiver are recovered from the line codes and processed in the receiver timing unit to generate the clock signals for processing the received signals. In this manner the receiver is kept synchronized to the transmitter. Synchronization between the transmitter and receiver is vital for TDM systems. The codes belonging to the speech signal are then converted to PAM signals by the D/A converter. Next, they are selected by their respective gates and sent to their own channels via the respective low-pass filters, which reconstruct the original analog speech patterns. The bits belonging to signaling are converted into signaling information by the receive signaling converter and sent to the respective telephone channels. The frame alignment word and service bits are processed in the frame alignment and alarm units. Frame alignment word (FAW) detection is done here, and if a FAW error is detected in four consecutive frames, the frame alignment loss (FAL) alarm is initiated. Some of the service bits are used to transmit and receive alarm conditions.

2.1.6   Formats for 30-channel PCM systems

The construction of the frame and multiframe is shown in Fig. 2.11. The details of a multiframe in time slot 0 and time slot 16 are shown in Fig. 2.12.

In time slot 0 of each frame number, the FAW (0 0 1 1 0 1 1) is sent on every *even* frame, and the service bits (Y 1 Z X X X X X) are sent on every *odd* frame.

In time slot 16 of frame number 0 only, the multiframe alignment word (0 0 0 0) is sent.

In time slot 16 of frame numbers 1 to 15, the signaling information of channel pairs 1/16, 2/17, etc., are sent in the respective frame order.

2.1.7   Frame alignment

As mentioned earlier, a time slot of 8 bits per frame is available for frame alignment. This means that 64 kb/s are reserved for this pur-

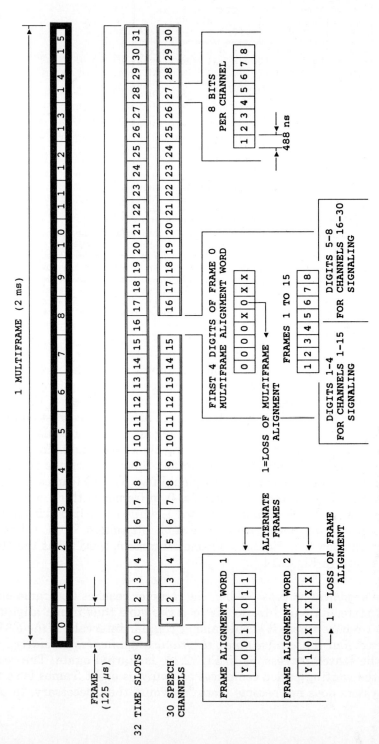

Figure 2.11 Thirty-channel PCM frame and multiframe details.

| FRAME | TIME SLOT 0 | TIME SLOT 16 |
| NO | BIT NO | BIT NO |
	1 2 3 4 5 6 7 8	1 2 3 4 5 6 7 8
0	Y 0 0 1 1 0 1 1	0 0 0 0 X Z X X
1	Y 1 Z X X X X X	SIG CH1 SIG CH16
2	Y 0 0 1 1 0 1 1	SIG CH2 SIG CH17
3	Y 1 Z X X X X X	SIG CH3 SIG CH18
4	Y 0 0 1 1 0 1 1	SIG CH4 SIG CH19
5	Y 1 Z X X X X X	SIG CH5 SIG CH20
6	Y 0 0 1 1 0 1 1	SIG CH6 SIG CH21
7	Y 1 Z X X X X X	SIG CH7 SIG CH22
8	Y 0 0 1 1 0 1 1	SIG CH8 SIG CH23
9	Y 1 Z X X X X X	SIG CH9 SIG CH24
10	Y 0 0 1 1 0 1 1	SIG CH10 SIG CH25
11	Y 1 Z X X X X X	SIG CH11 SIG CH26
12	Y 0 0 1 1 0 1 1	SIG CH12 SIG CH27
13	Y 1 Z X X X X X	SIG CH13 SIG CH28
14	Y 0 0 1 1 0 1 1	SIG CH14 SIG CH29
15	Y 1 Z X X X X X	SIG CH15 SIG CH30

Figure 2.12 Actual bits in time slot 0 of the multiframe for the no alarm condition. *Note 1:* (a) X bits not allocated for any purpose. Normally set to 1. (b) Y bits reserved for international use. Normally set to 1. (c) Z bits used to inform the distant end if frame alignment loss is detected: normal state—0; alarm state—1. *Note 2:* (a) Frame alignment signal (0011011) is sent during time slot 0 of even frames. (b) Multiframe alignment signal (0000) is sent only once per multiframe in time slot 16 of frame 0. This signal is inserted in the bit positions 1, 2, 3, and 4. Loss of multiframe alarm is sent to the distant end with bit 6 as 1.

pose. The basic principle of frame alignment is that the receiver identifies a fixed word and then checks its location at regular intervals. This makes it possible for the receiver to organize itself to the incoming bit flow and to distribute the correct bits to the correct channels. In addition to frame alignment, the assigned time slot is also used for transmission of information concerning the alarm states in the near-end terminal to the remote-end terminal. Spare capacity is also available for both national and international use. The 16 frames are numbered 0 to 15. The words in time slot 0 in frames with even numbers are often called *frame alignment word 1,* while those in odd frames are called *frame alignment word 2.* Frame alignment word 1 has the structure shown in Fig. 2.13. Frame alignment word 2 has the structure shown in Fig. 2.14.

Frame alignment process.   When the receiver reaches the frame alignment state, its only function is to make sure that frame alignment word 1 recurs where it should and at regular intervals. *If the FAW is incorrect four consecutive times, the frame alignment is considered lost,* and the search process, as in Fig. 2.15, is started again. The reason why the waiting period comprises as many as eight frames is so that the system does not become more "nervous" than necessary. In prac-

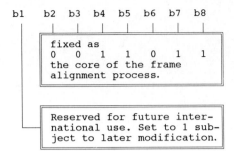

```
b1   b2  b3  b4  b5  b6  b7  b8
 |   |   |   |   |   |   |   |

    ┌──────────────────────────┐
    │ fixed as                 │
    │ 0   0   1   1   0   1   1 │
    │ the core of the frame    │
    │ alignment process.       │
    └──────────────────────────┘

    ┌──────────────────────────┐
    │ Reserved for future inter-│
    │ national use. Set to 1 sub-│
    │ ject to later modification.│
    └──────────────────────────┘
```

Figure 2.13 Structure of the frame alignment word 1.

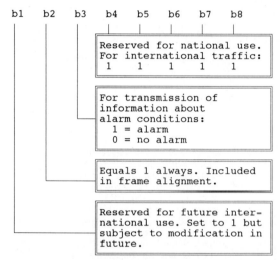

```
b1   b2   b3   b4   b5   b6   b7   b8
 |    |    |    |    |    |    |    |

        ┌──────────────────────────────┐
        │ Reserved for national use.   │
        │ For international traffic:    │
        │   1    1    1    1    1       │
        └──────────────────────────────┘

        ┌──────────────────────────────┐
        │ For transmission of          │
        │ information about            │
        │ alarm conditions:            │
        │   1 = alarm                  │
        │   0 = no alarm               │
        └──────────────────────────────┘

        ┌──────────────────────────────┐
        │ Equals 1 always. Included    │
        │ in frame alignment.          │
        └──────────────────────────────┘

        ┌──────────────────────────────┐
        │ Reserved for future inter-   │
        │ national use. Set to 1 but   │
        │ subject to modification in   │
        │ future.                      │
        └──────────────────────────────┘
```

Figure 2.14 Structure of the frame alignment word 2 for the no alarm condition.

tice it may happen that a bit in the frame alignment word becomes distorted in transmission, and it would be quite unnecessary to resynchronize the system every time this happens. By waiting for four consecutive incorrect alignment words before taking any action, we get a very stable synchronizing system with a high degree of insensitivity to disturbances. In fact, realignment will seldom be required in normal operation.

The strategy for the frame alignment alarm is shown in Fig. 2.16. This diagram illustrates the electronic decision-making process which results in the frame alignment alarm being indicated if four consecutive frame alignment errors (FAEs) exist. It also shows the decision process for the alarm state (FA) to be normalized only when three consecutive correct frame alignments (FAC) are detected. For example, if

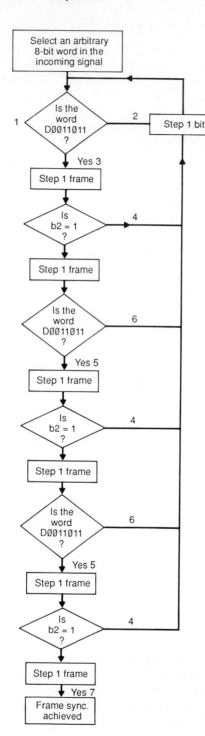

Figure 2.15 Flow chart for the frame alignment process. *Notes:* (1) D means the bit can be 0 or 1, i.e., don't care. (2) If the observed word is not D0011011, an attempt is made one bit length later. (3) When frame alignment word 1 is regarded as found, a check is made to ensure that the word is not an imitation. This check is done by studying frame alignment word 2. (4) If b2 = 0, then it was an imitation. The search then starts from the beginning. (5) If the word D0011011 and one frame later a word with b2 = 1 are found, a check is made to ensure these two events are not imitations. This means that one frame later we check whether frame alignment word 1 is where it should be. (6) If D0011011 is not found, the previous events were caused by imitations. The search starts again from the beginning. (7) If frame alignment word 1 has been found three times, and if frame alignment word 2 is found in between these two events, then frame alignment has been established.

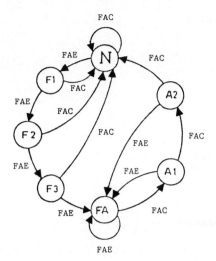

Figure 2.16 Frame alignment strategy. FAC = frame alignment correct; FAE = frame alignment error; N = normal state; FA = alarm state; F1, F2, F3 = prealarm state; A1, A2 = postalarm state.

the system is in the normal state N and one incorrectly received frame alignment word occurs, the system is in the prealarm state F1. If the next FAW is correct, the system is back to the normal state N, but if it is incorrect, the system is in the prealarm state F2. Another incorrect word takes it to F3. At this point a correct FAW would take the system back to the normal state N, but an incorrect word would be the fourth one, taking the system into the alarm state FA.

2.1.8 Multiframe alignment

Multiframe alignment may seem more complicated than frame alignment, since the multiframe alignment word occurs only once every 16 frames and should therefore be harder to find. However, the system first performs frame alignment and then multiframe alignment. The multiframe alignment logic receives information about the starting point of the frame from the frame alignment logic, via the so-called 64-kb/s interface. If the starting point of the frame is known, it is easy to establish the location of time slot 16 and then just wait for the frame that contains the multiframe alignment word (i.e., frame number 0). The structure of the multiframe alignment word is shown in Fig. 2.17.

Multiframe alignment (synchronization) process. The multiframe alignment process is quite simple. The system is multiframe aligned as soon as a multiframe alignment word is found (b1 b2 b3 b4 = 0 0 0 0). The reason for this is that the risk of imitation is practically nonexistent since the starting point of the frame is known and the combination 0000 never occurs in either the first half or the second half of time slot number 16 in any other frame except frame number 0. This leads to the requirement that the combination 0000 never be used for

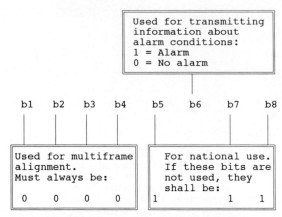

Figure 2.17 Structure of the multiframe alignment (synchronization) word.

signaling. *The multiframe alignment is considered as lost if two consecutive incorrect multiframe alignment words have occurred.* This means that we have an element of inertia, which makes it possible to avoid unnecessary realignment in the event of isolated bit errors.

2.1.9 Formats for 24-channel PCM systems

The 30-channel primary systems used in Europe and most of the developing world are not used in North America and Japan, where they have designed a system which has 24 channels for the primary PCM system, and the basic concept of A/D conversion is the same. The companding for each system achieves the same objective but uses slightly different companding curves. The frame structure is the major difference, as indicated in Fig. 2.18. The 24-channel frame is 125 μs long, just the same as the 30-channel frame. However, the 24-channel frame contains only 24 time slots each having 8 bits. The first 7 bits are always used for encoding, and the eighth bit is for encoding in all frames except the sixth frame, where it is used for signaling. At

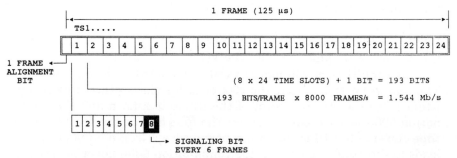

Figure 2.18 Twenty-four-channel frame structure.

the start of every frame, 1 bit is included for frame and multiframe alignment purposes. Each frame therefore contains $(24 \times 8) + 1 = 193$ bits. Since the sampling rate is 8 kHz, there are 8000 frames per second, giving $193 \times 8000 = 1.544$ Mb/s. The signaling bit rate is $(8000/6) \times 24 = 3200$ b/s.

The 24-channel system also has a 1.5-ms multiframe consisting of 12 frames. The frame and multiframe alignment words are transmitted sequentially, by transmitting 1 bit at the beginning of each frame. They are sent bit by bit on the odd and even frame cycles, and their transmission is completed only after each multiframe has been transmitted. The frame and multiframe alignment words are both 6-bit words (101010 and 001110, respectively).

A comparison of the construction of the frame and multiframe for the CCITT 30-channel and U.S./Japan 24-channel primary PCM systems is summarized as in Table 2.1.

2.2   Line Codes

The PCM signal is made up of bit sequences of 1s and 0s. Even when transmitting this information a meter or two from one multiplexer to the next higher-order multiplexer or to the microwave radio rack or to the optical fiber line terminal rack, the 1s and 0s can be detected incorrectly if they are not transmitted in the correct form. There are several types of pulse transmission, which are called *line codes*. The two main categories of line code are the *unipolar* and the *bipolar*. Also, the pulses in the line code are categorized as either *non return to zero (NRZ) or return to zero* (RZ). NRZ (as in Fig. 2.19) means that the 1 pulses return to 0 only at the end of one full clock period, whereas the RZ signal (Fig. 2.20) has 1 pulses that return to 0 during the clock period.

TABLE 2.1   Comparison of 24- and 30-Channel Systems

	24-CH system	30-CH system
Sampling frequency (kHz)	8	8
Duration of time slot (μs)	5.2	3.9
Bit width (μs)	0.65	0.49
Bit transfer rate (Mb/s)	1.544	2.048
Frame period (μs)	125	125
No. of bits per word	8	8
No. of frames per multiframe	12	16
Multiframe period (ms)	1.5	2
Frame alignment signal in	Odd frames	Even frames
Multiframe alignment signal in	Even frames	TS16 of frame
Frame alignment word	101010	0011011
Multiframe alignment word	001110	0000

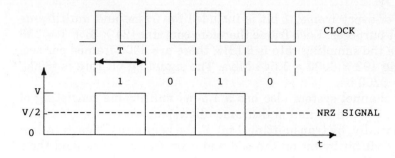

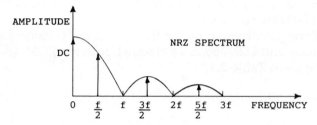

Figure 2.19   The NRZ code and its spectrum.

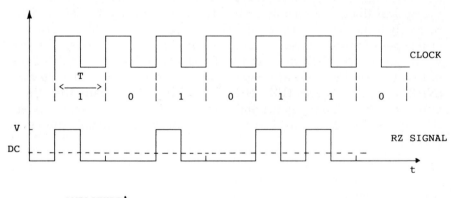

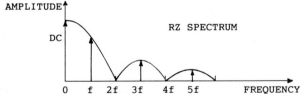

Figure 2.20   The RZ code and its spectrum.

It is usually necessary to convert from one code to another in the PCM transmission equipment. These conversions are done by appropriate circuits called *code converters*. The basic features of transmission coding must fulfill the following conditions:

1. There must be no significant direct current (dc) component, because an alternating current (ac) transformer or capacitive coupling is used in most wire-line transmission systems to eliminate ground loops.

2. The energy at low frequencies must be small; otherwise physically large components will be needed for the equalization circuitry.

3. A significant number of zero crossings must be available for timing recovery at the receiving end (i.e., for clock frequency recovery).

4. The coded signal must be capable of being uniquely decoded to produce the original binary information signal (i.e., with no ambiguity).

5. The code must provide low error multiplication.

6. Good coding efficiency is necessary to minimize bandwidth.

7. Error-detection or correction capability is necessary for high-quality performance.

The most common codes will be examined in the following subsections.

NRZ code (100 percent unipolar). From the circuitry point of view, the NRZ code is the most common form of digital signal because all logic circuits operate on the ON-OFF principle, and so the NRZ code is used inside the equipment (e.g., multiplexers, DMR, optical fiber line terminals, etc.). By observing the signal in Fig. 2.19, one can see that all 1 bits have a positive polarity, and so its spectrum has a dc component whose average value depends on the 1/0 ratio of the signal stream. If the signal consists of a 10101010 sequence, the dc component will be $V/2$. Depending on the signal, the dc component can assume any value from 0 (all 0) to V volts (all 1). The spectrum of the NRZ code is as shown in Fig. 2.19. The fundamental frequency occurs at half the clock frequency f, and only the odd harmonics are present. Furthermore, there is no signal amplitude at the clock frequency, so it is impossible to extract the clock frequency at the receiving end. Also, during transmission via wire-cable, if the noise peaks are summed up so that a 0 is simulated as 1, it is impossible to detect the error. These disadvantages make the use of the NRZ unsuitable for transmission via cable.

RZ code (50 percent unipolar). This is similar to the NRZ signal but with a pulse duration reduced to one-half. This code is also convenient for equipment circuitry since all logic circuits operate on the ON-OFF principle, and so the RZ is another code used inside the apparatus. One can see from Fig. 2.20 that it also produces a dc component in the spectrum. However, the fundamental frequency is now at the frequency of the clock signal with only the odd harmonics existing as in Fig. 2.19. This makes it possible to extract the clock frequency at the receiving end provided long sequences of 0s are not present. Detection of errors, as explained earlier, is not possible. This code is therefore rarely used at any stage of the system. However, a bipolar version of the RZ code is widely used, as discussed shortly.

Alternate mark inversion (AMI) code (bipolar code).   By observing the signal in Fig. 2.21, one can see that the 1s are alternately positive and negative, so there is no dc component in the spectrum. The apparent absence of the clock frequency in the spectrum can be overcome by simply rectifying the received signal to invert the negative 1s and making the resulting signal similar to the RZ signal. In this case, the received AMI signal has already passed through the line, and consequently the appearance of the dc component is of no interest, and the clock frequency can be extracted from the new spectrum of the signal. Another advantage of the AMI signal is that it can correct errors. If, during line transmission, noise peaks are summed up to simulate a 1 instead of 0, there would be a violation of the code which necessitates that the 1s are alternately positive and negative. Unfortunately, the recovery of the clock frequency is not easy with the AMI coding if a long sequence of 0s is present. The AMI code is recommended by the CCITT (G.703) for the 1.544-Mb/s interface. Its modified version, as discussed shortly, is used for 2.048 Mb/s.

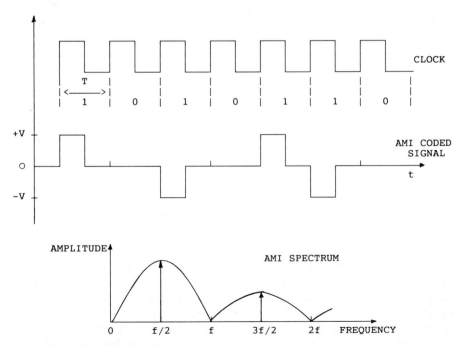

Figure 2.21   The AMI code and its spectrum.

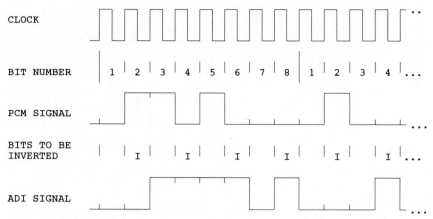

Figure 2.22 Code conversion by ADI.

Alternate digit inversion (ADI) code (unipolar 100 percent duty cycle). In this form of coding, every second, or alternate, digit or bit is inverted. The example in Fig. 2.22 shows how the 8-bit PCM words are coded in ADI code. The speech code from the A/D converter contains a relatively high number of 0s for speech levels close to zero, while the occurrence of 1s increases as the level is raised. The probability of the speech signal being near zero is great. First, when a given channel is not seized at all, the level is zero. Second, when a channel is seized, only one subscriber at a time speaks, which means that the other subscriber has zero level in his or her direction of transmission. Therefore, in general, the probability of a level near zero is great. ADI is very useful because even with large strings of 0s or 1s, the receiver extraction of the clock signal is possible when the transmit signal is ADI encoded.

HDB3 code. The purpose of the HDB3 code is to limit the number of zeros in a long sequence of zeros to three. This assures clock extraction in the regenerator of the receiver. This code is recommended by the CCITT (G.703) for the 2-, 8-, and 34-Mb/s systems. An example is shown in Fig. 2.23. Longer sequences of more than three zeros are avoided by the replacement of one or two zeros by pulses according to specified rules. These rules ensure that the receiver recognizes that these pulses are replacements for zeros and does not confuse them with code pulses. This is achieved by selecting the polarity of the pulse so as to violate the alternate mark inversion polarity of the AMI code. Also, the replacement pulses themselves must not introduce an appreciable dc component. The rules for HDB3 coding can be summarized as follows:

1. Invert every second *1* for as long as a maximum of three consecutive *zeros* appear.

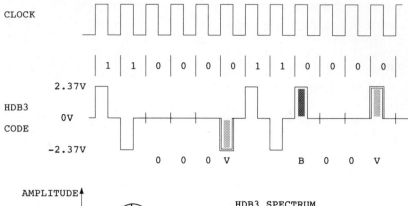

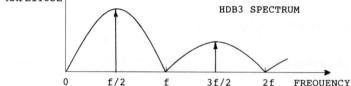

Figure 2.23   HDB3 code and its spectrum.

2. If the number of consecutive *zeros* exceeds three, set the *violation pulse* in the fourth position. The violation pulse purposely violates the AMI rule.

3. Every alternate violation pulse shall change polarity. If this rule cannot be applied, set a *1* according to the AMI rule in the position of the first *zero* in the sequence.

The HDB3 code can also be summarized as follows:

1. Apply the three rules step by step.
2. 000V and B00V generation:
   a. 000V is substituted if there is an odd number of 1s since the last violation pulse.
   b. B00V is substituted if there is an even number of 1s since the last violation pulse. The B-pulse follows the AMI rule for its polarity.
3. By observing the polarities of the preceding data pulse and the violation pulse.

Polarity of last V pulse	Polarity of preceding data	Substitution sequence
+	+	− 0 0 − B00V
−	−	+ 0 0 + B00V
+	−	0 0 0 − 000V
−	+	0 0 0 + 000V

Binary N zero substitution codes.   Binary N zero substitution line codes are used in North America. The B3ZS line code is specified by CCITT Recommendation G.703 for the 44.736-Mb/s (DS-3) interface. The binary three zero substitution (B3ZS) code is very similar to the HDB3 code. The only difference is that for B3ZS it is the third zero that is substituted, whereas for HDB3 it is the fourth.

Another code used in North America is the binary six zero substitution (B6ZS) used by the Bell system at the 6.312-Mb/s rate for *one symmetric pair*. Bipolar violations are introduced in the second and fifth bit positions of a six zero substitution, i.e., each block of six successive zeros is replaced by 0VB0VB:

Polarity of preceding data	Substitution sequence
−	0 − + 0 + −
+	0 + − 0 − +

The B8ZS line code is specified by CCITT Recommendation G.703 for *one coaxial pair* at 6.312 Mb/s. In this code, each block of eight successive zeros is replaced by 000VB0VB.

Coded mark inversion (CMI) encoding.   The CMI is a line code in which the 1 bits are represented alternately by a positive and a negative state, while the 0 bits are represented by a negative state in the first half of the bit interval and a positive state in the second half of the bit interval. The CMI code specifications are given in CCITT Recommendation G.703, as the interface code for signals transmitted at 139.264 Mb/s. An example of CMI encoding is shown in Fig. 2.24.

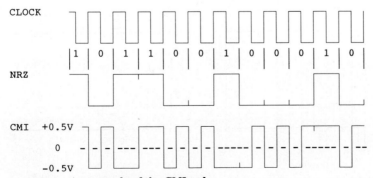

Figure 2.24   An example of the CMI code.

Comparison of codes. To summarize, Fig. 2.25 is a comparison of codes used for the transmission of digital signals at a clock rate of 34 Mb/s. This diagram provides an easy comparison of the way in which the codes described above differ from each other. It also gives a view of the energy at the clock frequency and at dc for each code.

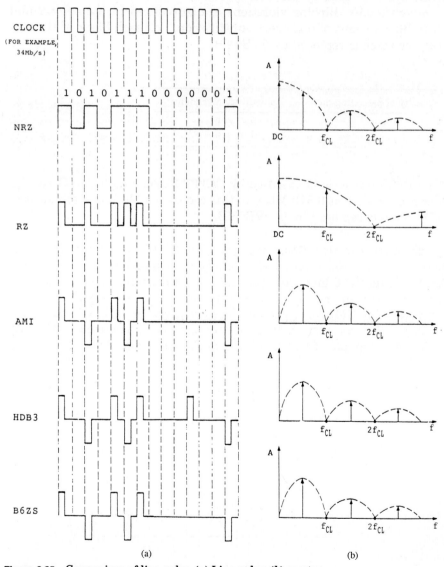

Figure 2.25 Comparison of line codes. (a) Line codes, (b) spectra.

2.3 Asynchronous Higher-Order Digital Multiplexing (CCITT)

2.3.1 The digital multiplexing hierarchy

The 30-channel PCM system is only the first, or primary, order of digital multiplexing as designated by the CCITT. If it is necessary to transmit more than 30 channels, the system is built up as in the hierarchy diagram of Fig. 2.26. Four primary systems are combined (multiplexed) to form an output having 120 channels. This is called the second order of multiplexing. Similarly, four 120-channel systems can be multiplexed to give an output of 480 channels (third order of multiplexing). Four 480-channel systems are multiplexed to give an output of 1920 channels (fourth order). Four 1920-channel systems are multiplexed to give an output of 7680 channels (fifth order). This is the highest level of multiplexing presently in service. Laboratory work is currently being done to produce higher orders of multiplexing, and these should be available from manufacturers very soon in the form of *synchronous* multiplexers. The approximate bit rate for each asynchronous multiplexer level is shown in the following table:

Level	No. of channels	Bit rate (Mb/s)
First	30	2.048
Second	120	8.448
Third	480	34.368
Fourth	1920	139.264
Fifth	7680	565.992

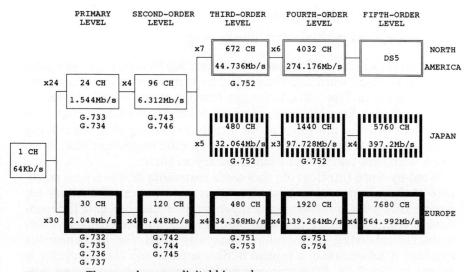

Figure 2.26 The asynchronous digital hierarchy.

2.3.2 Second-order multiplexing
(2 to 8 Mb/s)

The 8-Mb/s output of a second-order multiplexer is created by multiplexing *four* first-order (primary) multiplexer outputs. This is done by interleaving the bit streams of the four primary systems. Each individual bit stream is called a *tributary*. The main problem to overcome in this process is the organization of the four incoming tributaries. There are two categories of digital multiplexers:

1. *Synchronous* digital multiplexers

2. *Asynchronous* digital multiplexers

Synchronous digital multiplexers have tributaries with the same clock frequency, and they are all synchronized to a master clock. *Asynchronous digital multiplexers* have tributaries which have the same nominal frequency (that means there can be a small difference from one to another), but they are not synchronized to each other. The difference between the two types becomes apparent when one imagines the situation at a point where the four tributaries merge. For the synchronous case, the pulses in each tributary all rise and fall during the same time interval. For the asynchronous case the rise and fall times of the pulses in each tributary do not coincide with each other. The asynchronous designs are more widely used at present, but the synchronous systems will become prominent in the future. This will be explained in more detail in the following. The multiplexing of several tributaries can be achieved by:

1. Bit-by-bit multiplexing/interleaving

2. Word-by-word multiplexing/interleaving

Figure 2.27*a* and *b* illustrates the difference between the bit-by-bit and the word-by-word interleaving. The terms are self-explanatory. For example, in Fig. 2.27*a* there are four bit streams (tributaries) to be multiplexed. One bit is sequentially taken from each tributary so that the resulting multiplexed bit stream has every fifth bit coming from the same tributary. In Fig. 2.27*b* a specific number of bits, forming a word, are taken from each tributary in turn.

Word-by-word interleaving sets some restraints on the frame structure of the tributaries and requires a greater amount of memory capacity. Bit-by-bit interleaving is much simpler because it is independent of frame structure and also requires less memory capacity. Bit-by-bit interleaving is therefore generally used for asynchronous systems. A schematic of a typical 8-Mb/s asynchronous multiplexer is shown in Fig. 2.28. Four primary multiplexers, each having an out-

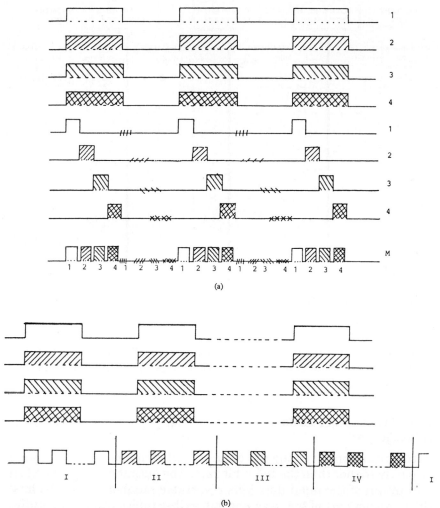

Figure 2.27 (*a*) Bit-by-bit multiplexing/interleaving; (*b*) word-by-word multiplexing/interleaving.

put of 2.048 Mb/s are bit-by-bit interleaved to form the next level in the hierarchy. Note that this output bit rate of 8.448 Mb/s is not exactly 4 times the tributary bit rates of 2.048 Mb/s. This is a result of the asynchronous nature of the system and will now be studied in detail. Every tributary has its own clock and is timed with what is called a *plesiochronous frequency,* that is, a nominal frequency about which the shifts around it lie within prefixed limits. For example, the primary multiplexer output is 2.048 Mb/s ± 50 ppm. To account for the small variations of the tributary frequencies about the nominal

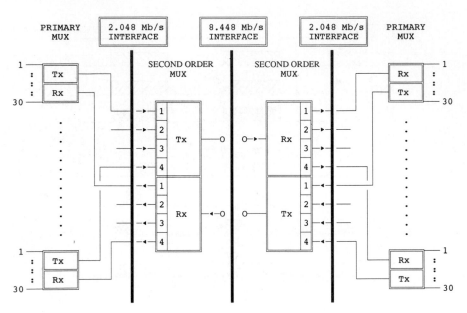

Figure 2.28 The 8-Mb/s multiplexer schematic.

value when multiplexing four tributaries to the next hierarchical level, a process known as *positive stuffing* (also known as *positive justification*) is used.

2.3.3 Positive pulse stuffing (or justification)

Pulse stuffing involves the use of an output channel whose bit rate is purposely higher than the input bit rate. The output channel therefore contains all of the input data plus a variable number of "stuffed bits" which are not part of the incoming subscriber information. The stuffed bits are inserted, at specific locations, to pad the input bit stream to the higher output bit rate. The stuffed bits must be identified at the receiving end so that "destuffing" can be done to recover the original bit stream. Pulse stuffing is used for higher-order multiplexing when each of the incoming lower-level tributary signals is unsynchronized and therefore bears no prefixed phase relationship to any of the others. The situation is shown by the simplified diagrams of the two-channel multiplexer in Fig. 2.29*a* and *b*. The input bit rates, in Fig. 2.29*a,* are *exactly* the same and the pulses arrive synchronized. The output subsequently has a perfect word-interleaved bit sequence. However, in Fig. 2.29*b*, the input bit rates are *not* identical, so the pulses do not arrive in a synchronized manner. The difference in bit

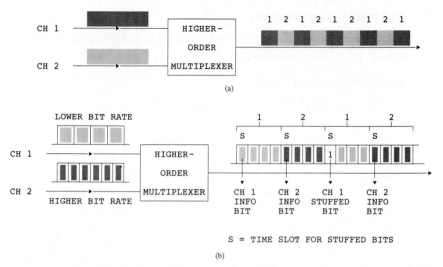

Figure 2.29 Simplified pulse stuffing example. (*a*) The input bit rates are equal. (*b*) The bit rates are not equal.

rates is exaggerated for the purpose of the example. This is the situation for an asynchronous multiplex system, where the input bit rates are not exactly the same, and there is a timing offset due to the unsynchronized arrival of the bit streams. The output of the system in Fig. 2.29*b* requires some additional (stuffed) pulses in order to make up the difference in input bit rates. Although bit-interleaving is the method used in asynchronous multiplexing, for the sake of comparison with the synchronous system, Fig. 2.29*b* shows a word-interleaved situation. Each word in the output signal in Fig. 2.29*b* contains a fixed number of time slots. Each word contains a stuffing control bit (C) and a stuffing bit (S). When a control bit is a 0, the (S) bit contains real data information. When a (C) bit is a 1, the respective (S) bit is a stuffed bit (i.e., a 1).

The higher-order asynchronously multiplexed outputs are composed of frames. The positive stuffing method involves the canceling of a clock pulse assigned to a particular tributary in some of the frames in order to coordinate the timing of the unsynchronized tributaries into a multiplexed output. Random spaces are therefore created in the frame, as are periodic spaces. In the periodic spaces FAW bits, service bits, and stuffing control bits are inserted. The tributary information bits are inserted in the random spaces in the absence of stuffing, or a logic 1 is used when stuffing takes place. Remember, the stuffing pulses carry no subscriber information.

2.3.4   The 8-Mb/s frame structure

The structure of the frame for the second-order multiplexer is shown in Fig. 2.30. The frame contains:

1. The frame alignment word
2. The stuffing control bits (stuffing message)
3. The stuffing bits
4. The tributary bit streams

It is important to notice that the output frame structure of a higher-order multiplexer is unrelated to the frame structure of the lower levels making up the input tributaries. As far as the higher-order multiplexer is concerned, each input tributary is merely a bit stream, within which frame alignment bits, multiframe alignment bits, and signaling bits are all transmitted along with the information bits

SUBFRAME	SIGNAL	NUMBER OF BITS	BIT NUMBERING
1	-Alignment word 1111010000	10	1 to 10
	-Service bits	2	11 to 12
	-Tributary bits	200	13 to 212
2	-Stuffing message	4	213 to 216
	-Tributary bits	208	217 to 424
3	-Stuffing message	4	425 to 428
	-Tributary bits	208	429 to 636
4	-Stuffing message	4	637 to 640
	-Stuffing bits	4	641 to 644
	-Tributary bits	204	645 to 848

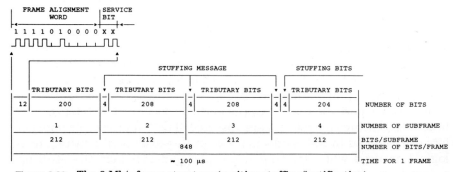

Figure 2.30   The 8-Mb/s frame structure (positive stuffing/justification).

(voice channels, data, or video). Each frame contains 848 bits and is divided into four subframes, each containing 212 bits. The first 12 bits in every frame contain 10 frame alignment bits (1111010000) and 2 service bits. The service bits relay information concerning alarms, synchronization errors, etc. A 3-bit stuffing control word for each tributary occupies a total of 12 bits (4 bits after each of the first three subframes). Finally, the stuffing bit for each tributary is inserted (if required) before the start of the fourth subframe. Each tributary signal is written into an *elastic memory* by the 2.048-Mb/s clock extracted from the incoming signal. An elastic memory expands the bit width to allow reading to take place during a greater time interval (Fig. 2.31). The data is read out by a multiplexing frequency clock at 2.112 Mb/s. Writing is inhibited at specific time slots for the inclusion of frame alignment bits, service bits, and stuffing control bits. The stuffing control message, requesting a stuffing bit, is generated when the memory reading and writing phases reach a predetermined threshold. The elastic memory is realized simply by a flip-flop circuit which maintains the 1 state until the following reading clock pulse arrives.

When a stuffing pulse is to be inserted, the stuffing control message is formed for the particular tributary concerned. This consists of setting the stuffing control message to 111 (i.e., a 1 in each of the three periodic spaces following each tributary information). The three sets of four periodic spaces allow the formation of the 3-bit stuffing control message for each of the four tributaries. When stuffing does not take place, the stuffing control message is 000. This is illustrated in Fig. 2.32 where the first C subscript denotes the *tributary* number and the second subscript denotes one of the 3 bits in the stuffing message word for that tributary. At the receiving end, if a majority of 1s (two out of three) is detected in the stuffing control message, the stuffing pulse is canceled by nulling the receive elastic memory writing clock for that pulse period (i.e., the stuffing bit is not part of the required received information). Conversely, if a majority of 0s (two out of three) is re-

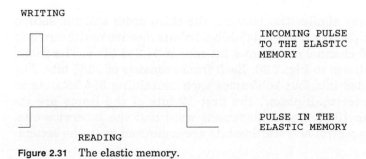

WRITING

INCOMING PULSE
TO THE ELASTIC
MEMORY

PULSE IN THE
ELASTIC MEMORY

READING

Figure 2.31   The elastic memory.

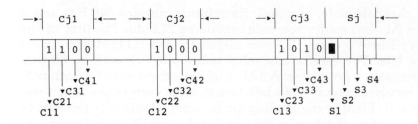

```
C11, C12, C13 = 1,1,1 ----▸ S1 IS A STUFFING BIT

C21, C22, C23 = 1,0,0 ----▸ S2 IS AN INFORMATION BIT

C31, C32, C33 = 0,0,1 ----▸ S3 IS AN INFORMATION BIT

C41, C42, C43 = 0,0,0 ----▸ S4 IS AN INFORMATION BIT
```

Figure 2.32   The formation of the stuffing message word.

ceived for the stuffing control message, the stuffing pulse time slot is not canceled, thus allowing a data bit (1 or 0) to be received.

The elastic memory of the receive writing clock has the same characteristics as that of the transmit reading clock. That is, it has a frequency, on average, the same as that of the tributary, but it presents periodic spaces for the frame structure and random spaces for the stuffing process. The elastic memory is read by a 2.048-Mb/s clock phase locked to the writing clock by a phase locked loop (PLL) circuit. The PLL circuit is able to reduce:

1. Jitter caused by the frame structure

2. High-frequency jitter components (waiting time) caused by stuffing

3. Tributary signal jitter

4. Jitter introduced by the 8.448-Mb/s link

2.3.5   Third-order multiplexing (8 to 34 Mb/s)

There are many similarities between the third order and the second order of multiplexing. Four 8.448-Mb/s tributaries are multiplexed to produce a 480-channel output at a bit rate of 34.368 Mb/s. The frame structure is shown in Fig. 2.33. Each frame consists of 1536 bits. The frame is divided into four subframes each containing 384 bits. As in the second-order multiplexer, the first 12 bits of the frame are reserved for the 10-bit frame alignment word and the 2 service bits. Also, stuffing control and stuffing bits are organized as in the second-order system.

SUBFRAME	SIGNAL	NUMBER OF BITS	BIT NUMBERING
1	-Alignment word 1111010000	10	1 to 10
	-Service bits	2	11 to 12
	-Tributary bits	372	13 to 384
2	-Stuffing message	4	385 to 388
	-Tributary bits	380	389 to 768
3	-Stuffing message	4	769 to 772
	-Tributary bits	380	773 to 1152
4	-Stuffing message	4	1153 to 1156
	-Stuffing bits	4	1157 to 1160
	-Tributary bits	376	1161 to 1536

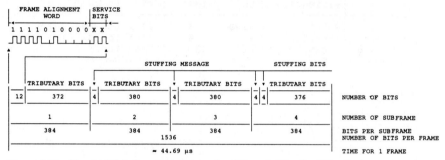

Figure 2.33   The 34-Mb/s frame structure (positive stuffing/justification).

2.3.6 Fourth-order multiplexing (34 to 140 Mb/s)

Again, the fourth-order multiplexer has a similar structure to the second- and third-order multiplexers, but it also has some important differences. In this case four third-order outputs at 34 Mb/s are multiplexed to give a 1920-channels output at a bit rate of 139.264 Mb/s. The frame structure is shown in Fig. 2.34. The 2928-bit frame is divided into *six* subframes each containing 488 bits. Note that this is different from the second- and third-order multiplexers which each had only *four* subframes. The FAW is different from the second- and third-order systems (i.e., 111110100000), occupying the first 12 bits in this case. Also the service bits occupy 4 bits here instead of the 2 bits in the other systems. The stuffing control message has 5 bits here instead of the 3 bits for the other systems. The actual stuffing bits are the same as the other systems (i.e., 1 bit for each tributary per frame). Finally, the line code used for the transmission of 140 Mb/s has

SUBFRAME	SIGNAL	NUMBER OF BITS	BIT NUMBERING
1	-Alignment word 111110100000	12	1 to 12
	-Service bits (1 for alarm)	4	13 to 16
	-Tributary bits	472	17 to 488
2	-Stuffing message	4	489 to 492
	-Tributary bits	484	493 to 976
3	-Stuffing message	4	977 to 980
	-Tributary bits	484	981 to 1464
4	-Stuffing message	4	1465 to 1468
	Tributary bits	484	1469 to 1952
5	-Stuffing message	4	1953 to 1956
	-Tributary bits	484	1957 to 2440
6	-Stuffing message	4	2441 to 2444
	-Stuffing bits	4	2445 to 2448
	-Tributary bits	480	2449 to 2928

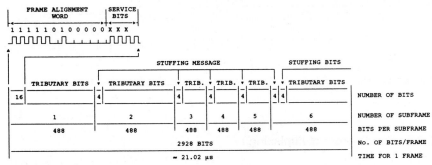

Figure 2.34 The 140-Mb/s frame structure (positive stuffing/justification).

been designated by the CCITT to be the CMI code instead of the HDB3 code used for the first-, second-, and third-order transmission codes.

2.3.7 Fifth-order multiplexing (140 to 565 Mb/s)

The fifth order of multiplexing is a natural progression from the previous levels. Four 139.264-Mb/s tributaries are multiplexed to give a 7680-channel output at 564.992 Mb/s. The frame structure is shown in Fig. 2.35. The 2688-bit frame is divided into *seven* subframes

SUBFRAME	SIGNAL	NUMBER OF BITS	BIT NUMBERING
1	-Alignment word -Tributary bits	12 372	1 to 12 13 to 384
2	-Stuffing message -Tributary bits	4 380	385 to 388 389 to 768
3	-Stuffing message -Tributary bits	4 380	769 to 772 773 to 1152
4	-Stuffing message Tributary bits	4 380	1153 to 1156 1157 to 1536
5	-Stuffing message -Tributary bits	4 380	1537 to 1540 1541 to 1920
6	-Stuffing message -Tributary bits	4 380	1921 to 1924 1925 to 2304
7	-Service bits -Stuffing message -Tributary bits	4 4 376	2305 2308 2309 2312 2313 2688

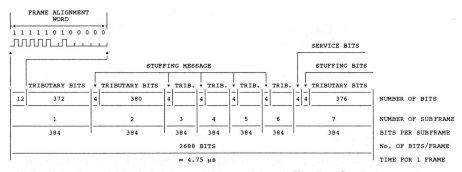

Figure 2.35 The 565-Mb/s frame structure (positive stuffing/justification).

each containing 384 bits. The frame alignment word occupies the first 12 bits. The service bits occupy 4 bits in the seventh subframe. The stuffing control message has 5 bits, composed of 1 bit at the beginning of subframes 2, 3, 4, 5, and 6, as in the case of the fourth level of multiplexing. The actual stuffing bits are the same as the other systems, that is, 1 bit for each tributary per frame. In this case the actual stuffing bits are in the seventh subframe. Finally, the line code used for the transmission of 565 Mb/s has been designated by the CCITT to be the CMI code.

2.4 Asynchronous Higher-Order Digital Multiplexing (North America)

2.4.1 The digital multiplexing hierarchy

The digital hierarchy used in North America is shown in Fig. 2.26 and is summarized as follows:

Level	No. of channels	Bit rate (Mb/s)	Line code
DS-1	24	1.544	Bipolar
DS-1C	48	3.152	Bipolar
DS-2	96	6.312	Bipolar (B6ZS)
DS-3	672	44.736	Bipolar (B3ZS)
DS-4	4032	274.176	Polar bipolar

The first level of multiplexing was described in Sec. 2.1.9. The line code used for transmission of this 1.544-Mb/s bit stream is the AMI code with 50 percent duty cycle (i.e., the pulse width is one-half of one bit interval). Notice that there is no simple relationship between the bit rates for this hierarchy and the CCITT hierarchy. This has been the source of considerable problems for interfacing international traffic between different regions of the world. The problem will be resolved satisfactorily only when there is a global adoption of the *synchronous* multiplexing hierarchy as described in Sec. 2.5. This new hierarchy has been created with a view to incorporating the existing asynchronous bit rates from both North American and CCITT hierarchies.

2.4.2 DS1-C multiplexing
(1.544 to 3.152 Mb/s)

The 3.152-Mb/s DS1-C output is obtained by multiplexing two DS1 input bit streams. As shown in Fig. 2.36, each DS1-C frame has 1272 bits and is composed of four subframes each having 318 bits. Each subframe has six microframes each containing 53 bits. Every microframe starts with an "overhead" (housekeeping) bit (e.g., frame alignment) followed by bits taken alternately from the two tributaries. Each frame has a 4-bit multiframe alignment word 011X, where X is an alarm service bit (X = 1 means no alarm). Each subframe contains the frame alignment signal 01, where the 0 is always the first bit in the third microframe and the 1 is always the first bit of the sixth microframe. Each subframe also has a 3-bit stuffing message (indicator) word, formed from the first bit of the second, fourth, and fifth microframes. The word 000 indicates no stuffing is required, whereas 111 denotes stuffing is required. The stuffing message indicator for

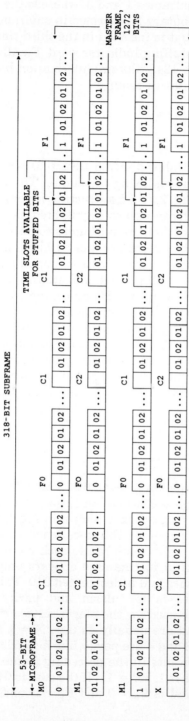

Figure 2.36 The 3.152-Mb/s frame format. *Notes:* (1) The frame alignment signal is F0 = 0 and F1 = 1. (2) M0, M1, M1, X is the multiframe alignment signal and is 011X, where X is an alarm service digit. The normal (no alarm) state is X = 1. (3) C1, C1, C1 and C2, C2, C2 are the stuffing (indicators) message words for DS-1 input channels 1 and 2, respectively, where 000 indicates no stuffing and 111 indicates stuffing is required.

59

tributary 1 is formed in subframes 1 and 3, whereas the stuffing message indicator word for tributary 2 is formed in subframes 2 and 4. If the stuffing bit is required, it is inserted in the third time slot following the completion of the stuffing indicator word. Finally, of the 1272 bits in each frame, 1244 to 1248 bits are information bits, depending upon how many stuffing bits are required.

2.4.3 DS2 multiplexing
(1.544 to 6.312 Mb/s)

Four DS1 signals are multiplexed into a DS2, 6.312-Mb/s bit stream using the asynchronous multiplexing technique. The composition of the frame is displayed in Fig. 2.37. The structure is very similar to the DS1-C frame, and the differences are highlighted as follows. Every frame has 1176 bits. Each of the four subframes contains 294 bits, and each microframe within the subframes has 49 bits. All microframes contain bits taken sequentially from the four input tributaries. The frame alignment, multiframe alignment, and stuffing indicator words are the same as for the DS1-C frame. However, the stuffing indicator words are distributed such that each subframe has a stuffing indicator word specific to one of the four tributaries (e.g., subframe 1 has the stuffing indicator word for tributary 1, etc.). The stuffing bits, if required, are inserted in the first time slot designated for each tributary following the frame alignment bit in the sixth microframe of each subframe. Note that the B6ZS line code is used at the output from this level of multiplexer.

2.4.4 DS3 multiplexing
(6.312 to 44.736 Mb/s)

For the next level of multiplexing, seven 6.312-Mb/s input tributaries are asynchronously multiplexed to give an output of 44.736 Mb/s having a B3ZS line code. The frame format shown in Fig. 2.38 indicates a 4760-bit frame containing seven subframes, each having 680 bits. Each subframe comprises eight microframes each having 85 bits. The multiframe alignment signal in this case appears only in the fifth, sixth, and seventh subframes and is designated 010. This time slot for the other four subframes is for alarm and parity information. Subframes 1 and 2 may be used for the alarm service channel, where XX must be identical, and are usually 1s (i.e., X = 1 for no alarm). The two Ps in subframes 3 and 4 are parity bits formed by considering parity over all *information* bits in the preceding frame. If the digital sum of all information bits is 1, PP = 11. If the sum is 0 then PP = 00. The frame alignment signal is $F_0 = 0$ and $F_1 = 1$, and each pair ap-

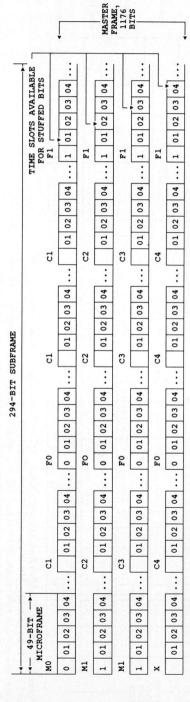

Figure 2.37 The 6.312-Mb/s frame format. *Notes:* (1) The frame alignment signal is F0 = 0 and F1 = 1. (2) M0, M1, M1, X is the multiframe alignment signal and is 011X, where X is an alarm service digit. The normal (no alarm) state is X = 1. (3) C1, C1, C1 and C2, C2, C2, and C3, C3, C3 and C4, C4, C4 are the stuffing (indicators) message words for DS-1 input channels 1, 2, 3, and 4, respectively, where 000 indicates no stuffing and 111 indicates stuffing is required.

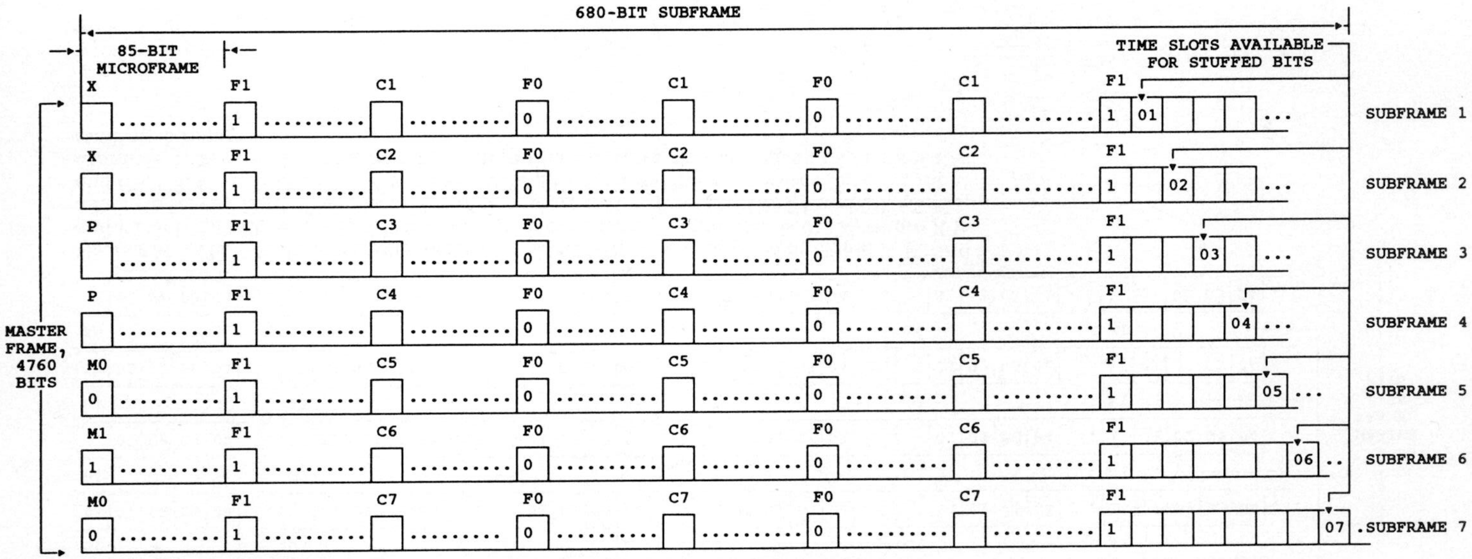

Figure 2.38 The 44.736-Mb/s frame format. *Notes:* (1) The frame alignment signal is F0 = 0 and F1 = 1. (2) M0, M1, M0 is the multiframe alignment signal and appears in the fifth, sixth, and seventh subframes. M0 = 0 and M1 = 1. (3) PP is parity information taken over all message time slots in the preceding M frame. PP = 1 if the digital sum of all message bits is 1 and PP = 0 if the sum is 0. These two parity bits are in the third and fourth M subframes. (4) XX is for alarm indication. In any one M frame the two bits must be identical. Presently XX = 1. (5) C1, C1, C1 to C7, C7, C7 are the stuffing (indicators) message words for DS-2 input channels 1 to 7, respectively, where 000 indicates no stuffing and 111 indicates stuffing is required.

pears twice in each subframe. The stuffing indicator words are the same as previously discussed except that here they appear in microframes 3, 5, and 7. The stuffing bits, if required, are placed in the final microframe of each subframe in the first time slots for their respective tributaries. The B3ZS line code is used at the output from this level of multiplexer.

2.4.5 DS4 multiplexing
(44.736 to 274.176 Mb/s)

For the next level of multiplexing, six 44.736-Mb/s input tributaries are asynchronously multiplexed to give an output of 274.176 Mb/s having a two-level binary line code. The frame format shown in Fig. 2.39 indicates a 4704-bit frame containing 24 subframes, each having 196 bits. Each subframe comprises two microframes each having 96 bits. In this scheme, the first two bits of each microframe are devoted to overhead (housekeeping). The multiframe alignment (M), alarm (X) and stuffing indicator (C) bits occur in complementary pairs (i.e., the second bit is the inverse of the first. The multiframe alignment bits appear in the first three subframes and follow the sequence 101. The alarm bits occur in the second three subframes. The stuffing indicator words occur over the remaining six sets of three subframes. As previously, the word 111 indicates stuffing is required, whereas 000 indicates stuffing is not required. The time slot for stuffing is the eighth slot for a given tributary after the last bit of the stuffing indicator word has appeared for that particular tributary. The P bits appear as identical pairs of bits, where P1 indicates even parity over the 192 previous *odd* numbered information bits and P2 indicates even parity over the 192 previous *even* numbered information bits.

2.5 Synchronous Digital Multiplexing

Asynchronous multiplexers have the benefit of operating independently without a master clock to control them. Each 2-Mb/s multiplexer has its own independent clock. This so-called plesiochronous transmission has small differences in frequency from one multiplexer to another, so when each provides a bit stream for the next hierarchical level, bit stuffing (justification) is necessary to adjust for these frequency differences. Despite the attractive aspects of asynchronous multiplexing, there is one major drawback. If, for example, a 140-Mb/s system is operating between two major cities, it is not possible to identify and gain access to individual channels at towns en route. In other words, drop and insert capability requires a complete demultiplexing procedure. The synchronous multiplexing technique does allow

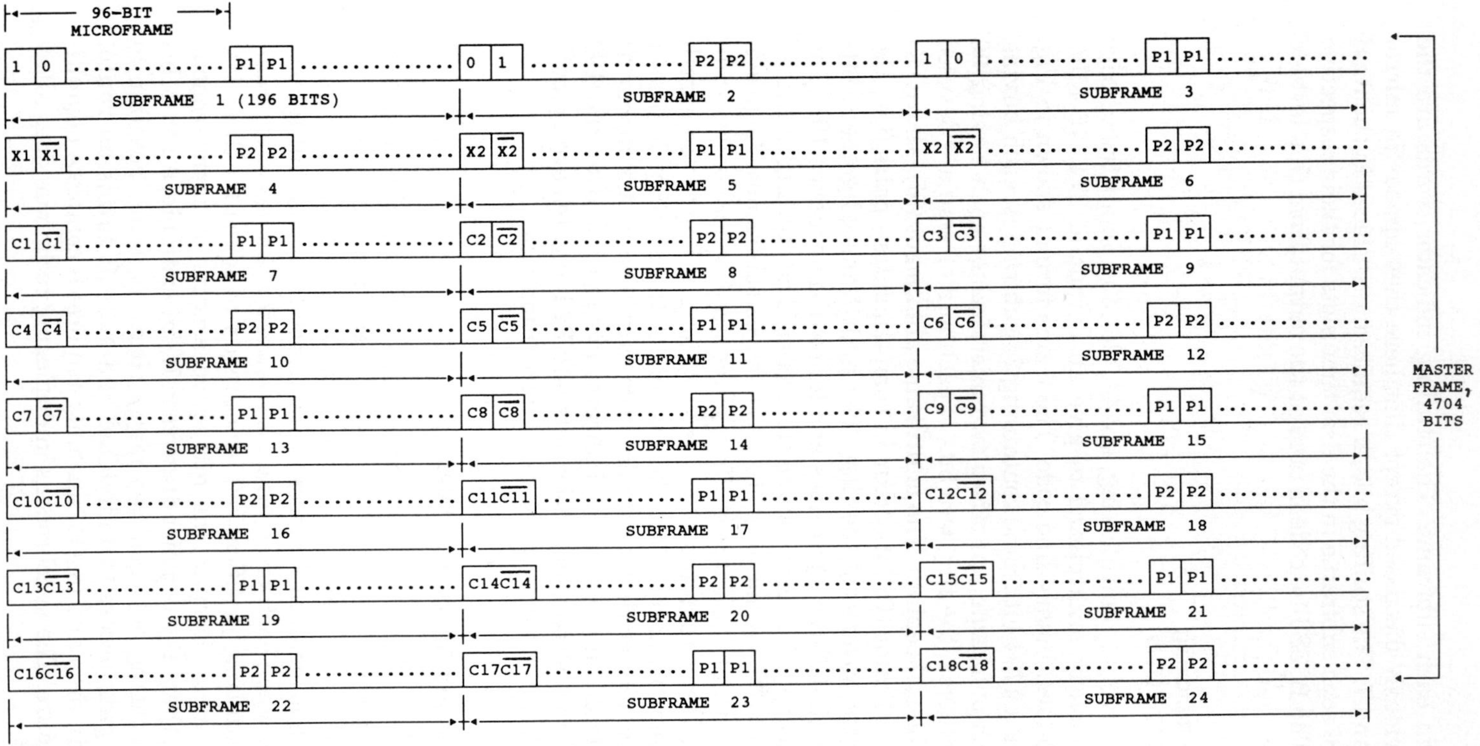

Figure 2.39 The 274.176-Mb/s frame format. *Note 1*: M, X, and C bits appear as complementary pairs. (a) Appears in the first three subframes and must be 10, 01, 10. (b) X appears in the fourth, fifth, and sixth subframes. (c) C1, C1, C1 to C6, C6, C6 are the stuffing (indicators) message words for DS-3 input channels 1 to 6, respectively, where 000 indicates no stuffing and 111 indicates stuffing is required. The time slot for stuffing is the eighth message bit position following the last C1 bit. *Note 2*: The P bits appear as identical pairs and are for even parity over the 192 previous odd-numbered message bits. *Note 3*: All other bits are message (information) bits.

this drop and insert facility. The synchronous multiplexing scheme also allows multiplexing of tributaries that have different bit rates.

In 1988, the CCITT reached an agreement on a worldwide standard for the *synchronous digital hierarchy* (SDH) in the form of Recommendations G707, 708, and 709. In addition to being a technical milestone, this agreement also unifies the bit rates so that this new synchronous system does not have the existing interface problems between North America and Japan and the rest of the world. The resulting Recommendations were intended for application to optical fiber transmission systems and they were originally called the synchronous optical network (SONET) standard. Although SDH now supersedes the SONET description, they both refer to the same subject matter and are used interchangeably in the literature. During the development stages of the new SDH, it was essential to establish a system which allowed both CCITT and North American hierarchical bit rates to be processed simultaneously. This led to extensive discussions on how the synchronous system should be constructed. Each negotiating team proposed solutions which favored their own existing systems, so eventually both sides had to make a compromise to ensure a successful outcome.

2.5.1 Network node interface

For international communications and the increasing demand for broadband services, it has become more and more important to specify a universal *network node interface* (NNI). The NNI is a term that is acquiring increasing importance with respect to digital telecommunication systems. Within a communication system a transport network has two elementary functions: first, a transmission facility, and second, a network node. The transmission facility has various media such as optical fiber, microwave radio, and satellite. The network node performs terminating, crossconnecting, multiplexing, and switching functions. There can be various types of node such as a 64-kb/s-based node or broadband nodes. The NNI is the point at which the transmission facility and the network node meet. The objectives or requirements for a NNI are that the network should have:

- Worldwide universal acceptance
- Unique interface for transmission, multiplexing, crossconnecting, and switching of various signals
- Improved operation and maintenance capabilities
- Easy interworking with existing interfaces
- The ability to easily provide future services and technologies
- Application to all transmission media

A typical example of NNIs within a network is presented in Fig. 2.40. This diagram shows typical communication between two network users. In this case, there are two switching exchanges. The input to or the output from an exchange is via a path termination (PT) or line or exchange termination. The equipment between the path terminations is referred to as the *path*. Each transmission line, which could be an optical fiber or PCM cable, is terminated by line termination (LT) equipment. At the end of a *section* of transmission line there may be add/drop multiplexers (ADMs) to enable local traffic to enter or leave the network, or there could be a digital crossconnect (DCC) between line terminals. In this example, it is the points between the LT/ADM, ADM/PT, PT/LT, and LT/DCC that are the NNIs. The implementation of the SDH is a very significant milestone in moving toward the achievement of a worldwide, universal NNI.

2.5.2 Synchronous transport signal frame

From the North American point of view, the basic building block and first level of the SDH is called the *synchronous transport signal—level 1* (STS-1) if electrical or *optical carrier—level 1* (OC-1) if optical. The STS-1 has a 51.84-Mb/s transmission rate and is synchronized to the network clock. The STS-1 frame structure has 90 columns and 9 rows (Fig. 2.41). Each column has an 8-bit byte, so now there is a departure from the asynchronous bit-by-bit multiplexing in favor of word-by-word multiplexing (byte interleaving). The 8-bit bytes are

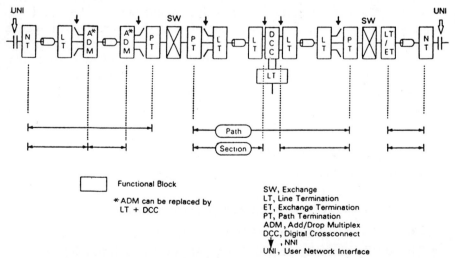

Figure 2.40 Network node interface (NNI) locations. (*Reproduced with permission from Ref. 3, © 1990 IEEE.*)

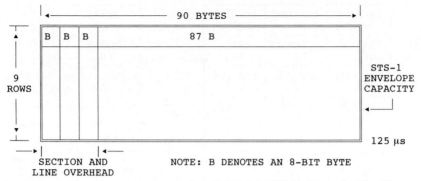

Figure 2.41 Synchronous transport signal—level 1 (STS-1). (*Reproduced with permission from Ref. 5, © 1989 IEEE.*)

transmitted row by row from left to right, and one complete frame is transmitted every 125 μs. The first three columns of the frame contain *section* and *line* overhead (housekeeping) bytes. The remaining 87 columns and 9 rows are used to carry the STS-1 *synchronous payload envelope* (SPE).

The SPE also includes 9 bytes of *path* overhead (see below). The STS-1 can carry a DS3 channel (44.736 Mb/s) or a variety of lower-order signals at DS1 (1.544 Mb/s), DS1C (3.152 Mb/s), and DS2 (6.312 Mb/s) rates.

Overhead. The overhead columns are divided into *section, line,* and *path* layers as indicated in Fig. 2.42. The overhead is structured in this manner so that each network element (terminal, repeater, ADM) needs only to access the necessary information. This enables cost savings to be made by designing each network element so that it accesses only the necessary overhead instead of all of it. Figure 2.43 shows the

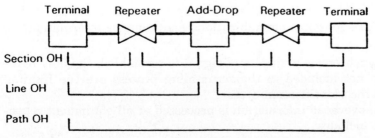

Figure 2.42 Allocation of section, line, and path overhead. OH = overhead. (*Reproduced with permission from Ref. 10, © 1990 IEEE.*)

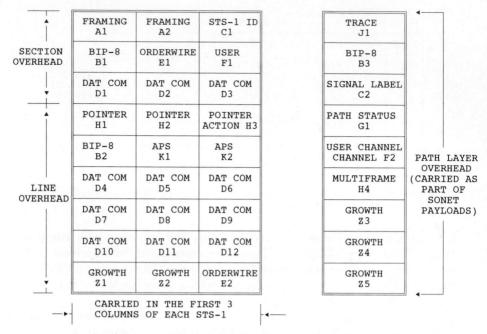

Figure 2.43 Synchronous multiplexer overhead bytes. (*Reproduced with permission from Ref. 5, © 1989 IEEE.*)

position of the overhead bytes in the STS-1 frame. The section overhead has:

- Two bytes which show the start of each frame
- A frame identification byte (STS-1 ID)
- An 8-bit bit-interleaved parity (BIP-8) check for monitoring section errors
- An order-wire channel for maintenance purposes
- A channel for operator applications
- Three bytes for data communications maintenance information

When a synchronous multiplexed signal is scrambled, the only bytes which are not included in the scrambling process are the framing bytes and the identification bytes.

The line overhead information is processed at all equipment regenerators. It includes:

- Pointer bytes (see below)
- A BIP-8 line error monitoring byte
- A 2-byte automatic protection switching (APS) message channel
- A 9-byte line data communications channel
- A line order-wire channel byte
- Bytes reserved for future growth

The path overhead bytes are processed at the terminal equipment. The path overhead includes:

- A BIP-8 for end-to-end payload error monitoring
- A byte to identify the type of payload being carried
- A byte to carry maintenance information signals
- A multiframe alignment byte

This very extensive array of overhead bits allows greatly enhanced maintenance, control, performance, and administrative capability of a network incorporating SDH equipment.

Synchronous digital hierarchy. Higher-bit-rate synchronously multiplexed signals are obtained by byte-interleaving N frame aligned STS-1s into an STS-N as in Fig. 2.44. In this manner the CCITT stan-

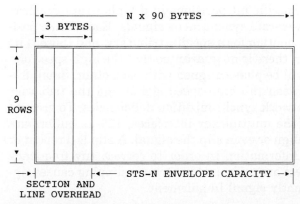

Figure 2.44 The STS-N frame. (*Reproduced with permission from Ref. 5, © 1989 IEEE.*)

dard level (155.52 Mb/s) or any other of the following synchronous hierarchical levels is constructed:

Level			Line rate (Mb/s)
OC-1	STS-1		51.84
OC-3	STS-3	STM-1	155.52
OC-9	STS-9	STM-3	466.56
OC-12	STS-12	STM-4	622.08
OC-18	STS-18	STM-6	933.12
OC-24	STS-24	STM-8	1244.16
OC-36	STS-36	STM-12	1866.24
OC-48	STS-48	STM-16	2488.32

Whereas all of the section and line overheads in the first STS-1 of an STS-N are used, many of the overheads in the remaining STS-1s are unused. Only the section overhead, framing, STS-1 ID, and BIP-8 channels and the line overhead pointer and BIP-8 channels are used in *all* STS-1s in an STS-N. The STS-N is then scrambled and converted to an optical carrier, whose line rate is one of the above hierarchical levels. Note: OC-N is exactly N times OC-1.

Fixed location mapping. The bit stuffing technique for asynchronously multiplexing several tributaries has been discussed previously. The *fixed location mapping technique* has been used in the early synchronous multiplexing equipment.

This technique uses specific bit positions in a higher-rate synchronous signal to carry lower-rate synchronous signals. Each frame position is dedicated to information for a specific tributary, and there is no pulse stuffing. However, there is no guarantee that the high-speed signal and the tributary will be phase aligned with each other. Small frequency differences between the high-speed signal and the tributary can occur because of network synchronization deficiencies. To account for such variations, at the multiplexer interfaces, 125-μs buffers are incorporated to phase align or even slip the signal. A slip is a repeat or deletion of a frame of information in order to correct any frequency differences. These buffers are undesirable because slipping causes signal delay and subsequently signal impairment.

Payload pointer. The SONET standard contains a very innovative technique known as the *payload pointer*. It is used to frame align STS-N or STM-N signals and also for multiplexing synchronization of plesiochronous (asynchronous) signals. The payload pointer allows easy access to synchronous payloads without requiring 125-μs buffers. The payload pointer is a number carried in each STS-1 line overhead which indicates the location of the first byte of the STS-1 SPE payload within the STS-1 frame. These are bytes H1 and H2 in Fig. 2.43. The payload is therefore not locked to the STS-1 frame structure, as in the case of fixed location mapping, but floats within the STS-1 frame. The positions of the STS-1 section and line overhead bytes define the STS-1 frame structure. Figure 2.45 shows how the 9-row by 87-column SPE payload fits into two 125-μs STS-1 frames. If the STS-1 payload has any small variations in frequency, the pointer value increases or decreases to account for the variations. Figures 2.46 and 2.47 illustrate the mechanism. Consider the payload bit rate to be high compared to

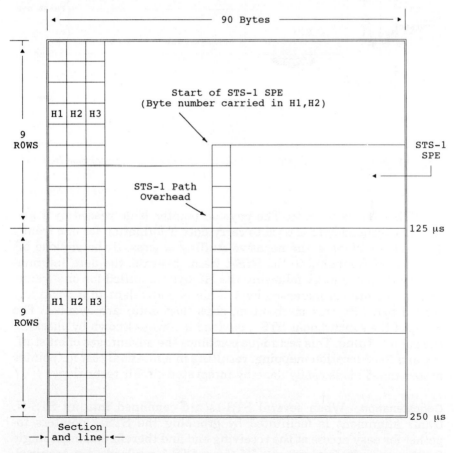

Figure 2.45 STS-1 SPE within a STS-1 frame. (*Reproduced with permission from Ref. 5, © 1989 IEEE.*)

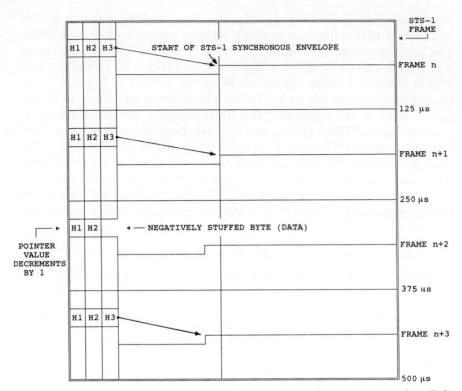

Figure 2.46 Negative STS-1 pointer adjustment. (*Reproduced with permission from Ref. 5, © 1989 IEEE.*)

the STS-1 frame bit rate. The payload pointer is decreased by 1 and the H3 overhead byte is used to carry data information for one frame. This is equivalent to the negative stuffing process. If the payload bit rate is slow compared to the STS-1 frame bit rate, the data information byte immediately following the H3 byte is nulled for one frame, and the pointer is increased by 1. This is equivalent to a positively stuffed byte. By this mechanism, slips (lost data) are avoided. The phase of the synchronous STS-1 payload is always known by checking the pointer value. This technique combines the advantages of bit stuffing and fixed location mapping, resulting in a minimal cost for pointer processing. This is easily done by integrated circuit technology.

Concatenation. When several STS-1s are combined into an STS-N, frame alignment is facilitated by grouping the STS-1 pointers together for easy access at the receiving end and therefore using a single framing circuit. For example, if three STS-1 payloads are required, the phase and frequency of the three STS-1s must be locked together,

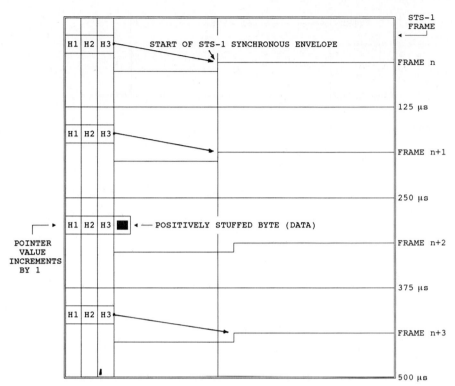

Figure 2.47 Positive STS-1 pointer adjustment. (*Reproduced with permission from Ref. 5, © 1989 IEEE.*)

and the three signals are considered as a single signal that is transported through the network. This is called *concatenation* and is achieved by using a *concatenation indication* in the second and third STS-1 pointers. The concatenation indication is a pointer value which notifies the STS-1 pointer processor that the present pointer being considered should have the same value as the previous STS-1 pointer. The STS-N signal that is locked by pointer concatenation is called an STS-Nc signal (c for concatenation). In North America an STS-3c signal would be considered as 3 STS-1s, but CCITT countries consider the STS-3c as the basic building block for the new synchronous digital hierarchy.

2.5.3 Synchronous transport module frame

Countries other than those in North America and Japan call the STS-3c the *synchronous transport module—level 1* (STM-1), and it has a bit rate of 155.52 Mb/s. The STM-1 has a 9-row by 270-column frame structure as shown in Fig. 2.48. The STM-1 is also known as the 150-

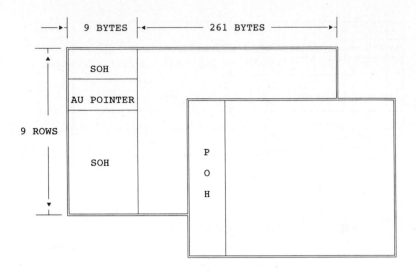

Main Functions of SOH and POH

SOH	- Framing - Error Check - Data Communication - Protection Switch Control - Maintenance
POH	- Error Check - Maintenance

Figure 2.48 The STM-1 frame structure. AU = administrative unit, SOH = section overhead, POH = path overhead. (*Reproduced with permission from Ref. 3, © 1990 IEEE.*)

Mb/s administrative unit (AU). The overhead and payload pointer for the STM-1 frame have a similar format to the STS-1 frame.

The diagrams in Fig. 2.49 compare the STM-1 frame format with that of the STS-1 frame. The compatibility, or equivalence, is self-evident. Notice how the pointers are used to indicate the beginning of the SPE. The SPS-1 is also known as a 50-Mb/s AU, and this is easily expanded to form the STS-3.

Nesting. As already stated with respect to international acceptance, it is essential that one type of AU (e.g., 155.52 Mb/s) in one country can be accepted by a country that has another type of AU (e.g., 51.84 Mb/s). This is where nesting is useful. A nested signal is a set of AUs contained within the STM-1 (i.e., 155.52-Mb/s AU), and this is

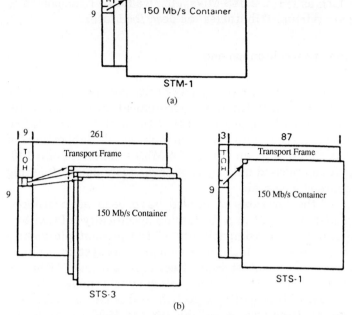

(a)

(b)

Figure 2.49 Comparison of the North American and European SDH frames. (*a*) Synchronous transport module (STM). (*b*) Synchronous transport signal (STS). (*Adapted with permission from Ref. 10, © 1990 IEEE.*)

then transported through the network. Figure 2.50 shows how a nested signal such as three AU-3s (at 45 Mb/s) can be carried in a European country in the STM-1 format without constructing a special network to manage it. Conversely, in North America, four European

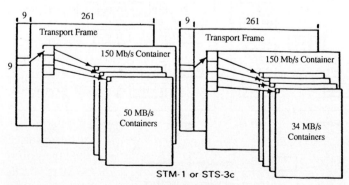

Figure 2.50 Nested signals. (*Reproduced with permission from Ref. 10, © 1990 IEEE.*)

34-Mb/s signals can be transported in an STS-3c (STM-1). Nested signals are used to carry bulk traffic through the network, whereas lower bit rate traffic such as DS1s, 2.048 Mb/s, etc., would be transported in virtual containers (virtual tributaries) as described later.

2.5.4 Comparison of asynchronous and synchronous interfaces

It is important to note that asynchronously multiplexed signals can be synchronously multiplexed to the next hierarchical level without difficulty. However, it must be appreciated that the drop and insert feature only applies to the complete asynchronous signal. Subportions of the asynchronous signal (e.g., one of the 34-Mb/s tributaries of a 140-Mb/s stream) can be accessed only by asynchronously demultiplexing the whole asynchronous signal.

So far, many of the bit rates discussed have been asynchronous ones. This may seem unusual in a discussion specifically related to a synchronous hierarchy, but compatibility of the present plesiochronous asynchronous system with the new synchronous system is one of the main successes of the latest SDH Recommendations. Although there will no doubt eventually be a completely synchronous network, a significant amount of time will elapse before this can be achieved. Present estimates indicate that it will be 1993 to 1996 in the developed world and much later in the developing world. The transition period will require the existing asynchronous network elements to be multiplexed into the synchronous format. Figure 2.51 illustrates the current interfaces compared to the new synchronous interfaces which will eventually be the worldwide, universal NNI. Note that even in

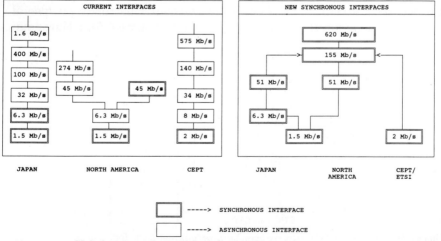

Figure 2.51 Global universal network node interfaces.

the asynchronous system the primary multiplexers at 1.544 and 2.048 Mb/s are described as synchronous. The asynchronous nature is only apparent when multiple primary multiplexers are grouped together to form a higher-order multiplexer and when these resulting bit streams are multiplexed to the next hierarchical level.

So, considering the new synchronous interfaces, the existing asynchronous networks of Japan and North America allow the basic primary 1.544 Mb/s to be synchronously multiplexed up to the STS-1 at 51.84 Mb/s and eventually to the STM-1 or STS-3c at 155.52 Mb/s. The European format is a direct synchronous multiplexing of 2.048 Mb/s up to 155.52 Mb/s. Any further higher-order synchronous multiplexing would be integer multiples of 155.52 Mb/s.

2.5.5 SDH multiplexing structure summary

Having discussed the major differences between synchronous and asynchronous multiplexing, particularly with respect to the frame format, the formal CCITT specifications for the synchronous scheme will now be summarized. The complete SDH multiplexing structure in Fig. 2.52 shows all of the possible ways of forming an STS-1 and subsequently an STM-1. To accompany this diagram there are several new definitions associated with SDH multiplexing which need to be clarified.

Administrative unit. An AU is simply a "chunk" of bandwidth which is used to manage a telecommunications network. In North America and

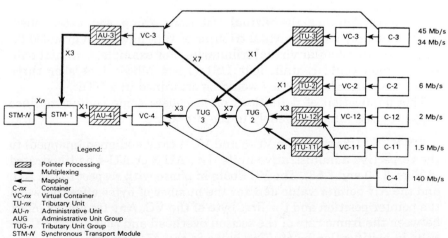

Figure 2.52 SDH multiplexing structure summary. (*Reproduced with permission from Ref. 40, © 1990 IEEE.*)

Japan, this value has been set at 51.84 Mb/s (DS3 size), whereas the rest of the world has the 155.52 Mb/s AU.

Container. The first block in Fig. 2.52 is called a *container* and is denoted as C-nx (where n = 1 to 4, x = 1 or 2); n refers to the asynchronous hierarchy level and x indicates the bit rate (i.e., x = 1 is for 1.544 Mb/s, and x = 2 is for 2.048 Mb/s). Note C-11 is stated as C-one-one and not C-eleven.

Virtual container. The next block is the *virtual container* denoted as VC-n (where n = 1 to 4). This consists of a single container or assembly of tributary units together with the path overhead so that the virtual container is a unit which establishes a path in the network. Each of the containers is said to be *mapped* into a virtual container.

Tributary units. The next block is the *tributary unit* denoted as TU-nx (where n = 1 to 3 and x = 1 or 2). The tributary unit consists of a VC together with a pointer and an AU. The pointer specifies the phase of the VC. The VCs are said to be *mapped* or *aligned* with respect to the TUs. The TUs and the AUs therefore contain sufficient information to enable crossconnecting and switching of the VC and its pointer.

In North America the VC is called a *virtual tributary* and sub-DS3 signals are placed in virtual tributary containers.

Tributary unit group. The next block is the *tributary unit group,* denoted as TUG-n (where n = 2 or 3). The TUG is a national grouping of TUs formed by the multiplexing process. For example, TUG-2 has four TU-11s, three TU-12s, or one TU-2. The TUG-2 can then be multiplexed to either a TUG-3 or VC-3, and the TUG-3 can be multiplexed to VC-4.

In North America the virtual tributary group was established, which is simply a set of virtual tributaries that have been grouped together to carry similar virtual tributaries. For example, a virtual tributary group could contain four DS1 (1.544 Mb/s) signals or three 2.048-Mb/s signals, and they would be contained in a VT6.

This duplication of definitions often causes some confusion since they are used interchangeably in the literature.

STM-N. At this stage the VC-3 and VC-4 can be aligned (mapped) to the respective administrative units (i.e., AU-3 or AU-4 as illustrated in Figs. 2.50 and 2.53). The VC floats in phase with respect to the AU, and the AU pointer value denotes the number of bytes offset between the pointer position and the first byte of the VC. Any frequency offset between the frame rate of the section overhead and that of the VC results in justification by stuffing bytes in the AU pointer.

The STM-1 is then formed by multiplexing either three AU-3s or an

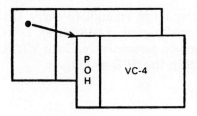

(a) VC-4 mapping into STM-1

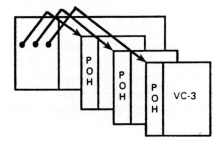

(b) VC-3 mapping into STM-1 for direct method

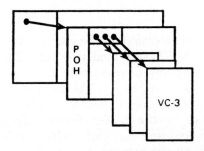

(c) VC-3 mapping into STM-1 for nested method

● : Pointer
POH: Path Overhead
VC-*n*: Virtual Container

Figure 2.53   Mapping methods. (*Reproduced with permission from Ref. 3, © 1990 IEEE.*)

AU-4 together with the *section* overhead information. The STM-1 can be multiplexed into an STM-N by synchronously byte interleaving N STM-1s.

Sub-STS-1 payloads.   The success of the synchronous hierarchy specifications in satisfying both North America and Japan and CCITT countries can be seen by observing the composition of the STS-1 signal. The STS-1 SPE is divided into VT (TU) payload structures. Four sizes of VT/TU have been specified: VT1.5/TU-11, VT2/TU-12, VT3, and VT6/TU-2. These have bandwidths large enough to carry DS1 (1.544 Mb/s), CCITT primary multiplexed signal (2.048 Mb/s), DS1C (3.152 Mb/s), and DS2 (6.312 Mb/s) signals, respectively. Each VT/TU occupies several nine-row columns within the SPE as follows:

VT1.5 occupies 3 columns (27 bytes)

VT2 occupies 4 columns (36 bytes)

VT3 occupies 6 columns (54 bytes)

VT6 occupies 12 columns (108 bytes)

A VT *group* (TUG-2) is defined as a 9-row by 12-column payload structure which can carry four VT1.5s, three VT2s, two VT3s, or one VT6. The STS-1 SPE therefore contains seven VT groups (84 columns), one path overhead column, and two unused columns.

There are two different methods of transporting the payloads within a VT/TU:

1. Floating mode

2. Locked mode

The floating mode uses a VT/TU pointer to establish the starting byte position of the VT/TU SPE within the VT/TU payload structure. This is the mode used for asynchronous mapping of nominally asynchronous signals. The locked mode does not use the VT/TU pointer. Instead, the VT/TU payload structure is directly locked to the STS-1 SPE (which still floats with respect to the STS-1 frame). Byte-synchronous mapping in both the locked and floating modes can be used for transporting unframed synchronous signals.

International compatibility is an important achievement. Also, on a network level there are many benefits that SDH offers in addition to the add and drop capability. Its ability to handle asynchronously multiplexed signals within the synchronous multiplexed hierarchy is a major achievement. The implementation and future implications of the SDH are discussed further in Chaps. 8 and 9.

2.6 Multiplexing Digital Television Signals

So far, the discussion has centered around the multiplexing of voice telephone channels into various hierarchical bit rates. In addition to voice traffic, data and video are also important media for incorporation into the multiplexing structures. The 24- or 30-channel primary multiplexers allocate voice channels for data transmission. The bandwidth required for television signals is relatively large, and a lot of work is currently being devoted to determining specifications for digital coding of TV signals. This subject is at present in a considerable state of flux, particularly as high-definition television (HDTV) is on the horizon, and specifications for the three main types of signal need to be harmonized. This is very difficult because the debate is still raging as to which type of HDTV system will evolve as the international standard. In the meantime, the CCITT is moving ahead with its introduction of Recommendation G.723, which refers to the transmission of component-coded digital television signals at the third hierarchical level, that is, 44.736 or 32.064 or 34.368 Mb/s.

The term *HDTV* is frequently used to describe the upcoming phenomenon that will sweep though our homes in the near future. Also, video telephones in the ISDN environment are just around the corner, and video conferencing is here already. The subject of video/TV is a very broad area of electronics which can be placed in the category of communications or broadcasting. Many books have been written solely on this subject. Here, it is intended to address the process of converting the analog video signals to digital bit streams. The resulting bit streams are multiplexed with voice and/or data traffic and transmitted over medium or long distance routes using microwave, optical fiber, or satellite media as required.

2.6.1 Digitization of TV signals

Until recently, TV signals have always been transmitted from the TV studio to the public in analog form. Approximately 6 MHz of bandwidth has traditionally been used for the TV signals. If the signal is translated to a digital bit stream, by an A/D process, the resulting digital signal is in excess of 140 Mb/s. For video telephone or even video conferencing this is an excessive bandwidth. The cost to the subscriber would be prohibitively large. For example, if a video conference were set up for 1 h between two cities at opposite sides of the country, the participants would have to pay for more than 1500 voice circuits for the duration of the conference. This might be acceptable in circumstances where air fares, hotels bills, etc., for several people would exceed the cost of the video conference. Of course, there is the personal, real-life aspect of public relations to consider, which might never be replaceable by the video conference. TV transmission for broadcast purposes would clearly be cheaper if the bandwidth could be reduced. For video telephones one might argue that the video bandwidth must not exceed the voice frequency range of 4 kHz. That presents a formidable problem to design engineers. Although it is not impossible, the question is: Can the minimum level of acceptable quality be achieved in a 4-kHz bandwidth? For international transmission, the problems are compounded by system incompatibilities. CCITT Recommendation H.261 specifies methods of video telephony communication.

The conclusion of this discussion is clear. For the digital transmission of video/TV signals, the bandwidth must be reduced to a minimum while maintaining high picture quality. It is this last proviso which makes this subject very interesting. There is a trade-off between picture quality and transmission bit rate. To appreciate this problem fully, first it is necessary to understand the basic principles of analog color television. The subject is large and complex, and only a very brief mention of the important features will now be highlighted.

2.6.2 Analog color TV

The color TV signal has two major components:

1. Black and white intensity, known as *luminance*
2. Color variation, called *chrominance*

The luminance signal (Y) can be expressed as a combination of varying intensities of the three primary colors—red, green, and blue:

$$Y = 0.299R + 0.587G + 0.114B$$

For the chrominance signal, instead of transmitting all three primary colors, the bandwidth is reduced by transmitting two color difference (chrominance) signals:

$$U = R - Y \quad \text{and} \quad V = B - Y$$

The picture to be transmitted is formed by allocating a certain number of vertical lines to the picture and horizontally scanning these lines from left to right and top to bottom. During the "flyback" time from the end of one line to the beginning of the next, and from the bottom of the picture back to the top, no picture information is obtained, and the signal is said to be *blanked*. This blanking time is wasted time in the analog system but can be used in the digital TV system for signal processing.

Two types of scanning have evolved: (1) interlaced and (2) progressive, also known as sequential. For the interlaced type, the video picture (called the frame or field) is divided into two fields where the scan lines of the first field fall between those of the second. In other words, every alternate (odd) line is scanned during the first half of a frame scan while the other alternate (even) lines are scanned during the second half of the frame scan. Two full field scans are therefore necessary to form a complete picture. The progressive type of scanning involves scanning the total number of lines contained in one field in a sequential manner. This technique provides higher vertical resolution but requires more transmission bandwidth than interlaced scanning.

One of the major problems for creating a global standard for HDTV is the fact that the NTSC system uses 525 lines for scanning and the power-line frequency of 60 Hz is used to define the rate of scanning, that is, 60 fields per second. The PAL and SECAM systems use 625 lines and 50 Hz for scanning parameters.

2.6.3 Video compression techniques

The standard A/D conversion for voice channels is also used as the basis for video digitization, after which compression is necessary to minimize the transmission bit rate. Compression is possible because there are redundancies in video transmission and also because the human visual perception has peculiar facets that can be exploited. For example, taking the analogy of the motion picture projection, it is well known that the motion is actually an illusion created by viewing a series of still pictures, each having a slightly different position of the moving components. This same principle is used in TV broadcasting. The number of still pictures required is determined by the human visual perception. If there are not enough stills, the picture appears to flicker. This occurs at about 16 stills per second or less.

When considering a motion picture, there is only a certain portion of the picture that changes from one still to the next. For a fixed camera position, only the parts of the picture which move need to be updated, since the background remains the same. This means a differential PCM (DPCM) type of processing will greatly reduce the required transmission bandwidth. In a video telephone application, it is mainly the facial movements (and to some extent the motion of whole head) that need to be periodically updated. DPCM is a very powerful tool when used together with some predictive signal processing so that past motion which causes differences in successive picture frames can be extrapolated to predict future frames.

Other aspects of the human visual system can be electronically exploited. For example, differences in brightness (luminance) are perceived much more prominently than differences in color (chrominance). This leads to the luminance being sampled at a higher spatial resolution. After sampling, each line in the frame contains a number of picture elements called pixels, or PELS. The luminance signal may have, for example, 720 × 480 pixels compared to 360 × 240 pixels for color signals. Also, the eye is more sensitive to energy with low spatial frequency than with high spatial frequency. This characteristic is exploited by using more bits to code the high-frequency coefficients than low-frequency coefficients. Combining all of these techniques results in bit rate compression ratios of 20:1 to 100:1. At 20:1 the reconstructed video signal is almost indistinguishable from the original signal. Even at a 100:1 compression ratio, the reconstructed signal is similar to analog videotape quality.

Inevitably, high compression ratios (greater than about 3:1) lose some information so that the received video signals are not exactly the same as those transmitted. The amount of compression possible and the effects of loss caused by compression vary depending on the application. For example, in a video telephone or conferencing situation there is relatively little motion. The amount of compression obtainable depends, to a large extent, on the video coding technique chosen. The two main categories of video coding are:

1. Source coding

2. Entropy coding

Source coding usually results in lossy, or degraded, picture quality and is classed as either intraframe or interframe coding. Intraframe coding is used for each new scene, whereas interframe coding is used for motion within each scene.

There are several techniques that are now being applied to video coding to achieve high compression ratios. The discrete cosine transform (DCT) has emerged as a powerful tool for intrafield, interfield, and interframe source coding. It is a process of mapping pixels of images from the time domain into the frequency domain. Images are separated into, for example, 8 × 8 blocks, meaning eight lines of eight samples. Each 8 × 8 block is transformed to produce another 8 × 8 block, whose coefficients are then quantized and coded. A large portion of the DCT coefficients are zero. A completely zero transform would mean no picture motion. Inverse DCTs are used to recover the original block.

Entropy coding is another valuable method which encodes frequent events with more bits than infrequent events and is theoretically lossless. Huffman coding is a form of entropy coding which uses predetermined variable code words. Since this subject is still rapidly evolving, no further details will be discussed here. Instead, a typical existing digital TV transmission system which uses some of the above compression techniques will be described.

2.6.4 A typical digital TV transmission system

While direct satellite broadcasting (DBS) and community antenna TV (CATV) via cable have become important modes of distributing existing TV signals (not HDTV), there is still the need to transport the TV signal from the studio to terrestrial transmitters or to link a TV switching center to a satellite earth station. Whether the telecommunications network or the broadcasting network organizes the link, the technical problems remain the same. High-quality transmission is essential. Digital transmission of high-quality TV signals at the 45- or 34-Mb/s interface has been devised to fit smoothly into the North American or European hierarchies.

A TV signal A/D converter can usually be used for digitizing either NTSC, PAL, or SECAM signals. Encoding is performed according to CCIR Recommendation 601-1, which refers to luminance signal sampling at 13.5 MHz and chrominance sampling each at 6.75 MHz with linear encoding into 8-bit words. This ratio is known as the 4:2:2 digital code studio standard. A ratio of 2:1:1 is not used because the Electronic News Gathering (ENG) organization already uses some frequencies which make 4:2:2 more appropriate. The 13.5- and 6.75-MHz rates are equivalent to 864/858 samples per line for the luminance signal and 432/429 samples per line for the chrominance signal. The number of active samples per line are 720 and 360 for luminance and chrominance, respectively. The number of active lines per frame is 485 for the 525-line system and 575 for the 625-line system.

Luminance processing.   The luminance signal can be processed to reduce the bit rate significantly. First, suppression of line and field blanking intervals reduces the bit rate by 25 percent. Adaptive DPCM coding and adaptive quantization reduces the number of bits per sample from 8 to 4. Figure 2.54 shows the feedback loops of intra- and interfield predictors used in the adaptive DPCM coding process. Switching between the two predictors is determined by evaluating the instantaneous picture movement inside the frame. Statistical coding then reduces the number of bits per sample to an average of 2.3.

Chrominance processing.   Coding and bit rate reduction of the chrominance signals U and V are achieved by a similar sequence of steps. U and V are sampled alternately at the transmit side, and the omitted color picture elements are reconstructed in the receiver by interpolation. The line and frame blanking intervals are suppressed. A 2:1 reduction is obtained by sampling only every other color picture element. DPCM coding with *fixed* prediction and adaptive quantizing gives an 8-to-4 reduction in the number of bits per sample. Statistical encoding then further reduces this to an average of 2.4 bits per sample.

Error correction.   Error correction circuits are included to compensate for the increased error sensitivity created by compression (bit rate re-

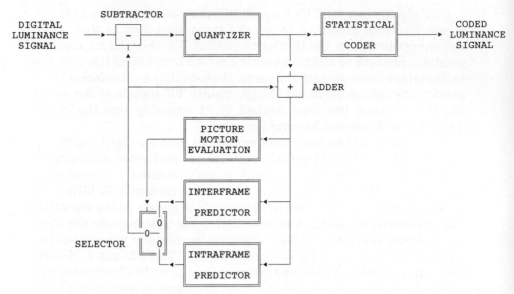

Figure 2.54 Luminance processing.

duction). A typical forward error correction (FEC) mechanism for correcting single errors and error bursts would have a 248-bit block length, containing 234 information bits, allowing the correction of an error burst up to 5 bits. Interleaving four blocks of 248 bits improves the correctable error burst length to 20 bits and individual errors to 5 bits. CCITT Recommendation 723 advises error protection by using a Reed-Solomon (255,239) code with an interlacing factor of 2.

2.6.5 High-definition television

The implementation of HDTV is such a huge commercial venture that the potential profits are enormous. Consequently, the major players are all trying to outmaneuver their competitors. As a result, the progress of HDTV has been plagued by the inability of the interested parties to converge on a global standard. The evolution of analog TV resulted in three main categories of system, namely, PAL, SECAM, and NTSC, with several variations within these categories. The necessity for HDTV to be compatible with the existing national analog systems has led to a divergence of research and a rather rigid stance of HDTV design from one group to another. At present, it appears inevitable that two studio standards will emerge, and much work is being done to minimize the differences between the two. The main debate has centered on the number of lines scanned per second and the proposed field rates of either 50 or 60 Hz.

There are some areas of universal acceptance. The picture aspect ratio (length to width) for HDTV will be 16:9 instead of the present TV value of 4:3. This will take advantage of the psychophysical characteristics of human visual perception. In other words, the picture will appear to be clearer and easier on the eye, particularly for programs that contain rapid movements.

CCIR Recommendation 601, which was initially created in 1982 for extended definition TV using digital technology, has become the basis for HDTV standards. If two systems do evolve, having field rates of 50 and 60 Hz, converters will be necessary to ensure compatibility. This will incur extra expense and a small deterioration of the converted signal. Even if a single system were designed to satisfy North American and European requirements, the resulting system would probably still be incompatible with the Japanese systems, which have already had substantial amounts of capital invested in development of the 1125/60 configuration. The number of lines is also a problem, with the NTSC group proposing 1050 lines, with 2:1 interlacing, and the European group proposing 1250 lines and 50-Hz progressive scanning. The Japanese are proceeding with their multiple sub-Nyquist encoding system (MUSE), which has 1125 lines at 60 fields per second.

Regardless of the incompatibility problems, concerning the transmission and broadcasting of HDTV, the "raw" bit rate will be approximately 1.2 Gb/s or higher. This high bit rate places HDTV in the category of broadband integrated services digital network, or BISDN, (see Chap. 9). The mode of transporting such high bit rates will be by the asynchronous transport mode (ATM) as described in Chap. 9. In the interim and at least until the middle to end of the 1990s, SDH will be used. The CCIR has defined three main categories of HDTV transmission:

1. *Contribution,* which will be for interstudio and intrastudio transmission. This will be at 622 Mb/s, which corresponds to the SDH level STS-12 or STM-4. The compression technique will probably be the lossless subband DPCM.

2. *Primary distribution,* for interconnecting broadcast stations, video theaters, and CATV suppliers. The transmission rate of 155 Mb/s will probably use subband DCT compression.

3. *Secondary distribution,* which will be for directly transmitting to customers at 30 to 50 Mb/s. The techniques necessary to achieve this high level of compression are still being studied.

Although the exact form of HDTV is still unknown, there is an emerging overlap between digital telecommunications and broadcasting. The customer benefits of this new and upcoming technology should be truly astounding.

3

Signal Processing
for Digital
Radio Communications

3.1 Modulation Schemes

The methods of modulation used in analog communication systems can be broadly categorized as amplitude modulation (AM) and frequency modulation (FM). Digital communication systems follow a different approach. The signal to be transmitted in a digital system is a stream of 1s and 0s. There are only two amplitude levels, ON or OFF. At first sight this appears to be a much simpler problem to solve than the transmission of an analog signal whose amplitude is varying in a very complex manner. Unfortunately, a pulse is composed of a fundamental tone plus an infinite number of harmonics. Theoretically, that requires an infinite bandwidth for the transmission of a single pulse. Any communication system is limited in available bandwidth, and it is this constraint that causes a considerable complexity in the design of digital modulators and demodulators (modems).

3.1.1 Bandwidth efficiency

AMR can transmit 1800 voice channels in a 30-MHz bandwidth in the 6-GHz band. In order to use the microwave spectrum efficiently, it is desirable, if not essential, for DMR systems to be able to transmit at least the same number of voice channels within the same analog radio RF bandwidth. In this respect, it is useful to define *bandwidth efficiency,* which is the number of transmitted *bits per second per hertz* (b/s/Hz). In the digital hierarchy, 140 Mb/s contain 1920 voice channels. If the bandwidth efficiency is 1 b/s/Hz, a transmission bandwidth of 140 MHz would be required. Since this is an excessive value, some system modifications must be applied to improve the bandwidth efficiency. This is done during the modulation process. There are several modulation techniques, which can be broadly categorized as follows:

1. Pulse amplitude modulation (PAM)

2. Frequency shift keying (FSK)

3. Phase shift keying (PSK)

4. A mixture of phase and amplitude modulation, called quadrature amplitude modulation (QAM)

Before discussing each type of modulation in detail, it is informative to briefly indicate the effects of *filters* used in the modulation process on the bandwidth efficiency of a digital signal. Since the bandwidth required to perfectly transmit pulses is infinite, if the available bandwidth is comparatively narrow, there will be a significant effect on the shape of pulses emerging from a bandwidth-limiting device such as a filter.

3.1.2 Pulse transmission through filters

The use of low-pass or bandpass filters in the modulation or up-conversion processes of a communication system is unavoidable. Passing the pulses through a low-pass or bandpass filter will eliminate some components of the pulses, resulting in output pulses having very "rounded" corners instead of sharp, right-angle corners. Eventually, if the cut-off frequency of a low-pass filter reaches a low enough value, the pulses become so rounded that they do not reach their full amplitude. *The Nyquist Theorem states that if pulses are transmitted at a rate of f_s b/s, they will attain the full amplitude value if passed through a low-pass filter having a bandwidth $f_s/2$ Hz.* This is the minimum filtering requirement for pulse transmission without performance degradation (i.e., *no intersymbol interference*). Figure 3.1 illustrates this *ideal* Nyquist filter, which allows pulses to reach their maximum amplitudes. Unfortunately, this type of ideal Nyquist filter does not exist. If it did, it would require an infinite number of filter sections, which would therefore have an infinite cost. The filter characteristics for transmission of impulses (approximately the same as for very narrow pulses) are shown in Fig. 3.2. The value of $\alpha = 0$ is the ideal filter case. A more practical value of $\alpha = 0.3$ requires a bandwidth of 30 percent in excess of the Nyquist bandwidth. This means that instead of transmission at an ideal bandwidth efficiency of 2 b/s/Hz, the value is $2/1.3 = 1.54$ b/s/Hz. Figure 3.1 also shows the output response

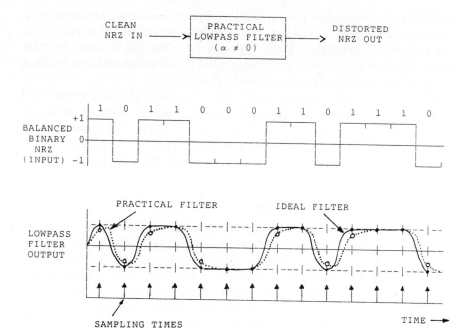

Figure 3.1   Pulse response for an ideal and practical low-pass filter.

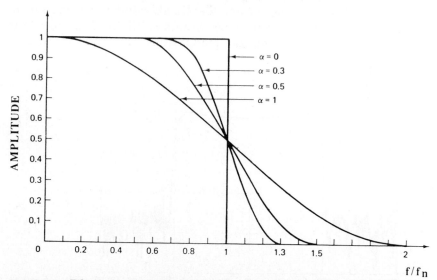

Figure 3.2   Filter characteristics. $f_n = f_s/2$; $f_s = 1/T_s$ = symbol transmission rate. (*From K. Feher, Digital Communications: Microwave Applications, © 1981, p. 48; reprinted by permission of Prentice-Hall, Englewood Cliffs, N.J.*)

for this type of nonideal filter. At the sampling instants, the signal does not always reach its maximum value, so the imperfect filter introduces intersymbol interference. Many types of complex filters have recently been designed to overcome the intersymbol interference problem without significantly reducing the bandwidth efficiency.

The surface acoustic wave (SAW) filter has recently been introduced into DMR equipment. It has some qualities which allow it to be designed very closely to the ideal Nyquist filter. SAW filters have cosine roll-off characteristics and "saddle"-shaped passband characteristics. Figure 3.3 shows a typical frequency response for a SAW filter.

3.1.3 Pulse amplitude modulation

There are several linear modulation techniques available to improve the bandwidth efficiency. The first, PAM, is a simple way of improving bandwidth efficiency. Figure 3.4 shows a conversion from binary NRZ to multilevel PAM signals. Here, a two-level NRZ signal is converted to four levels. Each four-level symbol contains 2 bits of information. Based on the Nyquist Theorem, it is theoretically possible to transmit, without intersymbol interference, two symbols per second per hertz. Since each symbol contains 2 bits of information, four-level PAM should ideally be able to transmit 4 b/s/Hz. Each eight-level PAM contains 3 bits of information, ideally allowing a transmission of 6 b/s/Hz. Unfortunately, the error performance (BER) of digital AM is inferior to other forms of digital modulation. However, there is a

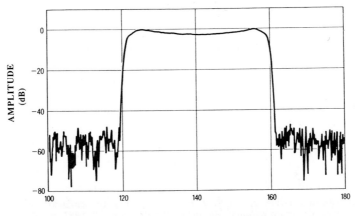

FREQUENCY   (MHz)

Figure 3.3 SAW filter passband characteristics. (*Reproduced with permission from Siemens Telecom Report: Special "Radio Communication," vol. 10, 1987, p. 242, Fig. 3.*)

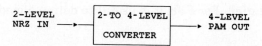

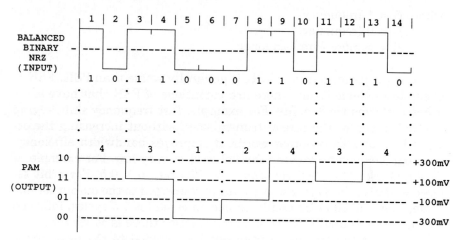

Figure 3.4   Two-level NRZ to four-level PAM conversion.

very important use of AM combined with phase modulation, QAM. Its use in DMR systems will be discussed in detail in Sec. 3.1.6.

3.1.4 Frequency shift keyed modulation, minimum shift keying (MSK), and Gaussian MSK

FSK is simply the allocation of one fixed frequency tone for 0s and another tone for 1s. The input data bit sequence is used to switch back and forth between these two frequencies in sympathy with the changes from 1 to 0 or 0 to 1 (Fig. 3.5). From a circuit point of view, this can be accomplished by feeding the input data into a voltage controllable oscillator (VCO) for the modulation process and using a PLL

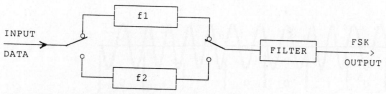

Figure 3.5   Binary frequency shift keying.

for demodulation. For FSK it is useful to define a modulation index m as

$$m = \frac{f_0 - f_1}{f_r}$$

where f_1 = 1 frequency
f_0 = 0 frequency
f_r = symbol rate frequency

   The error performance of FSK is generally worse than PSK. FSK is cheap to implement, and there are variations of FSK that have some other performance benefits. For example, fast frequency shift keying (FFSK) increases the rate of transmission without increasing the occupied bandwidth. In other words, it improves bandwidth efficiency. This can be done by making the value of m very low. For example, if f_1 = 1200 Hz, f_0 = 1800 Hz, and f_r = 1200 b/s, m = 0.5. For a bit sequence of 010, the 1200-Hz oscillator is connected to the output for one cycle, then the 1800-Hz oscillator is connected for 1.5 cycles, followed by the 1200-Hz oscillator for one cycle. Notice there is oscillator phase continuity at the transition from one bit interval to the next. From communications theory, it has been shown that there is equivalence between FFSK and MSK. MSK is binary FSK with the two frequencies selected to ensure that there is exactly a 180° phase shift difference between the two frequencies in one bit interval. MSK therefore produces a maximum phase difference at the end of the bit interval using a *minimum* difference in frequencies (Fig. 3.6) and maintains phase continuity at the bit transitions. MSK is attractive because it has a relatively compact spectrum with out-of-band characteristics that are better than FSK. Low out-of-band emission is necessary for low adjacent channel interference. One method of improving the out-of-band emission is to preshape the data stream with a filter prior to MSK modulation. Several types of filters have been used for this purpose. The classic Nyquist raised-cosine filter allows ISI-free transmission, but a Gaussian-shaped filter which accepts about 1 percent ISI

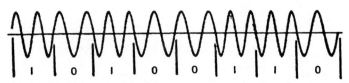

Figure 3.6   Minimum shift keying. f_0 = 1.5f_r, f_1 = 2f_r.

has considerably better out-of-band performance. This is consequently called Gaussian MSK, or GMSK, modulation.

3.1.5 Phase shift keyed modulation

PSK modulation is widely used in DMR technology today. There are several levels of PSK. The simplest is two-phase PSK as shown in Fig. 3.7*a*. In this type of modulation, the incoming bit stream is given a phase reversal of 180° every time a 1 changes to a 0 or vice versa. Figure 3.7*b* shows the waveform changing between 0 and 180°. The next level of PSK is four-phase PSK, otherwise known as quadrature or quaternary PSK (QPSK or 4-PSK). As indicated in Fig. 3.8*a*, the incoming bit stream is divided into two parallel bit streams by using a *serial-to-parallel converter*. These two bit streams are known as the in phase (or I) and the quadrature phase (or Q) bit streams. The transmit oscillator generates the unmodulated carrier frequency, which is passed through a 0 and 90° phase splitter. The I and Q baseband NRZ bit streams are time-domain multiplied by the carrier signals using a mixer. The summed output is then passed through a bandpass filter, resulting in the waveform of Fig. 3.8*b*. The phase diagram shows that there are four phase states, 90° apart from each other at 0, 90, 180, and 270°. The next level of PSK, as shown in Fig. 3.9, is 8-PSK. Here the incoming bit stream is divided into three before the carrier is modulated. The output waveform has eight states, spaced 45° apart. Each

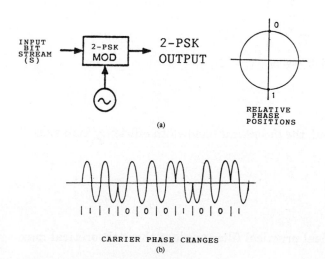

Figure 3.7 Two-phase shift keying.

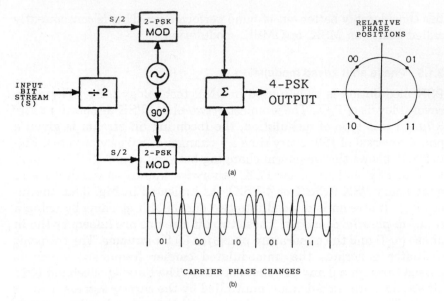

Figure 3.8   Four-phase shift keying.

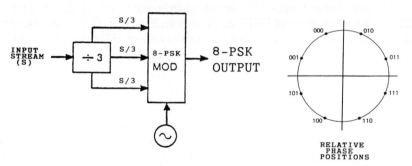

Figure 3.9   Eight-phase shift keying.

time the level is raised, the *theoretical* bandwidth efficiency is increased:

2-PSK → 1 b/s/Hz

4-PSK → 2 b/s/Hz

8-PSK → 3 b/s/Hz

In reality, the nonideal practical filters reduce these theoretical maximum values.

3.1.6 Quadrature amplitude modulation

The next level of PSK, 16-PSK, is not used very much. In preference, a modulation having both PSK and amplitude modulation has evolved—QAM. The reason for this is improved error performance. QAM can be viewed as an extension of PSK. In the special case of 4-QAM where two amplitude levels are used as inputs to a 2-PSK modulator, the system is identical to 4-PSK. However, higher-level QAM systems are distinctly different from the higher-level PSK systems. Figure 3.10*a* shows the generalized QAM modem. Figure 3.10*b* shows the modem for 16-QAM together with the respective waveforms. The bit stream to be modulated is split into two parallel bit streams which are then each converted into a four-level PAM signal. One of the signals is mixed with a carrier from a local oscillator (LO) while the other is mixed with the same LO after having undergone a 90° phase shift. The two signals are then added. The result is a 16-QAM signal;

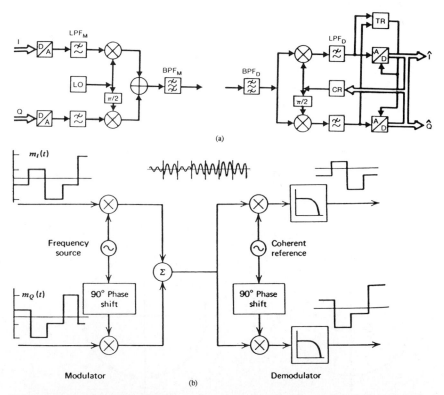

Figure 3.10   (*a*) Generalized QAM modem (*reproduced with permission from Ref. 59,* © *1986 IEEE*). (*b*) 16-QAM modem (*reproduced with permission from Bellamy, J.,* Digital Telephony, *Fig. 6.18,* © *1982, J. Wiley and Sons*).

Figure 3.11*a* shows the signal state-space diagram for 16-QAM. The *signal state-space diagram* is often referred to as the *constellation diagram*. Notice, by comparing this to the 16-PSK signal (Fig. 3.11*b*) that the 16-QAM signal does not have a constant envelope, whereas the 16-PSK does. This has some interesting noise implications, which are discussed later.

The prime motive for moving to higher levels of QAM is simply improvement in bandwidth efficiency. For example, a 64-QAM digital system has a bandwidth (spectral) efficiency which allows it to be a direct substitute for an existing analog system of similar capacity. It can operate in the same channel arrangement as the analog system. Also, analog and digital systems can operate on adjacent channels without mutual interference. Figure 3.12 is a comparison of QAM con-

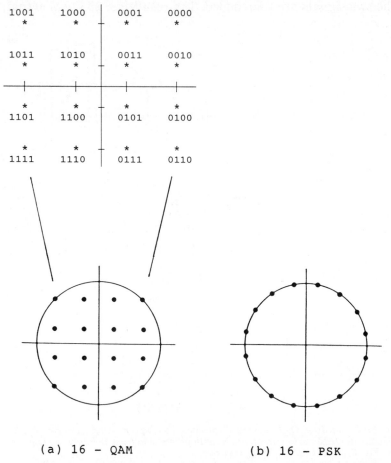

(a) 16 – QAM   (b) 16 – PSK

Figure 3.11   The signal state-space diagram for (*a*) 16-QAM and (*b*) 16-PSK.

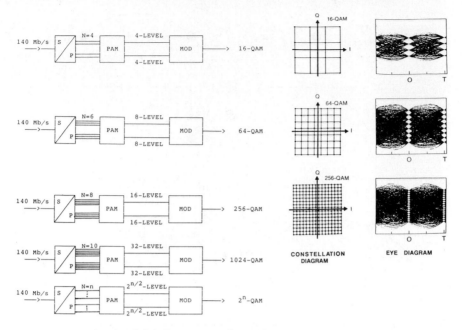

Figure 3.12   Higher-level QAM system configurations.

figurations for 16-, 64-, 256-, 1024-QAM, etc., together with their respective constellation diagrams. In each case, the input bit stream is split into four, six, eight, etc., bit streams by a series-to-parallel converter. A PAM process then provides two bit streams (I and Q) each with 4, 8, 16, etc., amplitude levels. Each I and Q bit stream is then mixed with the IF oscillator (directly and 90° phase shifted, respectively). The modulated outputs are then added to form the QAM signal.

3.1.7   Comparison of modulation techniques

The spectral shapes of PSK and QAM signals having the same number of states (e.g., 16-PSK and 16-QAM) are identical. The error performance of any digital modulation system is fundamentally related to the distance between points in the signal constellation diagram. Figure 3.13 illustrates how the distance between points is the same for 2-PSK as for 4-PSK. The next level, 8-PSK, has points more closely spaced, so the error performance is expected to be worse than 2- or 4-PSK. The reason for this is simply that, as the states become closer together, noise causes them to be located over a broader area around the required points instead of exactly at the points (as in Fig. 3.14).

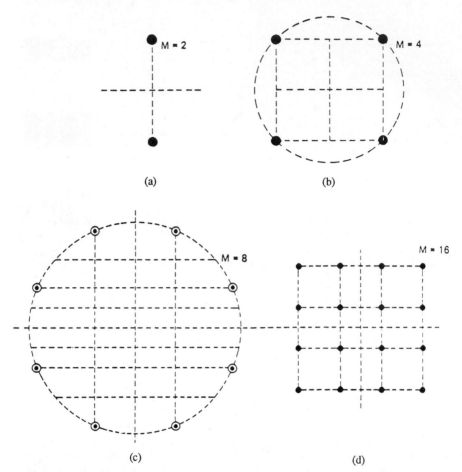

Figure 3.13 Signal state-space (constellation) diagram for several types of modulation. (*a*) 2-PSK: binary (DSB-SC-AM). (*b*) 4-PSK: 2-binary quadrature (DSB-SC-AM). (*c*) 8-PSK: eight selected states of two-level PAM quadrature (DSB-SC-AM). (*d*) 16-QAM. (*From K. Feher, Digital Communications: Microwave Applications, © 1981, p. 123; reprinted by permission of Prentice-Hall, Englewood Cliffs, N.J.*)

This simple result has important implications for coded modulation, as described in the next section. Overlapping of the points indicates interference, which causes errors. This constellation representation has a very useful application in DMR maintenance methods, as discussed in Chap. 5. It is quite evident that for 16-QAM the distance between the points is greater than for 16-PSK, thus providing better error performance. Notice, as indicated in Fig. 3.13, *PSK is a double sideband suppressed carrier (DSB-SC) modulation.* This is an important feature which must be taken into consideration when choosing demodulation circuitry.

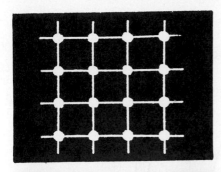

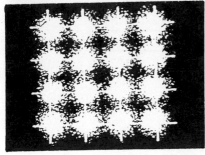

Figure 3.14 The constellation diagram showing noise.

Spectrum-power trade-off. Having established the desired value of bandwidth efficiency, it is important to assess how much received power is required to produce a specified BER (that is, to maximize power efficiency). Unfortunately, the higher levels of modulation require higher values of carrier-to-noise ratio (C/N) to achieve a given BER. Figure 3.15 is a comparison of the error rates for several PSK and QAM systems. For a required BER, the S/N must be increased significantly as the level of modulation increases. The probability of error is plotted against the C/N. This graph stresses the following important point: *Improving the bandwidth efficiency by employing a higher level of modulation requires a greater S/N to maintain a good BER.*

The spectral density (the same as bandwidth efficiency) η is simply the ratio of the bit rate to bandwidth W, the width of each channel. For M-PSK and M-QAM the transmitted bit rate is

$$\frac{1}{T}\log_2 M \quad \text{b/s}$$

where T is the symbol rate in bauds. Therefore

$$\eta = \frac{1}{WT}\log_2 M \quad \text{b/s/Hz}$$

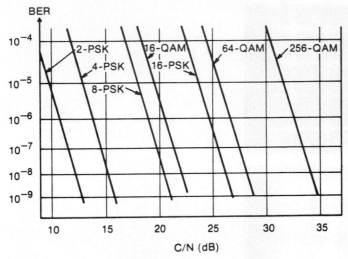

Figure 3.15 Error rates of QAM systems. (*Reproduced with permission from Ref. 59, © 1986 IEEE.*)

$WT = 1$ only for the case of $\alpha = 0$, which as already stated is the ideal (impractical) filter. For practical filtering in the transmitter and receiver, the total RF bandwidth is therefore $(1 + \alpha)/T$. This means that since $\alpha \neq 0$, the transmitted bandwidth exceeds the allotted channel bandwidth W. It is usual to organize filtering so that half of the cosine roll-off shaping is done in the transmitter and half in the receiver. Figure 3.16 shows the transmitted spectrum for cosine roll-off shaping for various values of α. Note that only $\alpha = 0$ keeps the transmitted energy within the channel region of ± 0.5. Studies by the FCC and Conference of Posts and Telecommunication (CEPT) have resulted in the transmitted spectrum emission masks of Fig. 3.17, which must not be exceeded by microwave radio operators. With $\alpha = 0.5$ and $1/WT = 0.75$, the FCC and CEPT masks for 4-, 6-, and 11-GHz operation are satisfied. The spectral efficiency in that case is

$$\eta = \frac{1}{(1 + \alpha)}\log_2 M$$

$$\eta = 0.75 \log_2 M$$

so for various levels of QAM

$$4\text{-QAM} \rightarrow \eta = 1.5 \text{ b/s/Hz}$$
$$16\text{-QAM} \rightarrow \eta = 3.0 \text{ b/s/Hz}$$
$$64\text{-QAM} \rightarrow \eta = 4.5 \text{ b/s/Hz}$$
$$256\text{-QAM} \rightarrow \eta = 6.0 \text{ b/s/Hz}$$

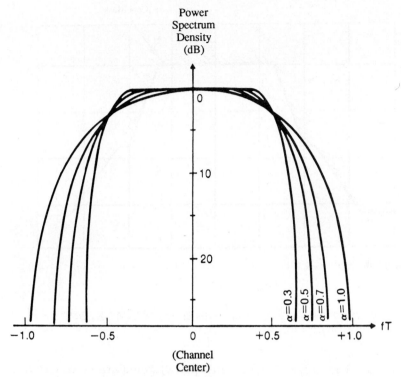

Figure 3.16 Frequency spectrum using a cosine roll-off filter. (*Reproduced with permission from Ref. 59, © 1986 IEEE.*)

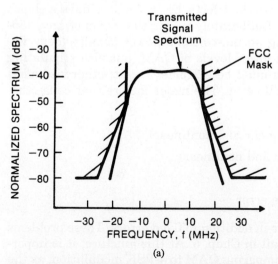

(a)

Figure 3.17 (*a*) FCC frequency spectrum limits for the lower 6-GHz DMR (*reproduced with permission from Ref. 59, © 1986 IEEE*).

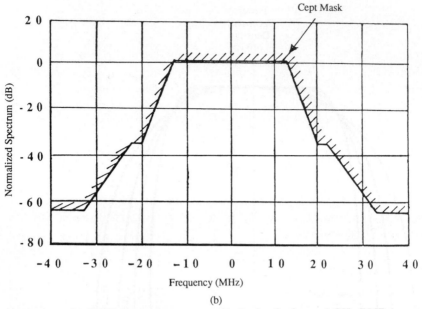

Figure 3.17 (*b*) CEPT frequency spectrum limits for the lower 6-GHz DMR (*reproduced with permission from Ref. 52, © 1989 Artech House, Inc., Norwood, MA*).

Impairments.   The higher-level modulation systems are more vulnerable to equipment and atmospheric propagation impairments. Both impairments cause degradation in transmission quality, but whereas the radio designer has no control over the propagation and can only take countermeasures to minimize the effects, he or she does have a choice concerning modem design. Deliberately taking the option of, say, 256-QAM is not only accepting an unavoidable power penalty (degradation in C/N), compared to lower levels of QAM, but also stipulating that the design tolerances must be extremely tight; otherwise additional C/N degradation will occur. The major factors that cause this further C/N degradation are:

1. Amplitude distortion (linear and nonlinear)
2. Delay distortion (linear and nonlinear)
3. Timing error
4. Recovered carrier phase error

Items 3 and 4 include jitter in addition to fixed errors. These problems are addressed in more detail in Chap. 5. At this juncture, it is important to stress that when comparing QAM to 4-PSK modulation, as the level of modulation increases (16-QAM, 64-QAM, etc.), any of the above impairments will very significantly increase the C/N required to achieve a specific BER. The C/N degradation is *in addition* to the values shown in Figure 3.15.

Conclusion.   Finally, the improvement of bandwidth efficiency at the expense of the C/N is shown in Figure 3.18. The choice of modulation scheme would appear to be simply a trade-off between bandwidth efficiency and power efficiency. There is also the constraint of staying within the required transmitted power spectrum mask. The tolerance to impairments is also important. As usual, cost is the prime deciding factor, and it is recognized that the cost per channel is lower for the higher levels of modulation. It may be that at 1024-QAM and above the system complexity required to overcome the technical impairments to ensure a satisfactory performance will be prohibitively expensive. Time will tell.

3.1.8 Demodulation

Demodulation is the inverse process of modulation. This rather obvious statement is evident by observing the demodulation schematic of Fig. 3.10. However, demodulation is not so simple in reality. The major difference between the modulation and demodulation diagrams lies in the carrier recovery and timing recovery circuits.

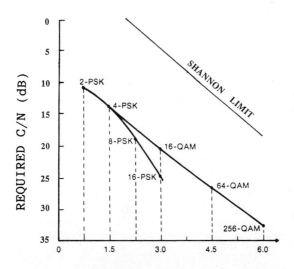

Figure 3.18   Graph of theoretical bandwidth efficiency against C/N for various types of modulation (BER = 10^{-6}; WT = 1.33).

A transmitted QAM signal, for example, is a suppressed carrier signal. The reinsertion of a locally generated carrier is necessary for recovery of the transmitted information. Furthermore, *coherent detection* is essential. This means that the inserted carrier must have the same phase and frequency as the transmitted carrier.

Two successful methods of carrier recovery are (1) the *Costas loop* and (2) the *decision directed* method, as described in the following.

The Costas loop. The Costas loop is an extension of the famous PLL method of carrier recovery and demodulation. The PLL has been used extensively in AMR circuits for establishing:

1. A stable oscillator frequency
2. Carrier recovery of a transmitted double sideband suppressed carrier (DSB-SC) signal
3. Demodulation

To refresh the memory, the PLL circuit of Figure 3.19 has an output that locks onto the input reference signal and then tracks any changes that occur in the input signal. The phase lock is done by using the phase comparator to compare the phase of the output signal with the phase of the input reference. The phase difference produces an error voltage which is used to modify the frequency (and therefore phase) of the VCO. As this error voltage tracks the phase or frequency of the input reference signal, it effectively demodulates the input reference signal. An equation can be derived for the error voltage required to maintain phase lock as follows:

$$V_f = K \sin(\theta_i - \theta_o)$$

where K is a constant and $\theta_i - \theta_o$ is the phase error.

The PLL circuit is applicable to both analog and digital demodulation. For an analog 2-PSK signal, the circuit for the Costas loop is shown in Fig. 3.20. One can see that this is merely an extension of the PLL concept. The loop filters are included to remove the harmonic fre-

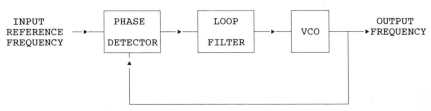

Figure 3.19 The basic phase lock loop.

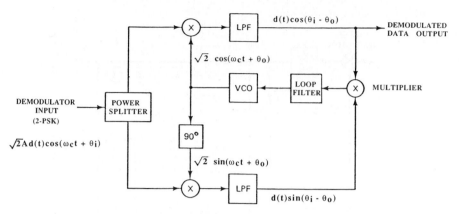

Figure 3.20   The 2-PSK analog Costas loop. θ_i = input signal phase, θ_o = VCO output phase, $(\theta_i - \theta_o)$ = phase error, $d(t)$ = data and HAS values ± 1, ω_c = angular carrier frequency, A = rms value of PSK signal. (*Adapted by permission from Ref. 54, p. 56,* © *1988, Prentice-Hall, Englewood Cliffs, N.J.*)

quency terms. The error voltage, which is the input to the loop filter, is found to be

$$e_f = C \sin 2(\theta_i - \theta_o)$$

where C is a constant. This result differs from the basic PLL equation only in the fact that the Costas loop error voltage is zero for a phase difference of 0 and 180° instead of just 0° as for the basic PLL. This means that the phase lock can occur for two different phase angles between the VCO and the input signal. Data can therefore appear in the upper or lower half of the circuit depending upon the angle onto which the loop locks. The data are a faithful reproduction of the modulated transmitted signal.

The differences between the analog and digital 2-PSK Costas loops are (1) the multiplier in the feedback loop is replaced by an EXCLUSIVE OR gate and (2) the low-pass filter outputs are followed by hard limiters for the digital 2-PSK circuit; 4-PSK requires a slightly more elaborate Costas loop for demodulation, as indicated in Fig. 3.21. Since a 16-QAM signal requires the demodulation of a 4-PSK prior to demodulation of the PAM signal, this type of Costas loop is widely used in 16-QAM demodulators.

The 4-PSK input in this case is

$$A_1 a(t)\sin\omega_c t + A_2 b(t)\cos\omega_c t$$

where $a(t)$ and $b(t)$ are the data signals, which can each have values of ± 1; A_1 and A_2 are the amplitudes of the PSK signals; and ω_c is the angular carrier frequency. The input to the upper phase detector is

$$2\sin(\omega_c t + \theta_e)$$

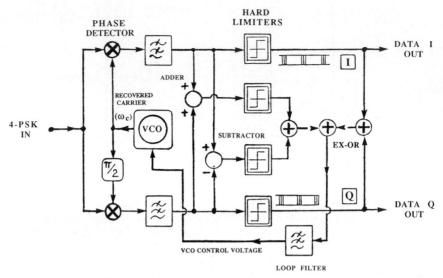

Figure 3.21   The 4-PSK digital Costas loop demodulator.

where θ_e is the phase error (i.e., $\theta_i - \theta_o$). The input to the lower phase detector is

$$2\cos(\omega_c t + \theta_e)$$

The data I output is

$$\text{sign}[a(t)\cos\theta_e + b(t)\sin\theta_e]$$

The data Q output is

$$\text{sign}[b(t)\cos\theta_e - a(t)\sin\theta_e]$$

The data outputs I and Q are subsequently processed at the baseband level to recover the originally transmitted bit stream at, for example, 140 Mb/s.

For a 140-Mb/s system, during the modulation process a 140-MHz VCO is used to produce the 16-QAM signal. This VCO can be very conveniently used as a means of transmitting a subbaseband (SBB) signal for service channel purposes. The SBB signal, which contains two channels of information, is transmitted by *frequency modulating* the 140-MHz VCO. At the receiving end, the SBB signal is recovered by a modified Costas loop circuit.

Other techniques can be used for carrier recovery. Remodulation is a favorable method, especially since it has a faster acquisition time than the 4-PSK Costas loop demodulator. The main purpose of the 4-PSK remodulation technique is to demodulate the signal into I and Q bit streams, after which the two I and Q signals are remodulated. The remodulated output is fed back into the loop filter together with the 4-PSK input signal to produce the loop error signal.

Decision-directed method. The decision-directed method of timing recovery is yet another variation of the PLL, and it has been applied successfully to high-level QAM demodulation circuitry. Over the past few years, so many variants of the original decision-directed loop have been proposed and successfully incorporated in DMR equipment, it is difficult to assess which circuit is the best because sufficient long-term field data is not yet available. Since this method is emerging as the overall best technique for carrier recovery of QAM signals, it is worthy of extensive coverage. However, this subject is unavoidably highly mathematical in its nature so, in this text, just a few details will be highlighted.

In the demodulation diagram of Fig. 3.10, the outputs of the A/D converters are the I and Q digital data streams. From these two streams, the constellation plot could be observed by feeding the I and Q streams into a constellation analyzer. The existence of noise in the DMR causes the constellation points to be spread as in Fig. 3.14. In addition, other impairments of the DMR system cause the constellation states to be offset from their precise locations.

Figure 3.22 is a 64-QAM constellation, superimposed on a template

Figure 3.22 The 64-QAM carrier recovery template.

of 256 small square regions (4 for each state). The coordinates of each of these 256 squares are stored in a memory in the carrier recovery circuit. Concentric circles are also drawn on this diagram. If noise and/or equipment impairment causes a constellation state to be moved from its ideal position to a position in a "black" square, the state needs to be rotated clockwise to correct its position. Counterclockwise rotation would be necessary for states displaced into the "gray" square areas. This information of control increments required to correct the position of the QAM states is applied to a VCO in a loop. The control increments derived from the template effectively replace the phase detector function of the familiar PLL.

More advanced decision-directed circuits incorporate the baseband adaptive equalizer (see Sec. 5.1.5) and automatic gain control (AGC) circuits within the loop. Sophisticated algorithms are applied in such circuits, which result in excellent carrier recovery even in the presence of very hostile propagation conditions.

3.2 Error Control (Detection and Correction)

The BER is perhaps the most important quality factor to observe when evaluating a digital transmission system. First, it is instructive to make a statement about the noise, which causes errors. The worst type of noise is random noise. Error calculations are often based on the type of random noise described as additive white Gaussian noise (AWGN). If the noise has a structure, that structure can be used to the system designer's advantage. Short error bursts are not random in the sense that there is a definite reason for and locatable source of such errors. The probability of receiving errors for a specific modulation scheme depends on the S/N at the receiver and the transmission bit rate. In practical systems, there is usually no opportunity to change those two factors, so one must resort to *error control coding,* also called *channel coding,* or broadly speaking, *FEC.* Recent developments in incorporating error control into the modulation process have produced some exciting results. This technique is called *trellis-coded modulation* (TCM).

A second category of error control technique is *automatic request for repeat* (ARQ). In this technique, redundancy is also added prior to transmission. The philosophy here is different from FEC, where the redundancy is used to enable a decoder at the receiver to correct errors. ARQ uses the redundancy to detect errors. Once detected, a request is made to the transmitter for a repeat of the transmission bit sequence. In this technique, a return path is essential.

3.2.1 Forward error correction

The objectives of error correctors are to reduce the residual BER by several orders of magnitude and also to increase the system gain. This is achieved by encoding the bit stream prior to modulation by a process of adding extra bits to the bit stream according to specific rules. The additional bits do not contribute information to the message transmitted. They serve only to allow the decoder in the receiver to recognize and therefore correct any errors which may arise during the transmission process. As a result, *error correction deliberately introduces redundancy.* In other words, the improvement in residual BER is achieved at the expense of an increase in the transmission bit rate. FEC does not need a return path. This is analogous to a person speaking over a noisy telephone line. If that person includes some repetitions from time to time (adding redundancy), the message is clarified without any need for the receiving party to ask for a repeat (say again). Increased bit rate implies that the bandwidth efficiency suffers some degradation. Although this is true for most types of coding, recent advances in TCM achieve improved noise immunity (low residual BER) with a minimal reduction in bandwidth efficiency.

Error correction codes can be broadly categorized into (1) *block* codes, (2) *convolution* codes, or (3) a combination of both block and convolution, called *concatenated* codes.

Block codes. In Fig. 3.23 the input bit stream is read into the FEC encoder in *blocks* of information bits containing k bits. The output from the FEC encoder has the k bits plus r check bits, so an n-bit *code word* is transmitted. This would be called an (n,k) encoder, having a *coding rate* of k/n. The FEC encoder adds check bits which are used in the receiver to identify and possibly correct any errors. In its simplest form, there is only 1 check bit, which is a parity check bit. The larger the number of check bits, the better the residual BER. Clearly there is a trade-off here between BER and excess bandwidth required because of the check bits increasing the bit rate. When comparing coded to

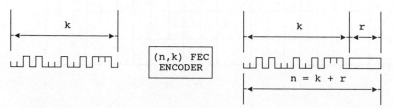

Figure 3.23 A block encoder. k = message bits, r = check bits, n = code word.

uncoded signals, the reduction in the value of coded signal C/N which is required to yield a specific residual BER (e.g., 1×10^{-10}) is called the *coding gain*. Using a block code having a coding rate of 81/84, which means there are only 3 check bits per code word, a coding gain in excess of 3 dB is easily achieved for a 64-QAM system at a BER of 10^{-6}

An important family of block codes is the *Reed-Solomon* (RS) codes. They are effective in correcting combinations of both random errors and error bursts. RS codes operate on multibit symbols instead of individual bits. For example, a RS(64,40) code with 6-bit symbols would have the information data grouped into blocks of 240 bits (6×40) in the encoder. The encoding process would then expand the 40 symbols into 64 symbols (384 bits), of which 24 symbols (144 bits) constitute redundancy. There is quite a lot of choice when considering RS codes. This is illustrated by the general inequality which states that RS(n,k) codes having b bits per symbol exist for all values of n and k, where $0 < k < n < 2^b + 2$. It is common to use $b = 8$, so each symbol is 1 byte. RS codes have a powerful correction capability for a given percentage of redundancy. A RS(64,62) code would have 3.2 percent redundancy, which is typical for a 64-QAM DMR system. As indicated in Figure 3.24, this would provide a coding gain of about 4.5 dB (at BER $= 10^{-8}$). The term *asymptotic coding gain* (ACG) is often used to

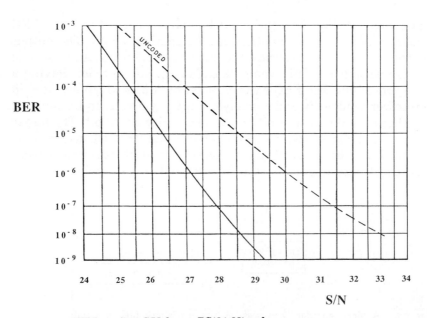

Figure 3.24   BER against S/N for an RS(64,62) code.

describe the coding gain at BER = $10^{-\alpha}$ (zero errors). Caution must be observed when using this term because it can be a little misleading. For example, the coding gain at BER = 10^{-3} for the above RS(64/62) code is only about 0.7 dB, which is considerably less than the ACG value.

The general corrective power is described as follows: If there are $r = n - k$ redundant symbols, t symbol errors can be corrected in a code word so long as t does not exceed $r/2$. For RS(64,62) this implies that only singly occurring symbol errors are correctable in each block of 61 symbols [i.e., $(64 - 62)/2 = 1$]. A highly redundant code such as RS(64,40) would allow every pattern of up to 12 symbol errors to be corrected in a block of 40 symbols.

The ability of a block code to control random errors is dependent upon the minimum number of positions in which any pair of encoded blocks differ. This number is called the *Hamming distance* for the code. For example, a received sequence of 110111 differs from the transmitted sequence of 111111 by a Hamming distance of 1, whereas 101101 would have a Hamming distance of 2.

Convolution codes.   The convolution code was the first type of coding to receive widespread acceptance, and it was initially used in satellite transmission applications.

A convolution code differs from a block code in that the convolution code word exiting the encoder depends on the block of message bits (k) *and* the previous ($n - 1$) blocks of message bits. This rather complex process is performed by shift registers and adders. Figure 3.25 illus-

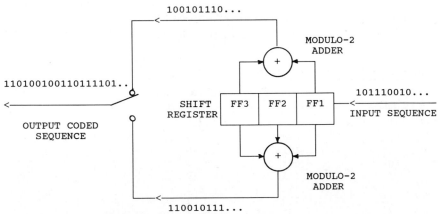

Figure 3.25   One-half rate, $K = 3$ convolution encoder.

trates a convolution encoder using only three shift registers. The number of shift registers forming the encoder is called the constraint length K, so Fig. 3.25 has $K = 3$. The input data is gradually shifted through the registers, and the outputs are fed into the two modulo-2 adders as indicated. A modulo-2 adder, by definition, produces a 1 output for an odd number of 1 inputs. A simple truth table check for two inputs to a modulo-2 adder confirms it to be the familiar EXCLUSIVE OR function. The switch moves back and forth to take alternate bits from each modulo-2 adder output to produce the encoded output. In this case there are 2 coded bits for every input data bit. This is called a 1/2-rate encoder. The code rate in general is $1/n$, where n is the number of coded output bits per input data bit. For example, consider an input data sequence of 101110010. The outputs of the modulo-2 adders are shown in Table 3.1 (considering the shift registers initially set at 000 and the sequence was followed by zeros). The switch then takes alternate bits to form the encoded output:

110100100110111101...

It is now instructive to represent the encoding process in the form of a tree (Fig. 3.26) and subsequently as the *trellis* of Fig. 3.27. This is done in most texts on convolution coding and will be presented here for completeness and as an introduction to TCM. The tree is a decision process diagram. Notice how the coded sequence can be derived from this tree diagram. The first 5 input bits to the coder were 10111. This would lead us through the path A-C-F-I-O-W, during which the coded sequence becomes 11 01 00 10 01. Notice that the tree repeats itself after three splits. LMNOLMNO is the same for the top half of the diagram as for the bottom half. This is because $K = 3$. The output is influenced only by the 3 bits that are currently in the three flip-flops of the shift register. As the fourth bit enters FF1, the first bit drops out of FF3 and no longer influences the output. Therefore, if $K = 4$, the

TABLE 3.1   Convolution Encoder Bit Sequences

Output from 3 I/P adder	Shift reg. values	Output from 2 I/P adder
1	001	1
1	010	0
0	101	0
0	011	1
1	111	0
0	110	1
1	100	1
1	001	1
1	010	0

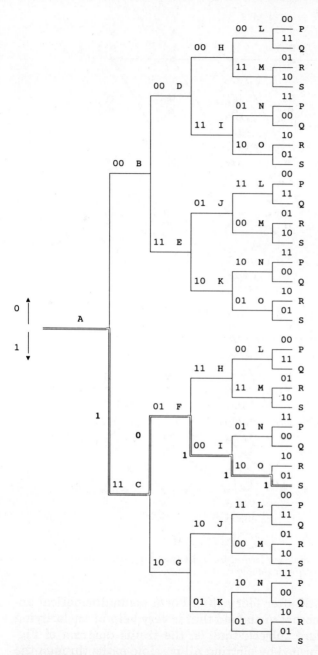

Figure 3.26   The tree structure for decoding.

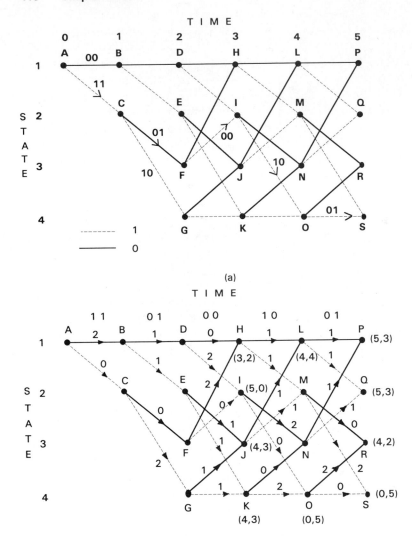

Figure 3.27 (a) Trellis code diagram formed from the tree; (b) trellis code diagram with a Hamming distance example.

tree would repeat itself after four splits. For a nonmathematical appreciation, there is another visual aid that is very helpful in clarifying this apparently obscure subject, that is, the trellis diagram of Fig. 3.27a. The trellis is formed by plotting all possible paths through the tree. Solid lines indicate a path due to a 0 and dotted lines indicate that of a 1. It is interesting to note that there are only four possible states in the trellis. State 1 is ABDHLP, state 2 is CEIMQ, state 3 is FJNR, and state 4 is GKOS.

The trellis diagram gives some insight into a means of decoding the convolution-coded sequence. The repetitive aspect of the trellis is the key to the operation of the decoding process known as the Viterbi algorithm. A block code has a formal structure, and this property is exploited to decode precisely the bit sequence in the receiver. Unfortunately, this is not the case for convolution decoding, and a considerably more complex mechanism is necessary.

First, the terms *hard* and *soft* need to be defined with respect to decoding. Previously, the output from a demodulator was considered to be a hard decision. For QAM, the output was a specific state in the constellation diagram. In the case of a soft decision, the output from the demodulator would be quantized into several levels (e.g., eight levels for a 3-bit quantization). A decoding algorithm can then operate on this output information to enhance the overall BER. Stated another way, a soft distance would be the Euclidean distance instead of the Hamming distance.

The Viterbi algorithm is known as a path maximum likelihood process. The paths through the trellis represent all possible transmitted sequences. The repetitive nature of the tree allows the number of computations required to home in on the correct sequence to be reduced to a manageable quantity. For example, consider decoding the received sequence 1101001001. Taking the digits in pairs (11 01 00 10 01), the path on the trellis is traced, noting the Hamming distance for all paths. Since 11 is the path AC, the Hamming distance for AC is 0, whereas the path AB is a Hamming distance of 2. This leads to the new trellis diagram of Fig. 3.27*b*, where the Hamming distances are noted for each branch of the tree. The cumulative distances are the important values and these are noted in brackets, the first number being the upper best route and the second number being the bottom route. At point E, the paths EK and EJ both have equal total distances of (4,3). One path is rejected while the other is the "survivor." This is the key to streamlining the calculation which is characteristic of the Viterbi algorithm. The least-distance path is ACFIOS, which is a coded sequence of 1101001001. The received sequence therefore has a Hamming distance of 0. The corresponding transmit sequence of 10111 is therefore the "most likely" sequence to have been transmitted.

All of the above analysis has been for a value of $K = 3$. The complexity clearly increases rapidly for higher values of K. Instead of the four states for $K = 3$, there would be eight for $K = 4$, sixteen for $K = 5$, etc., with correspondingly large increases in the number of possible paths in the trellis. The technological advancements of fast very large-scale integration (VLSI) circuits that can cope with such large numbers of computations have been essential to the success of convolution coding.

A convolution decoder requires a certain number of computations per decoded information bit. In general, the greater the number of computations the better the BER. This implies that computation time becomes a limiting factor. Indeed, this constraint is a serious limitation, and Viterbi algorithms are not widely used on high-data-rate systems (e.g., above 100 Mb/s).

The performance of convolution coders depends on the constraint length K. There is a trade-off between BER and bit rate. The higher the K value the better the BER performance, but the lower the highest operating bit rate. Typical design values for K range between 5 and 11, with 7 being very popular. A value of 11 would be more desirable from a BER improvement point of view, but this would limit the operation to very slow channels. For a data throughput of several hundred megabits per second, K would be restricted to about 5. In general, convolution codes tend to be superior to RS codes for *random error* correction.

Finally, the complexity of convolution coders decreases as the redundancy increases. Conversely, for block codes, the complexity increases as the redundancy increases.

Code interleaving.   Code interleaving is a countermeasure against short error bursts. Figure 3.28 illustrates how a simple block code deals with an error burst. Interleaving means that N blocks of data are transmitted bit by bit from each successive block instead of block

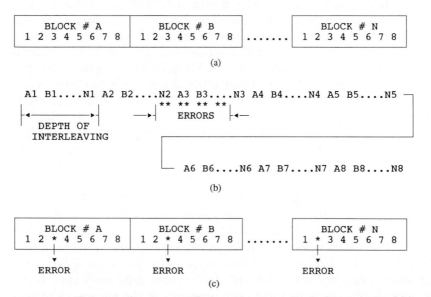

Figure 3.28   Code interleaving. (*a*) Blocks to be transmitted; (*b*) transmitted bit sequence; (*c*) blocks received.

by block. The number of blocks involved in the interleaving process is called the interleaving depth. Suppose the burst causes errors in N2, A3, B3,..., etc., of the transmitted bit stream. Because of the interleaving, these errors will show up in the receiver as isolated errors in blocks A, B, and N.

If the burst of errors is less than the interleaving depth, the errors can be corrected in the same manner as if they were random errors. The greater the interleaving depth, the longer the error burst that can be corrected but the greater the added redundancy and system delay. Interleaving is used on both RS and convolution codes, with the RS interleaving being preferable for long error burst (large interleaving depth) situations.

Interleaving tends to convert error bursts into semirandom errors, which is contrary to the philosophy that random errors are the most difficult to correct. Nevertheless, interleaving does have an important part to play in the practical implementation of error correction.

Concatenated codes.   Concatenated codes can provide the benefits of both RS codes and convolution codes to produce a performance that is synergistic in nature. In other words, the concatenated-coded BER is better than the sum of the two individual improvements of RS coding and convolution coding. The two coders and decoders are essentially in series, as indicated in Figure 3.29. If the first (inner) coder and decoder is used to improve the performance of a system degraded by Gaussian noise conditions, the second (outer) coder and decoder is then used to clean up short error bursts due to interference and fading. For relatively low bit rates (less than 100 Mb/s), the convolution coder with Viterbi decoder is a good option for the inner codec. Block coding would be more appropriate for higher bit rates and also if the noise is not Gaussian. The most suitable outer codec is usually the RS type. Figure 3.30 illustrates the dramatic performance improvement of a concatenated RS and convolution code compared with the convolution code alone and the uncoded signal. Admittedly, the redundancy of the RS(255,223) coder and (2,1), $K = 7$ convolution coder in this case is high, but even with a small percentage of redundancy, the performance is significantly improved.

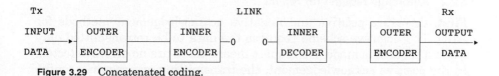

Figure 3.29   Concatenated coding.

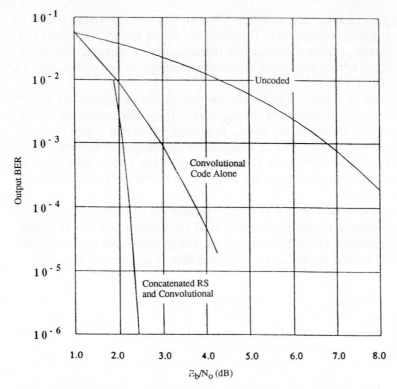

Figure 3.30 Performance of RS(255,223) and (2,1), $K = 7$ convolution code. (*Reproduced with permission from IEEE, © 1987, Berlekamp, E. R., et al., "The Application Error Control to Communications," IEEE Comms. Mag., Apr. 1987, Fig. 10.*)

Summary. To summarize, the following statements can be made about the use of coding to counter Gaussian noise:

1. Convolution coding is good for relatively low bit rate systems requiring BER values in the 10^{-3} to 10^{-7} region.

2. Concatenated convolution and block coding is applicable to low bit rate, excellent BER (10^{-10}) systems.

3. Block codes are useful for high bit rate, excellent BER requirement applications (better than BER = 10^{-10}).

3.2.2 Automatic request for repeat

First, there are positive and negative acknowledgement methods for ARQ. The negative scheme requires the data to be retransmitted *only* when a request is made. No request means there are no errors detected. In the positive acknowledgement, the transmitting end requires confirmation of every correctly received block of data. The negative method is clearly superior in terms of redundancy, but the positive method boasts superior data security. Because of the poor throughput of the positive method, it is used only in special circumstances, and the negative method is predominant. Two main types of ARQ are:

1. Go back N
2. Selective repeat

In the *go back N* system, negative acknowledgment is usually used when an error is received, in which case the receiver sends a request to the transmitter to repeat the most recent N blocks of data. N is typically 4 or 6. The advantage of this system is that the blocks do not have to be labeled, which minimizes redundancy. The main disadvantage is that the retransmission is greater than is desirable. This is necessary to ensure the error is eliminated. Furthermore, the maximum round-trip delay for the link must not exceed the time to transmit N blocks; otherwise the block containing errors may not be repeated.

As the name *selective repeat* suggests, the receiver only requests the transmitter to repeat specific blocks of data. While this method has the advantage that only the blocks containing errors are repeated (less redundancy than go back N), the blocks have to be individually numbered (more redundancy than go back N). Clearly, long blocks reduce the overhead, but they are also more likely to contain errors.

To summarize, ARQ is useful for channels that are mainly error-free but suffer rare error bursts of limited duration. While the output from an ARQ system has a predictable quality, the throughput of data is heavily dependent on the channel conditions. These characteristics make ARQ particularly well suited to packet-switched networks. For example, in a packet-switched network having selective ARQ, if one or more packets experience a noisy or failed path, the packet is repeated on a different path. The additional overhead inherent to the ARQ technique is easily absorbed by the packet-switched network.

Conversely, FEC systems have a constant throughput, and the data quality depends on the channel noise conditions. The complementary nature of ARQ and FEC leads one to believe that a combination of the two would be extremely effective for error control. In certain noise circumstances this is true. In a hybrid scheme where FEC is followed by ARQ, the idea is for the FEC to be dominant and for the ARQ to clear up any errors not detected by FEC. This combination is very effective for links that are almost uniformly noisy but have infrequent error bursts caused by interference or fading.

3.2.3 Trellis-coded modulation

There are several techniques for improving the BER of a digital communications system. As elaborated in more detail in Chap. 5, some techniques analyze and modify the received bit sequence, whereas others modify the received IF characteristics. Some techniques operate on the transmitter and receiver sides of the system. FEC, as just discussed, is successfully employed by sending some extra (redundant) bits to improve the BER. This has the disadvantage of increasing the bit rate (or using more bandwidth). Since higher levels of modulation have evolved to improve bandwidth efficiency, it seems a retrograde step to lose some of this valuable and hard earned bandwidth efficiency by using an error detecting code. *TCM was developed to improve the BER without increasing the bit rate.* Shannon's original work on information theory more than 30 years ago predicted the existence of such coded modulation schemes that could fulfill this function. Ungerboeck subsequently produced definitive papers (1976, 1982, etc.) outlining the implementation of TCM.

TCM can achieve a coding gain of up to 6 dB in the E_b/N_0, depending on the modulation scheme. Note that *asymptotic* coding gain is implied here, when the term *coding gain* is used, as is typical in literature on this subject. So far, it has been established that it is reasonable to relate the susceptibility of a modulation scheme to ISI by observing the distance between states in the signal state-space diagram. The trellis coding technique obtains its advantage by effectively maximizing the Euclidean (physical) distance between the states of a modulation scheme by expanding the signal set while avoiding bandwidth expansion. Notice this is distinctly different from convolution coding and Viterbi decoding which seek to minimize the Hamming distance. The term *trellis* refers to the state transition diagram, which is similar to the binary convolution coding trellis diagrams. The difference in the case of TCM is that the branches of the trellis represent redundant nonbinary modulation signals instead of binary code symbols.

Trellis coding for QAM. A technique which is central to TCM is called *set partitioning*. Consider the 16-QAM constellation diagram in Fig. 3.31. If this is now displayed as eight subsets each containing 2 of the 16 states, it is clear that the distance between each pair in any subset is larger than the minimum distance (between adjacent states) in the full 16-QAM constellation. Although this is a simplified explanation, it illustrates the essence of expanding the signal set to achieve coding gain while maintaining a fixed bandwidth efficiency.

This is called eight-state trellis coding, and if it were applied to higher levels of QAM, the number of states in the eight subsets would

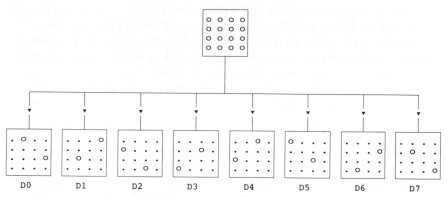

Figure 3.31 Set partitioning of the 16-QAM signal sets.

increase. For 2^{m+1}-QAM, where m is the number of bits per modulation interval, each subset would contain 2^{m-2} states, as follows:

m	2^{m+1}	2^{m-2}
3	16	2
4	32	4
5	64	8
6	128	16
7	256	32
8	512	64

The smallest distance between states, called the *free distance* (d_{free}), is an important parameter in establishing the coding gain. In a QAM signal this is simply the distance between adjacent states. The misinterpretation of any of the states in the constellation, caused by noise affecting the precise location of states, is referred to as the transition of one state to another. These transitions form the basis of the trellis diagram for TCM. In the 16-QAM diagram of Fig. 3.31, transitions are most likely to occur between adjacent states whose free distance is Δ_0. However, transitions between states other than between adjacent states can occur. In the eight subsets of Fig. 3.31, the free distance between states within each subset is $\sqrt{8\Delta_0}$. Taking into account the possibility of transitions occurring between states in different subsets, the free distance for this code is $\sqrt{5\Delta_0}$, which is still considerably better than Δ_0.

Figure 3.32 illustrates the eight-state trellis code for any eight-state QAM signal. For a specific code sequence of D0-D0-D3-D6, Fig. 3.32 traces four error paths at a distance of $\sqrt{5\Delta_0}$ from that code sequence. The paths all start at the same state and reconverge after three or four transitions. There will always be four error paths for this eight-state coding, regardless of the code sequence. Any error occurrences will most likely correspond to these paths, resulting in decision error bursts of length 3 or 4. It can be shown mathematically that the coding gain for eight-state coding for all relevant m values of QAM is around 4 dB. The coding gain can be increased to about 5 dB with 16 states and nearly 6 dB with 128 states. This is clearly a diminishing returns situation with considerable increase in circuit complexity as the number of states increases. Eight-state TCM is a good compromise between performance and complexity (and therefore cost).

Figure 3.33a and b graphically presents the theoretical performance curves for eight-state TCM. Figure 3.33a shows E_b/N_0 values plotted against BER for several levels of QAM. Figure 3.33b compares the E_b/N_0 values required to give a BER of 10^{-3} and 10^{-10} for coded and uncoded TCM for many QAM values. In fact the x axis plots the information rate (bits per symbol) which is equivalent to the bandwidth efficiency or is a measure of the QAM level. Notice the approximate asymptotic coding gain, denoted by BER = 10^{-10}, is nearly 4 dB.

3.3   Spread Spectrum Techniques

Initially, FM was used extensively for analog systems. Spread spectrum is now the subject of intense research and development for application to digital cellular mobile radio systems and fixed microwave radio systems. Spread spectrum techniques promise to improve the bandwidth efficiency and interference characteristics of future radio systems in general and, in particular, cellular radio systems.

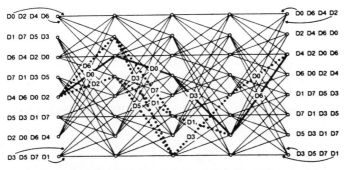

Figure 3.32   Eight-state trellis code for QAM. (*Reproduced with permission from Ref. 63, © 1987 IEEE.*)

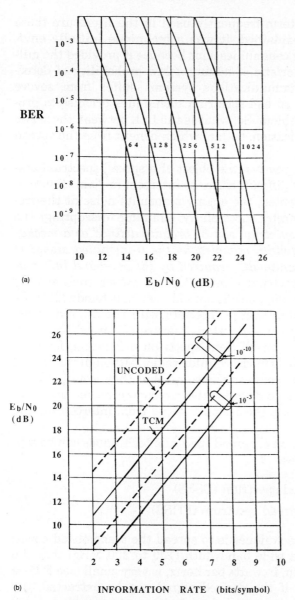

(a)

E_b/N_0 (dB)

(b)

INFORMATION   RATE   (bits/symbol)

Figure 3.33 (*a*) TCM performance curves (BER against E_b/N_0); (*b*) TCM performance curves (E_b/N_0 versus information rate).

Spread spectrum is a term frequently used in the literature these days, but it is seldom explained. It was a technique initially envisioned for use in military communications because it provided the militarily desirable characteristic of having a low probability of detection. Today's mobile communications systems suffer from severe spectral congestion, and spread spectrum techniques promise to provide a solution to this problem. So, what is the link between these two diverse types of communication, both of which benefit from the spread spectrum technique?

From the military perspective, the object of a spread spectrum system is to ensure that it is difficult either to intercept or "jam" the communication. Once intercepted, the enemy transmits noise at the frequency of operation, thereby successfully jamming (disrupting) the communication. *Spread spectrum can be broadly defined as a mechanism by which the bandwidth occupied by the transmitted signal is much greater than the bandwidth required by the baseband information signal.* A spread spectrum receiver operates over such a large bandwidth that the jamming noise is spread over that bandwidth in a manner that does not significantly interfere with the desired transmission. It is this interference-resistant feature of spread spectrum systems that is attractive for efficient spectrum utilization, for commercial applications such as micro-cellular radio. Although the spread spectrum signals are not secure in the sense that they would still require message encryption to disguise the message content, they provide enough security to stop the casual listener from intercepting the information.

The principle of operation of spread spectrum techniques can be separated into two categories:

1. Direct sequence spread spectrum (DSSS)

2. Frequency-hopping spread spectrum (FHSS)

The essence of the two techniques is to spread the transmitted power of each user over such a wide bandwidth (100 to 200 MHz) that the power per unit bandwidth, in watts per hertz, is very small (see Figure 3.34). This means that if the desired signal can be extracted, the power transmitted from unwanted, interfering sources and received by a typical narrowband receiver is a small percentage of the usual received power. The unwanted interference is therefore negligible.

In practice, a spread spectrum receiver spreads the energy of an interfering signal over its transmitted wide bandwidth while compressing the energy of the desired received signal to its original *prespread* bandwidth. For example, if the desired signal is compressed to its original bandwidth of 25 KHz, and all other interfering signals were

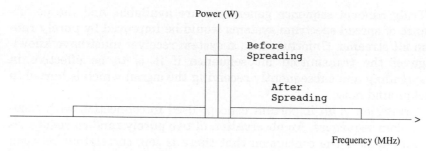

Figure 3.34 Spectrum spreading.

spread over 100 MHz in the receiver, the power in each interfering signal will be reduced by 100,000,000/25,000 = 4000, or 36 dB. This is a measure of the spectrum spreading and is defined as the processing gain N:

$$N = \frac{\text{bandwidth of the signal after spreading}}{\text{bandwidth of the unspread signal}}$$

As a background to the cellular mobile systems design discussed in Chap. 8, several associated techniques need to be clearly defined.

3.3.1 Pseudorandom noise generation

Both FHSS and DSSS use the principle of spreading the spectrum by using pseudorandom generated bit sequences. As the name suggests, these bit sequences are not truly random, but periodic. They are called pseudorandom because the periodicity is so large that usually more than 1000 bits occur before the sequence repeats itself, ad finitum. The simple pseudorandom noise generator circuit is shown in Figure 3.35. The periodicity depends on the number of flip-flops in the circuit n, with the length of the sequence being given by $2^n - 1$. This is often called a *maximal-length* sequence generation. In this case, as $n = 6$, the sequence repeats itself after every 63 bits.

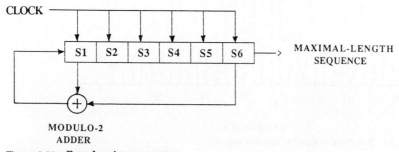

Figure 3.35 Pseudonoise generator.

Truly random sequence generators are available, and the performance of spread spectrum systems would be improved by purely random bit streams. Unfortunately, a system receiver must have knowledge of the transmitted bit sequence if it is to be effective in despreading and subsequently receiving the signal which is buried in background noise.

Correlation is an important concept used in connection with pseudorandom sequences. An observation of two purely random sequences would result in the conclusion that there is zero correlation between the two sequences. When comparing two sequences, the correlation can be defined as

$$R = \text{number of agreements} - \text{number of disagreements}$$

If a sequence is compared (correlated) to itself, the outcome is called autocorrelation, whereas correlation with another sequence results in cross-correlation. Comparison of two identical pseudorandom sequences would have perfect cross-correlation only for the in-phase or zero time shift condition and a very small correlation for all other phase relationships. Two sequences are said to be *orthogonal* if the degree of cross-correlation (the term "correlation" is more commonly used) is close to zero over the entire sequence comparison.

Figure 3.36 illustrates the line spectrum nature of a pseudorandom sequence. The null points occur at multiples of the shift-register clock frequency. The longer the sequence (defined by the number of shift registers), the more lines there are in the spectrum. For 10 shift registers, the sequence would be 1023 bits long, and the lines in the spectrum would be so close together that the spectrum would appear to be almost continuous.

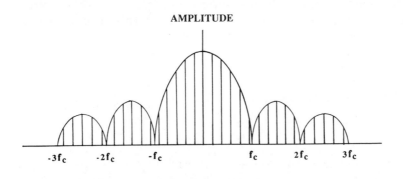

Figure 3.36 Practical spread spectrum signal.

3.3.2 Spread spectrum systems

The difference between FHSS and DSSS will now be addressed. The frequency-hopping systems achieve their processing gain by avoiding interference, whereas the direct sequence systems use an interference attenuation technique. In either technique, the objective of the receiver is to be able to pick out the transmitted signal from a wide received signal bandwidth in which the signal is below the background noise level. This sounds impossible, especially when considering the S/N are typically *minus* 15 to 30 dB. The receiver must have a certain minimum amount of information in order to perform this technological stunt. It must know the carrier signal frequency, type of modulation, pseudorandom noise code rate, and phase of the code. Only the phase of the code is a problem. The receiver must be able to establish the starting point of the code from the received signal. This process, which is referred to as *synchronization,* is essential to despreading the required signal while spreading all unwanted signals.

Frequency-hopping spread spectrum.   Frequency hopping, as the name suggests, periodically changes the frequency of the transmitted carrier signal so that it spends only a small percentage of its total time on any one frequency. If the total available bandwidth of, for example, 100 MHz, is partitioned so that carriers are spaced apart by 25 kHz, there are 4000 possible carrier frequencies to which the transmitted signal can hop. If a second user, occupying 25 MHz of bandwidth, also hops from one frequency to another, interference caused by the two simultaneously occupying the same frequency occurs only 1/4000 × 100 = 0.025 percent of the total time. It is evident that quite a number of users can simultaneously use such a system before the interference becomes even slightly noticeable. The main limitation of FHSS systems is the inability of the frequency synthesizer to change frequency quickly, without generating unwanted noise signals. As numerically controlled oscillator technology improves, tens of megahops per second within a 20-MHz bandwidth should be possible within the near future.

Figure 3.37 shows a schematic diagram of a frequency-hopping system. A pseudorandom noise code generator directly feeds a frequency synthesizer to produce the hopping. This allows the transmission of each bit or part of a bit (known as a *chip*) on a different carrier. The term *chip* arises because the frequency hopping rate is not equal to the transmission bit rate. The number of different carriers that can be used depends on the length of the pseudorandom bit sequence. At the receiving end the signal is dehopped by using a pseudorandom bit sequence identical to the one used in the transmitter. First, the RF sig-

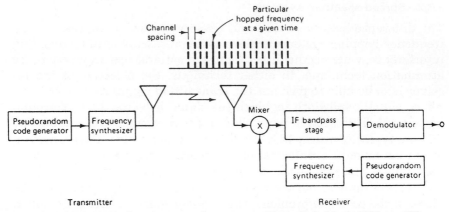

Figure 3.37   Frequency-hopping spread spectrum. (*From Ref. 54, © 1988, p. 238; reprinted by permission of Prentice-Hall, Englewood Cliffs, N.J.*)

nal is downconverted by mixing the received signal with the pseudorandom code-generated frequency. After passing through the IF bandpass filter, the signal is demodulated and the original bit stream recovered. A major advantage of frequency hopping is that the receiver does not need to have phase coherence with respect to the transmitter over the full spread-spectrum bandwidth. This means faster acquisition, because the receiver does not need to waste time searching and locking on to the correct phase of the transmitted signal. This noncoherent technique is, however, more susceptible to noise than a coherent method (e.g., direct sequence modulation).

3.3.3   Direct sequence spread spectrum

The DSSS technique acquires superior noise performance, compared to frequency hopping, at the expense of increased system complexity. The spectrum of a signal can be most easily spread by modulating (multiplying) it with a wideband pseudorandom code-generated signal (Figure 3.38). It is essential that the spreading signal be precisely known so that the receiver can demodulate (despread) the signal. Furthermore, it must lock onto and track the correct phase of the received signal within one chip time (partial or subinteger bit period). At the receiving end, a serial search circuit is shown. There are two feedback loops, one for locking onto the correct code phase and the other for tracking the carrier. For code phase locking, the code clock and carrier frequency generator in the receiver are adjusted so that the locally generated code moves back and forth in time relative to the incoming received code. At the point which produces a maximum at the correlator output, the two signals are synchronized, meaning the cor-

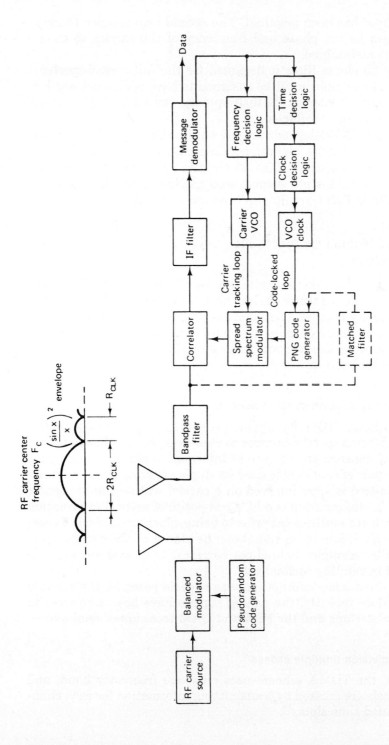

Figure 3.38 Direct sequence spread spectrum. (*Adapted by permission from Ref. 54, pp. 238 and 239, © 1988, Prentice-Hall, Englewood Cliffs, N.J.*)

rect code phase has been acquired. The second loop (carrier tracking loop) then tracks the phase and frequency of the carrier to ensure phase lock is maintained.

The input bandpass filter is designed for the full spread-spectrum bandwidth. The IF filter must be designed to have additional width to cope with small doppler shifts in frequency caused by high-velocity mobile stations (e.g., airplanes).

The success of DSSS depends largely on the chip rate. The spectrum should be spread at the highest possible chip rate. Complementary metal-oxide semiconductor (CMOS) circuits can operate at approximately 100 megachips per second, and gallium-arsenide field-effect transistor (GaAs FET) circuits, up to several gigachips per second.

3.4   Access Techniques for Mobile Communications

Both frequency division multiple access (FDMA) and TDMA are widely used for digital transmission, and these subjects are covered in most introductory texts on telecommunications. For completeness and to refresh the memory of those who may have forgotten, a very brief distinction between the two will now be clarified for the mobile communications application.

3.4.1   Frequency division multiple access

Broadly speaking, FDMA simply means splitting up an available frequency band into a specific number of channels, and the bandwidth of each channel depends on the type of information signals to be transmitted. One pair of channels is used for duplex operation. Information to be transmitted is superimposed on a carrier at the channel center frequency. The information can be a composite of several information signals, which are multiplexed prior to being superimposed on the carrier, or a single information signal can be placed on the carrier. This would be called a single channel per carrier (SCPC) system, which is widely used in satellite technology.

Initially, the analog information was superimposed on the carriers using FM. More recently, the analog signals have been converted to digital pulse streams and the PSK and QAM techniques employed.

3.4.2   Time division multiple access

By contrast, the TDMA scheme uses only one frequency band, and many channels are created by transmitting information for each channel in allocated time slots.

In a TDMA mobile radio system, each base station is allocated a 25- or 30-kHz channel, and users share this same channel on a time-allotted basis. The *maximum number* of users of each channel depends on how many bits per second are required to digitize the voice of each user. As indicated in Chap. 2, the conventional voice A/D conversion process requires 64 kb/s. ADPCM is already accepted as a means of reducing this to 32 kb/s. By incorporating several other digital processing techniques, such as linear predictive coding, 8.5 kb/s is possible without significantly noticeable degradation in quality, and 2.4 kb/s can be used if some noticeable degradation is tolerated. If a digital modulation technique such as 4-PSK is used, each voice channel can be digitized with a bandwidth efficiency of at least 1 b/s/Hz (i.e., 8.5 kHz of bandwidth is required for each digital voice signal). This means that for a 30-kHz TDMA channel, there can be three users per channel (or twelve users per channel with degradation).

3.4.3 Code division multiple access

Although the previous two access methods will already be known to most engineers and technicians, CDMA has only recently received widespread interest because of its potential benefits in mobile telecommunication systems. The objective of CDMA is to allow many users to occupy the exact same frequency band without interfering with each other. It sounds too good to be true. However, with our spread-spectrum knowledge, it is quite a reasonable statement. Each user is assigned a unique orthogonal code.

In a CDMA system all signals from all users will be received by each user. Each receiver is designed to listen to and recognize only one specific sequence. Having locked onto this sequence, the signal can be despread, so the message stands out above the other signals, which appear as noise in comparison. Interference does become a limiting factor because, eventually, as more users occupy the same frequency band, the noise level rises to a point where despreading does not provide an adequate S/N.

The asynchronous nature of CDMA compared to TDMA gives it the advantage that network synchronization is not required. Second, it is relatively easy to add users to a CDMA system. Third, CDMA is more tolerant to multipath fading than TDMA or FDMA. In conclusion, CDMA has the potential to enable a more efficient use of the frequency spectrum than other techniques, and when the remaining technical problems have been solved, it should be particularly well suited to cellular mobile radio systems (see Chap. 8).

The Microwave Link

In its simplest form the microwave link can be one *hop,* consisting of one pair of antennas spaced as little as one or two kilometers apart, or can be a *backbone,* including multiple hops, spanning several thousand kilometers. A single hop is typically 30 to 60 km in relatively flat regions for frequencies in the 2- to 8-GHz bands. When antennas are placed between mountain peaks, a very long hop length can be achieved. Hop distances in excess of 200 km are in existence. The "line-of-sight" nature of microwaves has some very attractive advantages over cable systems. *Line of sight* is a term which is only partially correct when describing microwave paths. Atmospheric conditions and terrain effects modify the propagation of microwaves so that even if the designer can see from point A to point B (true line of sight), it may not be possible to place antennas at those two points and achieve a satisfactory communication performance.

The objective of microwave communication systems is to transmit information from one place to another without interruption and clear reproduction at the receiver. Figure 4.1 indicates how this is achieved in its simplest form. The voice, video, or data channels are combined by a technique known as multiplexing to produce a BB signal. This signal is frequency modulated to an IF and then upconverted (heterodyned) to the RF for transmission through the atmosphere. The reverse process occurs at the receiver. The microwave transmission frequencies are within the approximate range 2 to 24 GHz. The frequency bands used for digital microwave radio are recommended by the CCIR and summarized in Fig. 4.2. Each Recommendation clearly defines the frequency range, the number of channels that can be used within that range, the channel spacing, the bit rate, and the polarization possibilities. There are also CCIR reports (indicated in Fig. 4.2)

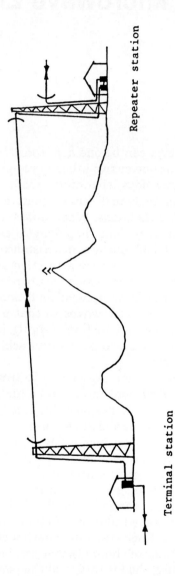

Figure 4.1 Simplified microwave link with schematic diagram.

CCIR ref.	Frequency range (GHz)	Digital			Analog	
		Bit rate (Mb/s)	No. of speech channels	No. of RF ch. pairs	No. of speech channels	No. of RF ch. pairs
Rec. 283	1.7–1.9	2, 2 × 2	30, 60	6	60, 120, 300	6
Rec. 701	1.9–2.1	2, 2 × 8, 8, 2 × 8	30, 60, 120, 240	6	60, 120, 300	6
Reps. 379,	2.1–2.3	2, 2 × 8, 8, 2 × 8	30, 60, 120, 240	6	60, 120, 300	6
934, 1055	2.5–2.7	2, 2 × 2	30, 60, 120, 240	24	60, 120, 300	—
Rec. 382	1.7–2.1	34	480	6		
Rep. 934	1.9–2.3	34	480	6		
	3.8–4.2	34, 2 × 34, 90	480, 960, 1344	6	1800, TV	—
Rec. 635		140, 200, 155	1920, 2880, 2016			
Rec. 383	5.925–6.425	34, 2 × 34	480, 960	8	1800, TV	
Rec. 934		140, 155	1920, 2016			—
Rec. 384	6.43–7.11	34, 2 × 34	480, 960	8	2700, TV	
Rep. 934		140, 155	1920, 2016			—
Rep. 934	7.1–7.75	34, 2 × 34	480, 960	10	300	20
Rep. 1055	7.725–8.275	34	480	12	960, TV	8
Rec. 386	7.725–8.275	34, 90	480, 1344	8	960/TV	8
Rep. 1055	8.275–8.5	34 (2 × 8)	480 (240)	6 (12)	960/TV	8
Rec. 387	10.7–11.7	Up to 140/155	1920/2016	12	1800	—
Rep. 782	10.7–11.7	140, 155	1920, 2016	6	1800/TV	12
	10.7–11.7	140, 155	1920, 2016	8		
Rec. 497	12.75–13.25	34	480	8 + 8	960	8
Rep. 607						
Rec. 636	14.4–15.35	70, 140	960, 1920	8 or 16	—	—
Rep. 607						
Rec. 595	17.7–19.7	140	1920	8 + 8	—	—
Rep. 936		34	480	8 + 8		
		280	3840/4032	7		
Rec. 637	21.2–23.6	—	—	—	—	—
Rep. 936						

Figure 4.2 Frequency bands for radio-relay systems.

that give additional technical detail associated with the Recommendations. Some AMR information is also given in Fig. 4.2 as a comparison to the digital radio. Factors influencing and leading to those Recommendations for point-to-point microwave radio transmission are discussed in the following.

4.1 Antennas

Since antennas play a central role in microwave communications, they will be considered first. There are several shapes of antenna available for transmitting microwaves. Telecommunication systems almost always use the parabolic type and sometimes the horn type. These antennas are highly directional. The microwave energy is focused into a very narrow beam by the transmitting antenna and aimed at the receiving antenna, which concentrates the received power by a mechanism analogous to the telescope. Figure 4.3 shows how the microwave energy is transmitted by a parabolic antenna, by placing the microwave guide opening at the focus of the parabola. For the simplest style of antenna feed, the waveguide opens in the form of an enlarging taper which is designed to match the impedance of the waveguide to that of free space. This system is analogous to the searchlight or flashlight beam at optical frequencies. Both light and microwaves are electromagnetic waves, so they have similar qualities. Since we can see light (or light allows us to see), it is often helpful to use the analogous optical mechanism to shed some light on the subject of microwaves. Parabolic antennas are available in sizes ranging from about 1 to 36 m in diameter.

4.1.1 Antenna gain

The most important characteristic of antennas is the gain. This is a measure of the antenna's ability to transmit the waves in a specific direction instead of in all directions. It is a measure of directionality. An antenna radiating energy equally in all directions is called an

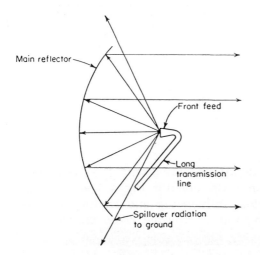

Main reflector

Front feed

Long transmission line

Spillover radiation to ground

Figure 4.3 Front feed paraboloid antenna. *Note:* Microwaves follow most of the rules of optics. (*Reproduced with permission from Ref. 95, Fig. 11-4, © 1972 McGraw-Hill.*)

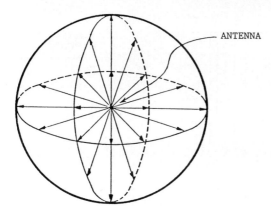

Figure 4.4 Radiation pattern—isotropic antenna.

omnidirectional, or *isotropic,* antenna (Fig. 4.4). For a point-to-point system, as in microwave communication systems, it is desirable to have a high degree of directionality. In other words, the isotropic antenna is not efficient because energy is wasted. The gain of an antenna describes the extent to which an amount of isotropically radiated energy can be directed into a beam. The narrower the beam, the more highly directional the antenna and therefore the higher the gain.

Mathematically,

$$\text{Gain } (G) = 10 \log_{10}\left(\frac{4\pi Ae}{\lambda^2}\right) \quad \text{dB} \tag{4.1}$$

where A = effective area of the antenna aperture
  e = efficiency
  λ = wavelength

An isotropic antenna, by definition, has a gain of 1 (or 0 dB). For a parabolic antenna the efficiency is not 100 percent because some power is lost by "spillover" at the edges of the antenna when it is illuminated by the waveguide fixed at the focus. Also, the antenna dish is not fabricated perfectly parabolic in shape. The waveguide feed at the focus causes some reduction of the transmitted or received power since it is a partial blockage to the microwaves. Commercially available parabolic antennas have efficiencies in the region of 50 to 70 percent. For a conservative efficiency of approximately 50 percent, Eq. (4.1) can be rewritten

$$G = 20 \log_{10}(7.4 \, Df) \quad \text{dB} \tag{4.2}$$

where D = antenna diameter (m)
  f = frequency (GHz)

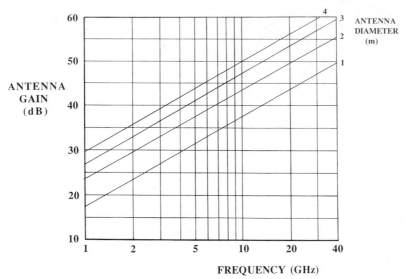

Figure 4.5 Variation of antenna gain with frequency.

The graph in Figure 4.5 shows the relationship between antenna diameter, frequency, and gain for an antenna efficiency of 50 percent.
As the equation suggests, the gain increases with frequency and also
with antenna diameter. The most frequently used antenna for the 4-
to 6-GHz band is the 3-m-diameter antenna, which has a gain in the
region of 40 dB. The diagram in Fig. 4.6 illustrates the manner in
which the microwave power is radiated from a parabolic antenna. The
measured radiated power level is mapped over a full 360° range. The
maximum power is transmitted in the direction called the *boresight.*
The boresight field strength is designated as 0 dB, so the field
strength in any other direction is referred to that maximum value.
The major lobe (main beam) is quite narrow, and the sidelobes and
backward radiation are more than 25 dB below the major lobe. Ideally, the transmitted energy should be only in the main beam, but imperfect illumination of the parabolic antenna and irregularities in its
reflecting surface cause the sidelobes.

An important characteristic of antennas is the *front-to-back ratio.* It
is defined as the ratio of maximum gain in the forward direction to the
maximum gain in the backward direction (backlobes). To illustrate
the point, Fig. 4.6 shows a rather poorly designed antenna because it
has a front-to-back ratio of only 32 dB. This backward radiation has
great significance with respect to noise and interference. Signals

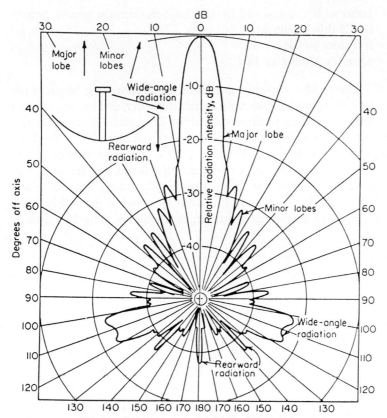

Figure 4.6 Radiation pattern for a microwave antenna. (*Reproduced with permission from Ref. 95, Fig. 11-1, © 1972 McGraw-Hill.*)

reach the receiver after reflection from the ground behind the antenna. This introduces unwanted noise. Secondly, at repeaters where the same frequency is used for incoming and outgoing signals, coupling occurs between the receiver and transmitter antennas. Since transmitter power levels are usually at least 60 dB higher than receiver levels, it is evident that there must be a high degree of isolation between antennas in order to avoid severe interference.

4.1.2 Beamwidth

The beamwidth is another important characteristic of antennas, and for parabolic antennas the beamwidth is

$$\phi = \frac{21.3}{fD} \qquad (4.3)$$

where φ = beamwidth measured at the half-maximum power points
(3-dB down points)

f = frequency (GHz)

D = antenna diameter (m)

Figure 4.7 shows the antenna gain and corresponding beamwidth plotted against the parabolic antenna dish area divided by the square of the wavelength, which produces straight-line plots. Note that the beamwidth for a 3-m antenna is very narrow, less than 2° in the 4- to 6-GHz range (i.e., for A/λ^2 = 1.3 to 2.9 × 10³). If larger antennas are used, the beamwidth is reduced further. Interference from external sources and adjacent antennas is minimized by using narrow beam antennas. Although large-diameter antennas provide desirably high gain, the decrease in beamwidth can cause problems. The two anten-

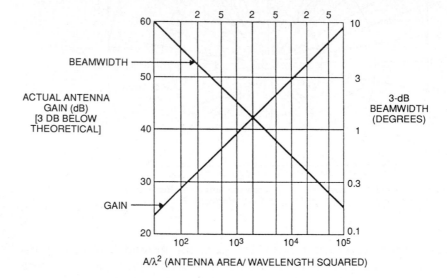

ANTENNAS	A/λ^2		
	4 GHz	6 GHz	11 GHz
10 ft (3.05 m) PARABOLOID	1.3 x 10³	2.9 x 10³	9.8 x 10³
8 ft (2.44 m) PARABOLOID	8.3 x 10²	1.9 x 10³	5.6 x 10³
4 ft (1.22 m) PARABOLOID	2.1 x 10²	4.1 x 10²	1.6 x 10³
8 ft SQUARE	1.1 x 10³	2.4 x 10³	8.0 x 10³

Figure 4.7 Approximate antenna gain and beamwidth.

nas in each hop must be aligned very precisely; the narrower the beamwidth, the higher the alignment precision required. A very small movement in either antenna will cause degradation of the received signal level. This problem can be serious, particularly when large antennas are used on high towers in very windy regions.

4.1.3 Polarization

If a horn antenna is used to illuminate the antenna, it is orientated in one of two positions. The narrow dimension of the rectangular opening is placed either horizontally or vertically with respect to the ground. If it is vertical, the electric field is in the horizontal plane and the antenna is said to be *horizontally polarized* (Fig. 4.8a). If the narrow dimension of the horn is horizontal, the electric field is in the vertical plane and the antenna is *vertically polarized* (Fig. 4.8b). If the signal is transmitted, for example, in the horizontal polarization, there is unavoidably a small amount of power transmitted in the vertical polarization. The signal in the unwanted polarization is typically 30 to 40 dB lower than the intended polarization. This is referred to as a *cross-polarization discrimination* (or **XPD**) of 30 to 40 dB. Often two radio channels having opposite polarizations are placed on the same

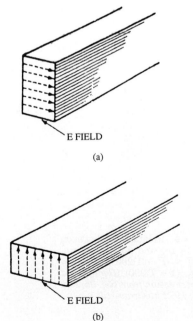

E FIELD

(a)

E FIELD

(b)

Figure 4.8 Waveguide representation of horizontal (*a*) and vertical (*b*) polarization.

antenna at the same frequency. This is known as frequency reuse, or *frequency diversity,* and is discussed in detail in Sec. 4.7 and Chap. 5.

4.1.4 Antenna noise

Noise entering a system through the antenna can set a limit to the performance of the system. A communication system performance is designed to have the lowest possible noise figure, often stated in terms of noise temperature. These parameters are used interchangeably in most texts, so the relationship between noise figure and noise temperature is given in the graph of Fig. 4.9. It is useful to become familiar with both types of notation and to memorize the approximate equivalent values for, say, 1-, 3-, 5-, and 10-dB noise figures (i.e., 1 dB $\equiv$ 75 K, 3 dB $\equiv$ 290 K, 5 dB $\equiv$ 627 K, and 10 dB $\equiv$ 2610 K, respectively). The noise temperature of the antenna, T_a, depends on:

- The loss between the antenna and the receiver input (i.e., waveguide run)
- Sky noise from the galaxy, sun, and moon
- Absorption by atmospheric gases and precipitation
- Radiation from the earth into the backlobes of the antenna
- Interference from artificial radio sources

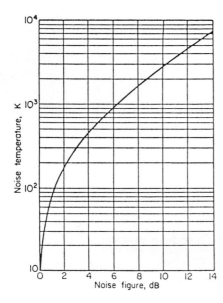

Figure 4.9 Relationship between noise figure and noise temperature. $T = (F - 1)290$ or $NF_{dB} = 10 \log_{10} (1 + T/290)$. (*Reproduced with permission from Ref. 95, Fig. 6-6, © 1972 McGraw-Hill.*)

Suppression of the sidelobes is an important aspect of reducing antenna noise. Although the main beam may not be directed toward a "hot" part of the sky (e.g., the sun), one of the sidelobes may be, causing T_a to be large. A countermeasure used to minimize sidelobe problems is to place a shroud (shield) around the edge of the parabolic dish. This is in the form of a metallic rim, as shown in Fig. 4.10. This improves the front-to-back ratio of a 3-m antenna operating at 6 GHz from the unshielded value of approximately 50 dB to greater than 70 dB. This is sufficient to allow transmission of the same frequency in both directions of a back-to-back repeater station with negligible interference between signals. A thin, negligible attenuation, plastic

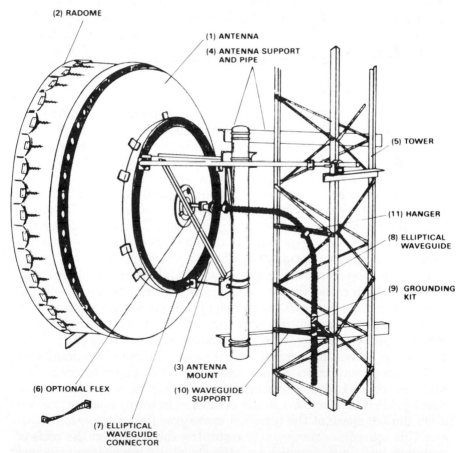

Figure 4.10 Parabolic antenna with shroud and radome connected to an elliptical waveguide system. (*Reproduced with permission from Andrew Corporation.*)

cover called a *radome* is often attached to the shroud to provide weather protection.

4.1.5 High-performance antennas

An even higher performance can be obtained from a *horn reflector* antenna (Fig. 4.11). This design has a section of a very large parabola placed at an angle so that the energy fed from the waveguide is both focused and reflected from the parabolic surface. This antenna has excellent sidelobe suppression and a 3-m antenna at 6 GHz has a front-to-back ratio of about 90 dB. It also has very low voltage standing-wave ratio (VSWR) characteristics, low noise temperature, and substantially wider bandwidth than ordinary parabolic antennas. Although the electrical characteristics are superior to the usual style of parabolic antenna, the horn reflector type is more expensive, larger, heavier, and more difficult to mount. Nevertheless, in circumstances where very high performance is warranted, it is indispensable.

4.1.6 Antenna towers

The towers that are built for the microwave antennas significantly affect the cost of the microwave link. The higher the towers are built, the longer the hop and therefore the cheaper the overall link. Calculations show that for a 48-km hop, over relatively flat terrain, towers of about 76 m are required at each end. If there are obstacles such as trees or hills at some points on the path, towers of 107 m or more may be necessary to provide adequate clearance. Towers of this height require guy wires to support them. This is because the cost of self-supporting towers is too expensive since the cost increases almost exponentially with height, whereas the cost of guyed towers (having constant cross section) increases linearly with height. In both cases, the amount of land required for the towers increases considerably with tower height. As an example, using Fig. 4.12, the land area required for a guyed tower is determined by the area occupied by the guy wires. They extend outward from the tower a distance of about 80 percent of the tower height. The minimum ground area required for an approximately 91-m guyed tower is 1.47 hectares, or 3.64 acres. Although this may not be a problem in the countryside where land is usually available, in towns or cities there is often limited space at the terminal exchanges for building high towers. This sometimes necessitates mounting the towers on the roofs of exchange buildings, in which case the structural adequacy of the roof must be carefully evaluated. In addition, local building codes or

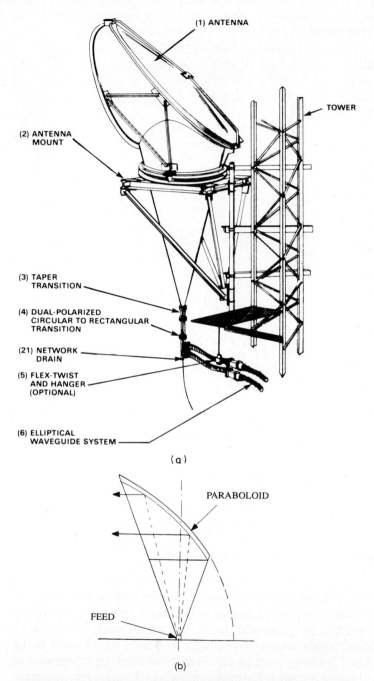

Figure 4.11 Horn reflector antenna. (*a*) Horn antenna connected to an elliptical waveguide system (*reproduced with permission from Andrew Corporation*). (*b*) Schematic diagram of the microwave beam propagating from the antenna.

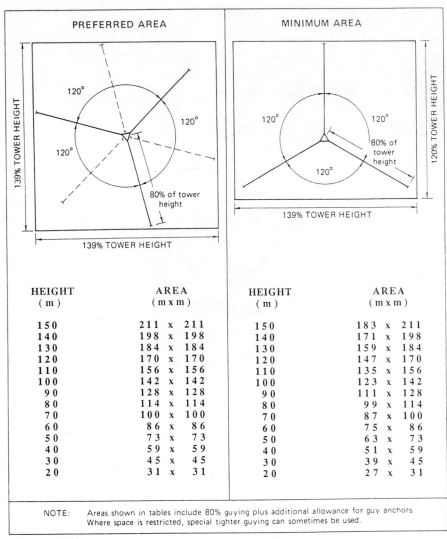

HEIGHT (m)	AREA (m x m)	HEIGHT (m)	AREA (m x m)
150	211 x 211	150	183 x 211
140	198 x 198	140	171 x 198
130	184 x 184	130	159 x 184
120	170 x 170	120	147 x 170
110	156 x 156	110	135 x 156
100	142 x 142	100	123 x 142
90	128 x 128	90	111 x 128
80	114 x 114	80	99 x 114
70	100 x 100	70	87 x 100
60	86 x 86	60	75 x 86
50	73 x 73	50	63 x 73
40	59 x 59	40	51 x 59
30	45 x 45	30	39 x 45
20	31 x 31	20	27 x 31

NOTE: Areas shown in tables include 80% guying plus additional allowance for guy anchors. Where space is restricted, special tighter guying can sometimes be used.

Figure 4.12 Approximate area required for guyed towers.

air-traffic control regulations can impose restrictions on the height of towers. Local soil conditions must also be taken into account. Extra cost can be incurred in areas with hard rock which must be moved or in very soft soil areas where extra-large concrete bases need to be built. Also, wind loading must be taken into account, or movement of the tower will cause outages. Even the antennas themselves can cause problems if there are too many placed on one tower in an imbalanced configuration.

4.2 Free Space Propagation

Since microwave energy is electromagnetic energy, it will pass through space (a vacuum) in a manner similar to that of light. Nobody knows how it propagates through "nothingness," but we know it does propagate. The atmosphere and terrain have modifying effects on the loss of microwave energy, but first only the loss as a result of free space will be considered. Free space loss can be defined as the loss between two isotropic antennas in free space, where there are no ground or atmospheric influences. The isotropic antenna by definition radiates energy equally in all directions. Although this is a hypothetical ideal, which cannot be realized physically, it is a useful concept for calculations. As stated previously, it allows the directivity (or gain) of an antenna to be described relative to this omnidirectional reference, the isotropic antenna.

When imagining how energy emanates from an isotropic antenna, it is easy to appreciate how a fixed amount of energy emitted by the antenna will spread out over an increasing area as it moves away from the antenna. It is analogous to considering the spherical surface area of an expanding balloon. The energy loss due to the spreading of the wavefront as it travels through space is according to the inverse-square law. Since the receiving antenna occupies only a small portion of the total sphere of radiation, it is reasonable to assume only a minute fraction of the total emitted radiation is collected. Mathematically, the free space loss has been derived as

$$\alpha_{fs} = \left(\frac{\lambda}{4\pi d}\right)^2 \tag{4.4}$$

where λ = wavelength
d = distance from source to the receiver

If the path distance is in kilometers and the frequency of the radiation is in gigahertz, then the previous equation can be written as

$$\alpha_{fs} = 92.4 + 20 \log_{10}f + 20 \log_{10}d \tag{4.5}$$

(If d is in miles, 92.4 changes to 96.6.) This equation is illustrated in the form of a graph (Fig. 4.13) where free space loss is plotted against distance for several frequencies. For a typical repeater spacing of approximately 48 km, the free space loss is 132 dB at 2 GHz and 148 dB at 12 GHz. This is an extremely large value of attenuation. If only 1 W of microwave power is transmitted, only 10^{-13} W of power are received. These figures are for the case of isotropic antennas at both the transmitter and receiver. Fortunately, the gain of each high-performance directional antenna used in microwave communication systems is ap-

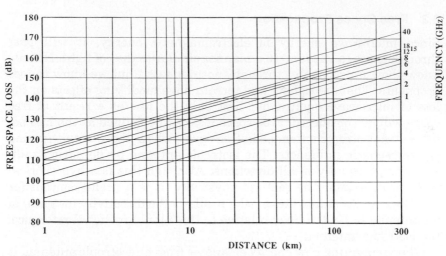

Figure 4.13 Free space loss between isotropic radiators.

proximately 40 dB. The received power is therefore improved from 10^{-13} W to about 10^{-5} W. Equation (4.5) is presented in a different form in Fig. 4.14 by plotting free space loss against frequency for various distances. This graph gives a good feel for how the free space loss increases with distance for a particular operating frequency.

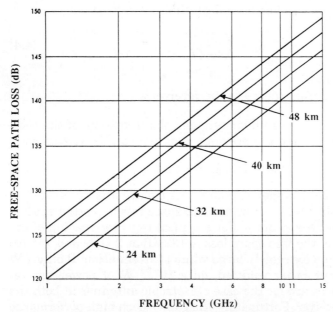

Figure 4.14 Free space loss versus frequency and path length.

4.3 Atmospheric Effects

4.3.1 Absorption

True free space propagation exists for a portion of a satellite communication path, but terrestrial microwave communication paths always require propagation through the atmosphere. The oxygen in the atmosphere absorbs some microwave energy, as shown in Fig. 4.15. Fortunately, this attenuation is relatively small in the frequency range used for microwave communication. It is approximately 0.01 dB/km at 2 GHz and increases to 0.02 dB/km at 26 GHz.

The effect of rain on microwave radio propagation is quite significant, especially at the higher frequencies. As the graph in Fig. 4.15 shows, the attenuation increases rapidly as the water content of the atmosphere in the microwave path increases. At 6 GHz the attenuation due to water vapor in the air is only 0.001 dB/km. As the water content increases to fog then light rain, the attenuation increases to 0.01 dB/km, and for a cloudburst (very heavy rain) the attenuation is about 1 dB/km. The microwave energy is absorbed and scattered by the raindrops. For a 40-km hop, this would cause a 40-dB increase in the attenuation, which is enough to cause problems with the quality of transmission. Usually, cloudbursts do not cover a distance of 40 km, so the attenuation would not be as high as 40 dB. It can be concluded that rain is not a serious problem below 6 GHz, although it can impair quality in regions that frequently have very heavy rainfall.

At higher frequencies, especially above 10 GHz, rainfall can cause severe transmission problems. For example, at 12 GHz, the attenuation can reach almost 10 dB/km. An extensive cloudburst can cause a temporary break in transmission. In a DMR system, that may cause several hundred calls to be dropped. In circumstances where very high reliability is required, it may be necessary either to reduce the length of hops or to use a lower transmission frequency.

4.3.2 Refraction

In addition to the attenuation effects of the atmosphere, there is also the problem of refraction. The effect of refraction is to cause the microwave beam to deviate from its line-of-sight straight path. Its effects on transmission can be very serious, causing outages of a fraction of a second up to several hours. Predicting its occurrence can be done only on a statistical basis. Methods of protecting a link against this problem are discussed later.

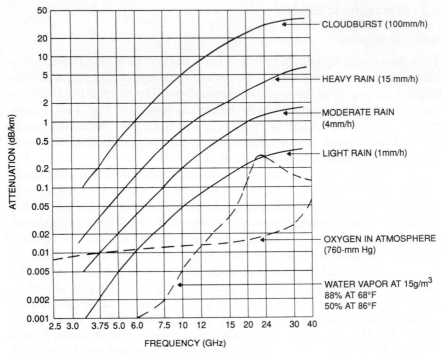

Figure 4.15   Estimated atmospheric absorption.

Refraction is a bending of the radio waves due to changes in the characteristics of the atmosphere. The atmosphere changes in temperature, density, and humidity as the altitude from the earth's surface increases. The change in density affects the velocity of microwaves traveling through the atmosphere:

$$\text{Velocity } (v) = \frac{c}{n} = \frac{c}{\sqrt{\epsilon_r}} \tag{4.6}$$

where n = refractive index
c = speed of light
ϵ_r = relative permittivity

Figure 4.16 recalls to the memory the simple physics diagram of how the high velocities at the top of a wavefront, in thinner air, cause the microwave beam to bend. This is analogous to light rays bending on entry into water from air. The bending in these circumstances is downward, toward the earth, which allows the microwaves to be transmitted further than the direct straight-line path by tilting the

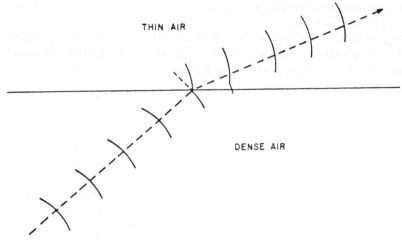

Figure 4.16 Refraction of a microwave beam.

antennas slightly upward (Fig. 4.17). This is equivalent to increasing the radius of the earth's curvature. On average, atmospheric conditions cause the propagation path to have a radius of curvature that is approximately 1.33 times the true earth radius. In practice, this allows a propagation path length that is approximately 15 percent longer than the line-of-sight path. It is common practice to plot the radio path on profile paper that is corrected to four-thirds the earth's radius so that the microwave beam can be plotted as a straight line. The change of earth curvature caused by refraction is denoted by the *k-factor,* which is defined as the ratio of the effective earth radius to the true earth radius:

$$k = \frac{\text{effective earth radius}}{\text{true earth radius}} \qquad (4.7)$$

The effective earth radius is often misunderstood. It is not the radius of the microwave beam. For a given atmospheric condition, it is the

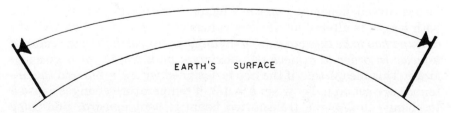

Figure 4.17 Transmission distance increased by refraction.

radius of a fictitious earth which allows the microwave beam to be drawn as a straight line.

A more mathematically precise definition of the effective earth radius is $1/[(1/a) + (dn/dh)]$, where a is the true earth radius and dn/dh is the gradient of the refractive index n, with respect to the altitude h. So,

$$k = \frac{1}{1 + (a\,dn/dh)} \tag{4.8}$$

The standard value of k is taken to be 4/3. Variations in atmospheric conditions occur daily and also hourly, depending upon geographical location. On a larger time scale, significant seasonal changes can affect transmission. In general, k is larger than 4/3 in warm temperature areas and less than 4/3 in cold temperature areas, perhaps lying between 1.1 and 1.6 for most countries, depending upon latitude and season.

In addition, k can have values of less than 1 to ∞ or even can be negative. For example, when a temperature inversion occurs, a warm layer of air traps a cooler surface layer of air. The k-factor in this case is less than 0. Other conditions, such as a steep change in temperature of the air from the earth's surface to an altitude of several hundred meters can produce $k = \infty$. This special case denotes that a microwave beam follows the exact same curvature as that of the earth. If this condition were present around the complete surface of the earth, it would be possible to transmit across the Atlantic Ocean with only one hop. Of course, this situation never occurs, and even if it did, the received signal level would be too low to be useful. Nevertheless, $k = \infty$ does occur over short distances under the appropriate atmospheric conditions. Further downward bending beyond the $k = \infty$ condition produces a negative k-factor. Figures 4.18 and 4.19 illustrate the k-factor effects on radio paths. Figure 4.18 shows a straight-line microwave beam and how the effective earth radius causes a "bulging" effect which can enhance or reduce the path clearance, depending on the k-factor. Figure 4.19a shows the true situation of the real earth curvature and how a microwave beam transmitted at 90° to the earth's surface is curved, depending on the k-factor. Figure 4.19b shows how the path length is altered for various values of the k-factor. *Note that the antenna has to be tilted to a different angle to the earth for each value of k-factor in order to achieve the maximum path length for a given k-factor.* This means that if the hop is designed for a $k = 4/3$ and the antennas are set on a day when $k = 4/3$, if temperature changes cause k to change to 2/3, the transmitted beam is bent *upward,* and some power will be lost since the reception will be off the peak of the major

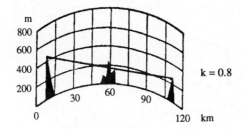

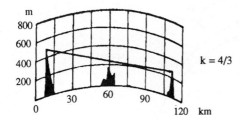

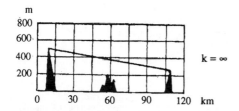

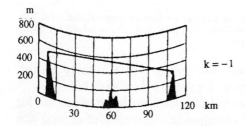

Figure 4.18 Effective earth profile for several *k*-factors showing relative obstacle clearance.

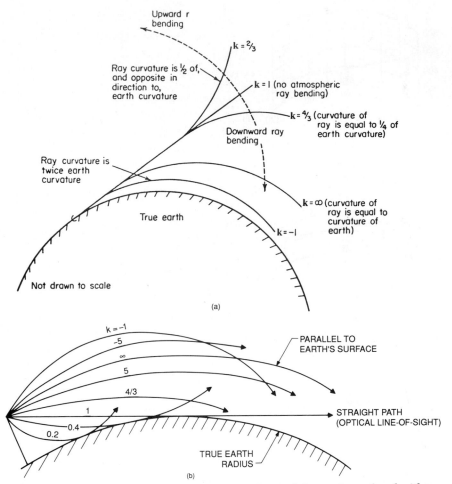

Figure 4.19 (a) Microwave beam bending for different k-factors (*reproduced with permission from Ref. 95, Fig. 13-4, © 1972 McGraw-Hill*). (b) Antenna oriented for maximum path length for various k-factor values.

lobe (off boresight). Maximum received power will similarly be unachievable if the temperature causes k to change to, say, 2.0.

In this case, the beam will be bent *downward,* with antennas set for $k = 4/3$. These are very extreme changes in the value of k, and the power loss would still only amount to about 1 dB for a 50-km-length link using 3-m-diameter antennas. For longer links and larger antennas the loss is worse. Furthermore, the downward bending of the $k = 2.0$ case could lead to loss problems due to inadequate clearance from obstacles.

When designing a microwave hop, it is often more convenient to plot a profile of the path using a straight-line microwave path rather than a curved line. For this purpose, an equivalent earth profile template is available, as in Fig. 4.20. This is derived using the equation

$$h = \frac{0.079 \, d_1 d_2}{k}$$

(4.9)

where h is the vertical distance (in meters) between the flat earth ($k = \infty$) and the effective earth at any given point, and d_1 and d_2 are the distances (in km) from a given point to each end of the path. The main disadvantage of this technique is that if several values of the k-factor are to be considered, multiple plots are necessary, each on different profile paper. Conversely, if the flat earth method is used, several microwave beams (one for each k-factor) can be plotted on the same diagram.

4.3.3 Ducting

Atmospheric refraction can, under certain conditions, cause the microwave beam to be trapped in an atmospheric waveguide called a *duct*, resulting in severe transmission disruption. Ducting is usually caused by low-altitude, high-density atmospheric layers, most frequently occurring near or over large expanses of water or in climates where temperature or humidity inversions occur. Figure 4.21 gives an example of how the transmitted beam becomes trapped in a duct. When the beam enters the duct and it reaches the other interface between the two different density layers, the critical angle is exceeded so that internal reflection occurs. Subsequently, the beam bounces back and forth as it travels along the duct, and the receiving antenna loses the signal.

4.4 Terrain Effects

Propagation of microwave energy is affected by obstacles placed in its path. As described previously, the earth curvature (or effective curvature) is a dominant factor in determining hop length. Previously, a smooth earth was considered. Now the effects of obstacles such as rocks, trees, and buildings will be discussed. The shape and material content of any obstruction must be taken into account when surveying a microwave path. Unfortunately, objects close to the direct-line path can cause problems even though they are not obstructing the line of sight.

4.4.1 Reflections

The raylike beam that has been drawn in previous illustrations is a convenient tool for describing several physical concepts. However, it

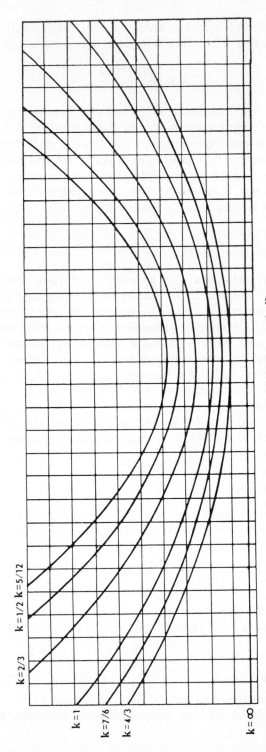

Figure 4.20 Equivalent earth profile template. Scale: Each square equals 30.5 m vertically and 3.2 km horizontally; or, each square equals 122 m vertically and 6.4 km horizontally.

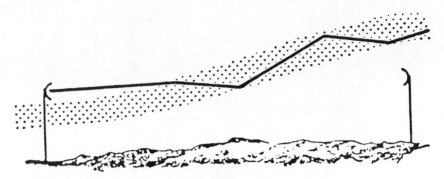

Figure 4.21 Microwave beam trapped in a duct.

must be remembered that, although a microwave beam may be only 1 or 2° in half-power width, this still represents a large area of energy spread at a distance of 40 km from the transmitter. Simple geometry indicates the half-power cone to have enlarged to a circle of approximately 1.4 km in diameter for a 2° beam or 0.7 km for a 1° beam. This means that some of the energy off boresight will be reflected from the ground (Fig. 4.22) or other nearby objects at either side of the direct line. Figure 4.23 shows how a wave undergoes a 180° phase reversal when it is reflected. At the receiving antenna, energy arrives from the direct and reflected paths. If the two waves are in phase, there is an enhancement of the signal, but if the waves are out of phase, a cancellation occurs which can disrupt transmission. This 180° phase shift occurs for horizontally polarized waves at microwave radio operating frequencies, whereas for vertically polarized waves the phase shift can be between 0 and 180° depending on the ground conditions and the angle of incidence.

4.4.2 Fresnel zones

The microwave energy that arrives at the receiving antenna 180° (or $\lambda/2$) out of phase with the direct wave determines the boundary of what is called the *first Fresnel zone,* as illustrated in Fig. 4.24. For a specific frequency, all points within a microwave link from which a

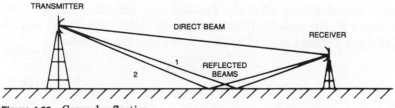

Figure 4.22 Ground reflection.

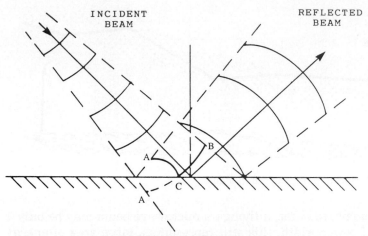

Figure 4.23   Phase reversal due to a reflection.

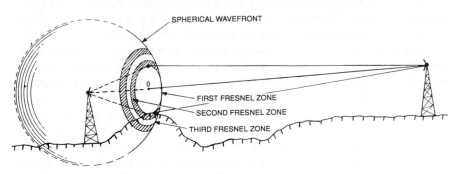

Figure 4.24   Fresnel zones.

wave could be reflected with a total additional path length of one half-wavelength (180°) form an ellipse which defines the first Fresnel zone radius at every point along the path (Fig. 4.26). The second and third Fresnel zones are defined as the boundary consisting of all points from which the additional path length is two half-wavelengths and three half-wavelengths, respectively. So, at any point along the path, there is a set of concentric circles whose centers are all on the direct line-of-sight path line, denoting all of the Fresnel zone boundaries (Fig. 4.24). The distance F_n (in meters) from the line-of-sight path to the boundary of the nth Fresnel zone is approximated by the equation

$$F_n = 17.3 \sqrt{\frac{nd_1d_2}{fD}} \qquad (4.10)$$

where d_1 = distance from one end of the path to the reflection point (km)

d_2 = distance from the other end of the path to the reflection point (km)

$D = d_1 + d_2$

f = frequency (GHz)

n = number of Fresnel zone (1st, 2d, etc.)

Figure 4.25a is a graph formed from Eq. (4.10), and it indicates the first Fresnel zone radius at specific distances from either end of a link of known total length. This family of curves is drawn for a frequency of 6.175 GHz. This is a very popular frequency band for microwave link operation. Figure 4.25b is a table which can be used to convert the values for the first Fresnel zone radius calculated at 6.175 GHz to values at other frequency bands. For example, a link of 40 km in total length would have a first Fresnel zone radius of about 12 m at a distance of 3.2 km from either end antenna at 6.175 GHz, but this increases to about 21 m at 12.8 km.

These Fresnel zone radii have significant consequences when obstacles such as trees or hills within the microwave path approach the first Fresnel zone radius. For determining the higher-order, nth, Fresnel zone radii when the first zone radius is known, the equation $F_n = F_1 \sqrt{n}$ is used.

When designing a microwave link, it is useful to obtain information of all obstacles in the region between the transmitter and receiver and draw them on the profile plot (as in Fig. 4.26). This diagram also has the first Fresnel zone plotted for 100-MHz and 10-GHz waves (note the ellipses defining these zones). This figure gives an indication of how the size of these zones varies with frequency. The radius of the first Fresnel zone at the center of the path in this case is about 17 m at 10 GHz, but it is about 170 m at 100 MHz. The rocky peak (point C) is outside the first Fresnel zone at 10 GHz, but at some lower frequency, it will be within the first Fresnel zone. However, an important point arises here. When the wave is reflected from point C for a small incident angle, a 180° phase reversal occurs (for horizontal polarization). Therefore, when AB and ACB differ by an odd multiple of a half-wavelength, the received signals add instead of canceling. Also, if the two paths differ by an even number of half-wavelengths, cancellation occurs. This means that signal energy at the second Fresnel zone distance from the reflection point and the direct signal produce cancellation. The loss in signal strength due to this cancellation can be very large. Atmospheric changes can cause refraction variations which make

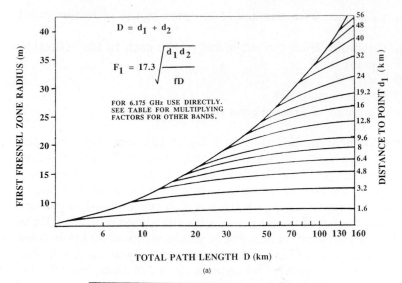

TOTAL PATH LENGTH D (km)

(a)

BAND, GHz	CENTER FREQUENCY	MULTIPLY BY
1.850 − 1.990	1.920	1.793
1.990 − 2.110	2.050	1.735
2.110 − 2.130 2.160 − 2.180	2.145	1.697
2.130 − 2.150 2.180 − 2.200	2.165	1.688
2.450 − 2.500	2.475	1.580
3.700 − 4.200	3.950	1.250
4.400 − 5.000	4.700	1.146
5.925 − 6.425	6.175	1.000
6.575 − 6.875	6.725	0.9582
6.875 − 7.125	7.000	0.9392
7.125 − 8.400	7.437	0.9112
	7.750	0.8926
	8.063	0.8751
10.700 − 11.700	11.200	0.7425
12.200 −· 12.700	12.450	0.7043
12.700 − 12.950	12.825	0.6939
12.700 − 13.250	12.975	0.6899

(b)

Figure 4.25   (a) First Fresnel zone radius (6.175 GHz); (b) multiplying factor to convert Fresnel zone radius calculated for 6.175 GHz to other bands.

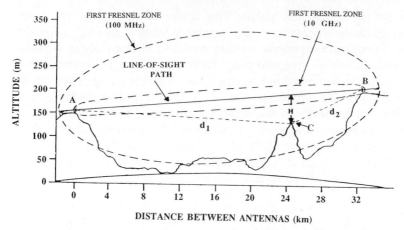

Figure 4.26 Typical profile plot showing first Fresnel zones for 100 MHz and 10 GHz.

the reflected waves intermittently pass through the cancellation conditions. This is one form of fading called *multipath fading,* and it can occur over short periods of a few seconds to longer periods of many hours. Considering reinforcement of waves arriving in phase, a signal level up to 6 dB higher than that obtained by direct free space loss can be achieved. This should be considered a bonus, and link and hop designs never include this as an expected signal enhancement.

Since the vertical polarized wave can have a 0° phase shift on reflection, this represents a worst-case situation because reflection from an obstacle at the first Fresnel zone distance causes cancellation. Hop design experience has shown that to achieve a transmission unaffected by the presence of obstacles, the transmission path should have a clearance from these obstacles of at least 0.6 times the first Fresnel zone radius, $H1$ for $k = 2/3$. Note that the $k = 2/3$ obstacle clearance condition is worse than the normal condition of $k = 4/3$ (*provided the antennas are set on a day when $k = 2/3$*). For conservative design, one or more first Fresnel zone distances are chosen for clearance of obstacles, particularly when operating in the higher-frequency bands. This clearance condition applies to objects at the side of the path as well as below the path (ground reflections).

4.4.3 Diffraction

So far, discussion has been confined to perfectly reflecting surfaces. In practice, this applies only to paths that pass over surfaces such as water or desert. Such highly efficient reflective surfaces are often labeled *smooth sphere diffraction* paths. The majority of microwave paths have an obstacle clearance in the category known as *knife-edge diffraction*. These paths traverse terrain which is moderately to severely rough with brush or tree covering. Diffraction is a characteristic of electromagnetic waves which occurs when a beam passes over an obstacle with *grazing incidence* (i.e., just touching the obstacle; Fig. 4.27). The beam energy is dispersed by an amount which depends on the size and shape of the obstacle. *Shadow loss* is a term often used to describe the loss in an area behind an obstacle. This loss is dependent on frequency. The high-frequency waves tend to follow a straight line of sight and not be diffracted into shadow area behind the obstacle. At

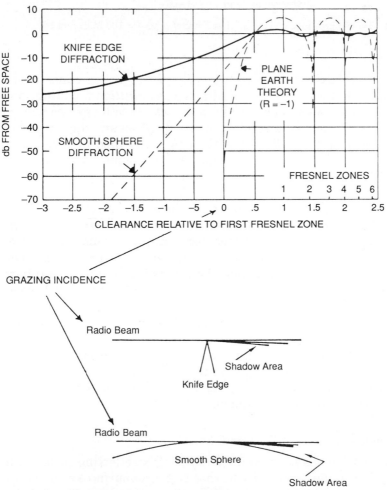

Figure 4.27   Effect of path clearance on radio wave propagation.

lower frequencies, more diffraction occurs, producing higher shadow loss since stronger signals exist in the shadow area. If the microwave hop antennas are placed at a low height compared to the Fresnel zone clearance, there will be a small angle of incidence for an obstacle near the line-of-sight path, and there will be some shadow loss (or diffraction loss) due to the grazing incidence. Figure 4.27 shows the loss compared to the expected free space value plotted against clearance from the obstacle, for the two terrain extremes: (1) the smooth sphere case and (2) the knife-edge case. The worst loss situation is for smooth sphere grazing, which can be up to 15 dB depending on clearance. The knife-edge grazing causes approximately 6 dB of loss. This loss is not small. Most microwave paths have reflections occurring at one or several points along the path. The height of antennas must be sufficient to prevent the reflected signals from causing high losses due to varying propagation conditions. Particular attention should be paid to this matter for paths crossing water or desert.

Figure 4.28 shows the resulting computer plot of a microwave hop with a beam plotted for a k-factor of 4/3. The plot also includes a one Fresnel zone radius factored into the calculations; 85-m antennas at each end of the hop are needed for the beam to clear the hill peak at the 10-km distance and the trees at the 27-km point. The question must be asked: Is this configuration adequate? The answer is yes, pro-

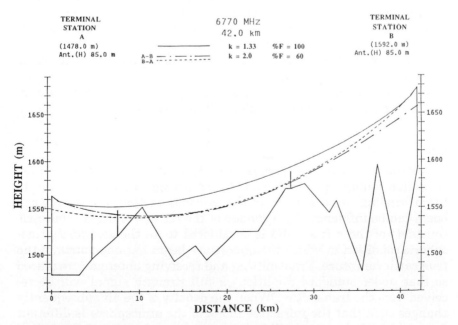

Figure 4.28 A typical hop using the flat earth presentation.

vided there are no significant atmospheric changes. In these circumstances it is useful to plot the situation on the same diagram for a k-factor of 2.0 to see if there is still adequate obstacle clearance in the event that the atmospheric conditions change for the antennas set at $k = 4/3$. As one can see, the clearance is now inadequate, so it would be preferable to increase the antenna height at one or both ends of the hop so that there is a little more clearance margin to allow for any atmospheric changes. The problem, in reality, is more complicated because the curves in Fig. 4.28 consider a homogeneous atmosphere, which is only an approximation of reality. Nevertheless, these graphs do provide a good engineering starting point from which modifications can be made to optimize each individual situation.

4.5   Fading

There are two main categories of fading:

1. Flat fading (frequency independent)
2. Frequency selective fading

Neither type of fading can be predicted accurately because each is caused by variations in atmospheric conditions. Experience has shown that some climates and terrain surfaces are more likely to cause fading than others, but in all circumstances fading can only be defined statistically. In other words, one can only say that based on probability theory the microwave system will be inoperative for a certain percentage of the year because of fading. In some regions this percentage is too large to be tolerated. Fortunately, there are techniques which can be used to improve the outage time. Before discussing the statistics, a more detailed view of fading will be examined.

4.5.1   Flat fading

Two forms of flat fading were indicated earlier: ducting and rain attenuation fading. A more frequently occurring flat fading is due to beam bending. As discussed in the previous section, the microwave beam can be influenced by a change of the refractive index (dielectric constant) of the air; $k = 4/3$ is considered to be the *standard* atmospheric condition in which the microwave beam has one-fourth of the true earth curvature. Transmitting and receiving antennas are placed so that under standard conditions a full-strength signal will be received from the transmitter. When the density of the air subsequently changes such that the refractive index of the atmosphere is different from standard, the beam will be bent upward or downward, depending on the k-factor. When k is less than 4/3, often called a *subrefractive* or

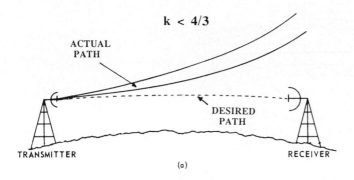

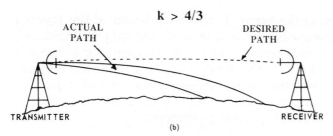

Figure 4.29 (a) Inverse bending, substandard conditions; (b) positive bending, superstandard conditions.

substandard condition, it causes upward bending, and when k is greater than 4/3, known as *superrefractive* or *superstandard* conditions, it causes downward bending (Fig. 4.29). Depending on the severity of the bending, either type can cause a considerable reduction in the received signal strength to the point of disrupting service. The most commonly occurring type is upward bending, above the receiving antenna.

For downward bending, provided the beam is not bent so far that some energy from the beam is reflected from an obstacle, the fading is wideband compared to the relatively narrow microwave frequency band (i.e., flat fading or non-frequency-selective fading). However, if some energy is reflected from an obstacle and it interferes with the direct path energy, the fading is frequency selective. Similarly, for upward bent beams, provided there is no energy reaching the receiver other than the direct path, as the beam is bent away from the receiver, flat fading occurs.

4.5.2 Frequency selective fading

Atmospheric multipath fading. When the atmospheric conditions are such that layers or stratifications of different densities exist, as indi-

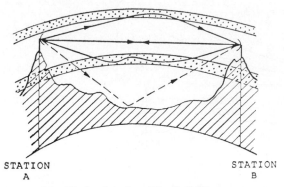

Figure 4.30 Mechanics of multipath fading.

cated earlier, ducting can occur. If the composition of the layers is such that the microwave beam is not trapped, but only deflected, as in Fig. 4.30, the microwave energy can reach the receiving antenna by paths that are different from the direct path. This multipath reception produces fading because the two waves are rarely received in phase. If they arrive in complete antiphase, for a few seconds a drop in received power, which can be 30 dB or more, is observed.

Ground reflection multipath fading.   As indicated in Sec. 4.4.1, ground reflection can cause a multipath reception which will be observed as fading if the waves are received in antiphase. When ground reflection and atmospheric multipath fading occur simultaneously, short-term fades as deep as 40 dB can occur. If corrective action is not taken, this will cause service disruption. Multipath fading is frequency selective because, for antiphase cancellation, the different waves must reach the receiver after traveling distances that differ by one half-wavelength. Because the size of one half-wavelength varies significantly from 1 to 12 GHz, fading conditions that exist at one frequency may not exist at another.

4.5.3 Factors affecting multipath fading

Experience has shown that all paths longer than 40 km can be subject to multipath fading for frequencies of operation above 890 MHz. Atmospheric multipath fading is most pronounced during the summer months or, more specifically, when the weather is hot, humid, and wind-free. It has been found that fading activity most frequently occurs after sundown and shortly after sunrise. At mid-day, the thermal air currents usually disturb the atmosphere so that layers do not form and fading is not a problem. In some regions, there are certain times of the year when the atmospheric conditions produce multipath fading outages every day. In general, frequency selective fading is "fast fading." The average duration of a 20-dB fade is about 40 s, and the average duration of a 40-dB fade is about 4 s.

As the length of a microwave path is increased, there is a rapid increase in the number of possible indirect paths by which the signal may be received. For microwave hops in desert regions or over water, it is often necessary to reduce the hop path length to 35 km or less to avoid serious ground reflection multipath fading.

Fortunately, multipath fading does not occur during periods of heavy rainfall, so fading is usually flat fading or frequency selective fading.

4.6 Availability

The measure of system reliability is usually referred to as its *availability*. Ideally all systems should have an availability of 100 percent. Since this is not possible, the engineer strives to ensure the availability is as high as possible. Table 4.1 shows the average outage time expected as the availability (reliability) degrades from 100 percent. The system becomes unavailable for two main reasons. First, *person-made faults* are caused during maintenance, or failures occur because of inadequate equipment design or fabrication. Failure due to old age can also be included in this category, because such equipment should be taken out of service before its rated lifetime expires. The second category can be called unavoidable or non-person-made faults, and these are primarily caused by *changing atmospheric conditions*. This can be controlled to some extent by equipment and route design. Usually the service interruption due to fading can account for up to about half of

TABLE 4.1   Relationship between System Reliability and Outage Time

Availability or reliability (%)	Outage time (%)	Outage time per		
		Year	Month (avg.)	Day (avg.)
0	100	8760 h	720 h	24 h
50	50	4380 h	360 h	12 h
80	20	1752 h	144 h	4.8 h
90	10	876 h	72 h	2.4 h
95	5	438 h	36 h	1.2 h
98	2	175 h	14 h	29 min
99	1	88 h	7 h	14.4 min
99.9	0.1	8.8 h	43 min	1.44 min
99.99	0.01	53 min	4.3 min	8.6 s
99.999	0.001	5.3 min	26 s	0.86 s
99.9999	0.0001	32 s	2.6 s	0.086 s

the total outage. There is a third category, which includes *disasters* such as earthquakes, fire, terrorism, etc., but fortunately this amounts to a very small percentage of the total unavailability. At the time of occurrence the effects can be temporarily devastating.

Since the fading effect on availability is closely linked to route design, a close look at this subject will now be made. Since the atmospheric changes occur over a period of time, the depth of multipath fades occurs on a statistical basis. Fortunately, 40-dB fades occur for only a very small portion of the total operating time (approximately 0.01 percent). It is therefore not possible to state exactly how much time a system will be interrupted each month or year, but one can calculate the statistically derived average times. As one can see from the table of availabilities in Table 4.1, a value of 99 percent reliability may sound impressive, but a closer evaluation shows that this is equivalent to a system average outage of 14.4 min each day.

For an analog system carrying only voice traffic this may not be too serious. However, it depends on how this 14.4 min is distributed throughout the day. It is statistically improbable that the 14.4 min will occur as a continuous time interval. At the other extreme (which is more probable) each subscriber could suffer a break in conversation for 1 percent of the time of the call. In reality certain times of the day have higher incidences of fading than others, and during these times subscribers would experience several instances where the speaker's voice fades to a level that is inaudible. For a digital system 99 percent reliability would be nothing short of disastrous for a telephone company. An outage on a digital system means that all calls are dropped for as much as several minutes until the system resets itself. This happens if the deep fade duration is as short as a few hundred milliseconds. Table 4.2 shows the order of magnitude of the maximum tolerable interrupt times for several types of information transmission. Video and data transmission are the most intolerant, but even 100 ms is enough to drop voice traffic. Obviously, if numerous short fades occur throughout the day, the system could be out of service for a large part of the day. A reliability of 99.999 percent would be an excellent

TABLE 4.2   Interrupt Time for Various Types of Traffic

Type of traffic	Maximum tolerable interrupt time	Effect if tolerance is exceeded
Video	100 μs	Loss of synchronization (rolling)
Telegraph (50 baud)	1 ms	Error
Data	10 μs	Error
Voice circuits	100 ms	Seizure of exchange switching equipment

target to achieve, and even this high value causes an average outage of about 0.86 s per day, which can cause problems when data transmission is required. A realistic value to try to obtain for a several thousand kilometer link is 99.9 percent. Long national and international links must be subdivided into separate categories or sections for the purpose of availability allocation. This has been done by the CCIR as follows.

4.6.1 Performance objectives

There is considerable variation in length, terrain, climate, etc., from one microwave link to another depending on the country in question. The CCIR has outlined a *hypothetical reference digital path* (HRDP) to characterize a typical realistic link within national boundaries. Although this concept is not currently adopted in North America, it is used in the rest of the world, so there is a possibility that North America will do so in the future. The HRDP as given in CCIR Rec. 556-1 is shown in Fig. 4.31. It is a 2500-km link which has nine radio sections. Each radio section has several hops incorporating frequency translating repeaters. At the ends of each section there are, for each direction of transmission, multiplexers and demultiplexers. Note that between sections 3 and 4 and 6 and 7, the multiplexing and demultiplexing is down to the single voice channel bit rate of 64 kb/s.

In addition, the CCITT has outlined Recommendations for acceptable digital transmission quality for the *hypothetical reference connection,* HRX (Rec. G.801, G.821 and G.921). The HRX can also include satellite connections as well as switching equipment, and it characterizes the international connection which incorporates the HRDP. Because of the importance of this aspect of microwave links, it will be treated in detail in the following. A national transmission network is

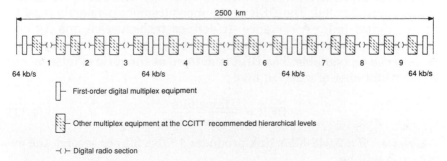

Figure 4.31 Hypothetical reference digital path for radio relay systems with capacity above 8 Mb/s. (*Reproduced by permission from the ITU, CCIR Rec. 556-1; XVIIth Plenary Assembly, Dusseldorf, 1990, vol. IX—Part 1, Fixed Service Using Radio-Relay Systems, p. 21, Fig. 1.*)

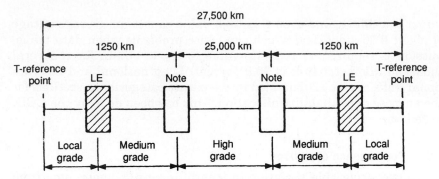

Figure 4.32 Circuit quality demarcation of longest HRX. *Note:* It is not possible to provide a definition of the location of the boundary between the medium- and high-grade portions of the HRX. (*Reproduced by permission from the ITU CCIR XVIIth Plenary Assembly, Dusseldorf, 1990, vol. IX—Part 1, Fixed Service Using Radio-Relay Systems, Rep. 1051-1, p. 21, Fig. 1.*)

subdivided so that each digital microwave link is denoted as being a high, medium, or local grade of reference circuit, depending upon its position within the network (Fig. 4.32).

High grade.   This category includes long-haul national and international connections operating mainly at high bit rates.

Medium grade.   These systems operate between local exchanges within the national network. The length of these connections is up to a maximum of 1250 km.

Local grade.   Systems in this grade operate between subscribers' premises and local exchanges and usually have a bit rate of 2 Mb/s or less.

In any of the above connections, the result of equipment failure or atmospheric effects is observed as the reception of "false bits," or errors, at the receiving end. The BER is the measurement primarily used to assess digital link performance. It defines transmission quality and acceptability. If the BER is too high, the system is declared to have failed or be unavailable. The BER is defined as the ratio of false bits to the total number of received bits:

$$\text{BER} = \frac{\text{false bits}}{\text{received bits}} \tag{4.11}$$

Example.   If a 2.048-Mb/s link produces 1 false bit per second, the error ratio will be

$$\text{BER} = \frac{1}{2.048 \times 10^6} = 4.9 \times 10^{-7}$$

The concept of unavailability (link failure) of an HRDP is defined in CCIR Rec. 557-2 as follows. The period of unavailable time begins when, in at least one direction of transmission, one or both of the following conditions occurs for 10 consecutive seconds:

- The digital signal is interrupted (i.e., alignment or timing is lost).
- The BER measured over a time period of 1 s (integration time) remains worse than 1×10^{-3}.

Those 10 s are declared as *unavailable time*.

The period of unavailable time terminates when for both directions of transmission, both of the following conditions occur for 10 consecutive seconds:

- The digital signal is restored (i.e., alignment or timing is recovered).
- The BER in each second is better than 1×10^{-3}.

Those 10 s are considered to be *available time*.

Further study by the CCIR is continuing on several aspects of this subject, particularly with respect to intermittent interruptions of duration less than 10 s. It is usual, in the field, to consider the moment the BER goes beyond the 1×10^{-3} value to be the failure point.

The BER performance objectives for a connection depending upon its length and grade are summarized in CCIR Report 1052-1 as indicated in Fig. 4.33. Regardless of these objectives, a link is designed to provide an availability as high as possible. Even at 99.99 percent availability, the link will fail for approximately 262 s per month. This may not seem to be many seconds, but if the equipment is designed to drop calls every time the BER increases above 1×10^{-3} for a few milliseconds, there could be calls dropped *thousands* of times each month. This is why the term *errored seconds* is an important parameter in performance specification. Errored seconds gives an indication of the distribution of errors (i.e., bursts or isolated events). BER measurements are usually made in the DMR station terminal equipment. Some manufacturers use the BER measurements to provide a *maintenance required* alarm if the BER becomes worse than, say, 1×10^{-5} or 1×10^{-6} and an *out of service* alarm for worse than 1×10^{-3}.

To expand on Fig. 4.33, each section of the 2500-km reference path which may form part of an ISDN network has its own performance objectives as follows.

High-grade circuits.   First, the overall availability objective for a 2500-km HRDP has been recommended by the CCIR (Rec. 557-2) to be 99.7 percent in 1 year for the HRDP. In addition:

Digital Section Quality Classifications for Error Performance

Section quality classification	HRDS length* (km)	Allocation† (%)	To be used in circuit classification
1	280	0.45	High grade‡
2	280	2	Medium grade
3	50	2	Medium grade
4	50	5	Medium grade

*The indicated values of length should be understood to correspond to maximum lengths of real digital sections. If a real digital section is shorter, there will be no reduction of the bit error allocation (i.e., percentage value in the third column). This takes into account that, in the interests of economy, short-haul systems may be designed with a greater per kilometer error ratio than long-haul systems.

†The values in this column are percentages of the overall degradation (at 64 kb/s specified in Recommendation G.821; that is, of the 8 percent errored seconds, of the 10 percent degraded minutes, and of the 0.1 percent severely errored seconds which are allocated according to the same rules as the two other parameters.

‡High-grade systems may also be used within the medium-grade portion of the connection.

Allocation of Error Performance Objectives to the Constituent Parts of the Longest HRX at 64 kb/s (Notes 1, 2, and 3)

Error performance criteria	Portion of HRX	Total (%)	High grade — Whole portion (%)	High grade — 2500-km HRDP (%)	Medium grade (Note 4) (%)	Local grade (Note 4) (%)
Degraded minutes	Minutes with $BER > 10^{-6}$	10	4	0.4	1.5	1.5
Severely eroded seconds	Seconds with $BER > 10^{-3}$	0.2	0.04	0.004 + 0.05 (Note 5)	0.015 + 0.025 (Note 5)	0.015
Errored seconds	Seconds with at least one error	8	3.2	0.32	1.2	1.2

Note 1: Each time percentage is the permissible threshold for the worst month.

Note 2: In accordance with CCITT Recommendation G.821, administrations may allocate the block allowance for the local-grade and medium-grade portions of the connection as necessary within the total allowance of 30 percent for any one end of the connection. In this case, the percentages of time given in the table for the medium-grade and the local-grade portion (besides the additional allowance for adverse propagation conditions) could be different.

Note 3: The figures quoted in this table apply only when the system is considered to be available in accordance with Annex A to Recommendation G.821.

Note 4: The allocated values are the objectives for one side of the HRX.

Note 5: Additional allowance for the adverse propagation conditions.

Figure 4.33 A summary of CCIR Report 1052-1 (1990) on error performance. (*Reproduced by permission from the ITU CCIR XVIIth Plenary Assembly, Dusseldorf, 1990, Annex to vol. IX—Part 1, Fixed Service Using Radio-Relay Systems Rep. 1052-1, p. 23, Table II.*)

1. The BER should not exceed 1×10^{-6} for more than 0.4 percent of any month (integration time of 1 min). These are declared minutes of degraded performance. The seconds during which the BER exceeds 1×10^{-3} should not be included in the integration time (CCIR Rec. 594-2).

2. The total number of errored seconds should not exceed 0.32 percent of any month (CCIR Rec. 594-2).

3. The residual BER, defined as the BER at the nominal received signal level (i.e., in the absence of fading and short-term interference), should not exceed 5×10^{-9}. This measurement is made over the period of 1 month using a 15-min integration time (CCIR Rec. 634-1 and Rep. 930-2).

CCIR Rec. 634 also specifies objectives for real high-grade digital radio links of length L, between 280 and 2500 km, within an ISDN as follows:

BER $\geq 1 \times 10^{-3}$ for no more than $(L/2500) \times 0.054$ percent of any month (integration time: 1 s)

BER $\geq 1 \times 10^{-6}$ for no more than $(L/2500) \times 0.4$ percent of any month (integration time: 1 min)

Errored seconds for no more than $(L/2500) \times 0.32$ percent of any month

Residual BER $\leq (L \times 5 \times 10^{-9})/2500$ (measurement over 1 month; integration time: 15 min)

CCIR Rec. 695 specifies the availability objective for real *high-grade* digital radio link of length L, between 280 and 2500 km, within an ISDN as follows:

$$\text{Availability} = 100 - \frac{0.3 \times L}{2500} \qquad (4.12)$$

Performance objectives for circuits shorter than 280 km are still under study.

Medium-grade circuits. The boundary between the high- and medium-grade portions of the HRX is not always clearly defined, but it can be assumed that all of the medium-grade portion is composed of several hypothetical reference digital sections (HRDSs), each with different quality and length classifications as given by CCIR Report 1052-1.

1. The error performance objectives of Table 4.3 apply to each direction and to each 64-kb/s channel of an HRDS of quality classifi-

TABLE 4.3   Performance Objectives (Medium-Grade Portion)

	Percentage of any month			
Performance parameter	Class 1 (280 km)	Class 2 (280 km)	Class 3 (50 km)	Class 4 (50 km)
BER > 1×10^{-3}; integration time: 1 s	0.060	0.0075	0.002*	0.005*
BER > 1×10^{-6}; integration time: 1 min	0.045	0.2000	0.200	0.500
Errored seconds	0.036	0.1600	0.160	0.400

*Allowance for adverse propagation not included.
SOURCE: ITU, CCIR Rec. 696, XVIIth Plenary Assembly, Dusseldorf, 1990, vol. IX—Part I, Fixed Service Using Radio-Relay Systems, p. 27, Table I; used by permission.

cations 1 to 4 using digital radio systems and forming part of the medium-grade portion of an ISDN connection. These objectives take into account fading, short- and long-term interference, and all other sources of performance degradation during periods for which the system is considered to be available.

2. The following summarizes the CCIR Recommendation 696 performance objectives that apply to each direction and to each 64-kb/s channel for the total medium-grade portion at each end of an HRX when it is realized entirely with digital radio-relay systems.

    BER should not exceed 1×10^{-3} for more than 0.04 percent of any month (integration time: 1 s)

    BER should not exceed 1×10^{-6} for more than 1.5 percent of any month (integration time: 1 min)

    Total errored seconds should not exceed 1.2 percent of any month

3. The total bidirectional unavailability due to all causes for the HRDS classes 1 to 4 using digital radio systems and forming part of the medium-grade portion of an ISDN connection should not exceed the following values (in percentages):

    - Class 1: 0.033
    - Class 2: 0.05
    - Class 3: 0.005
    - Class 4: 0.1

These percentages are to be determined over a long time period in order to be statistically valid. Although this time period is still under study, it will probably be specified to be greater than 1 year.

Local-grade circuits.   CCIR Rec. 697 outlines the following error performance objectives for each direction and to each 64-kb/s channel of a digital radio system used to form the local-grade portion at each end of an ISDN connection. These objectives take into account fading, short- and long-term interference and all other sources of performance degradation during periods for which the system is considered to be available.

1. In any month, the BER should not exceed:
   - 1×10^{-3} during more than 0.015 percent of any month, using 1-s integration time (severely errored seconds)
   - 1×10^{-6} during more than 1.5 percent of any month, using 1-min integration time (minutes of degraded performance)

2. The total errored seconds should not exceed 1.2 percent of any month (i.e., 98.8 percent error-free seconds).

3. The residual BER is still under study (CCIR Rep. 930-2).

4. The unavailability of a bidirectional link is still under study. CCIR Rep. 1053 indicates the present status of this subject, as follows. For local-grade radio systems, the unavailability is determined by (a) equipment unreliability and (b) adverse propagation, mainly due to rain attenuation.

In the above Recommendations, the BER, errored seconds, and degraded minutes are referred to the 64-kb/s interface. The difference between these values and those that would be obtained at the digital radio system bit rate is still under study. However, CCIR Report 930-2 briefly addresses this subject, and the main features are as follows:

1. The translation of severely errored seconds (SES) at 64 kb/s to the value at the system bit rate is accurate to within a few percent (i.e., % $SES_{64} \approx$ % $SES_{\text{system bit rate}}$). The percentage of severely errored seconds normalized to 64 kb/s can be assessed from measurements made at the system bit rate as follows:

$$\% \text{ SES at 64 kb/s} + Y \text{ percent} + Z \text{ percent}$$

where Y = % SES at the system bit rate.
Z = % of non-SES at the system bit rate containing one or more losses of frame alignment at the system bit rate. This term allows for error bursts during demultiplexing between the system bit rate and 64 kb/s.

2. The relationship between the errored seconds (ES) of a 64-kb/s channel and the corresponding parameter which may be measured at the system bit rate is presently denoted as follows:

$$\% \text{ ES at 64 kb/s} = \frac{1}{J} = \sum_{i=1}^{i=J} \left(\frac{n}{N}\right)_i \times 100\%$$

where n is the number of errors in the ith second at the system bit rate; N is the system bit rate divided by 64 kb/s; J is the integer number of second periods (excluding unavailable time) within the total measurement period; and $(n/N)_i$ for the ith s is n/N if $0 < n < N$, or 1 if $n > N$. This is a conservative estimate and in practice the actual result at 64 kb/s is better than the result calculated from the above equation due to the nonuniform distribution of errors.

3. The percentage of degraded minutes can be taken directly at the primary rate or above.

See CCIR Report 930-2 for more details.

As one can now appreciate, this is a rather complex subject, and the performance objectives have only recently started to evolve in a satisfactory form. This is because it has taken a long time to accumulate empirical data on the causes and effects of digital microwave radio outages. Further modifications will no doubt be made in the future. For example, a study in Australia (CCIR Rep. 930-2) for 140-Mb/s, 16-QAM digital radio systems indicated that although the CCIR Recommendations for errored seconds and severely errored seconds could easily be met, the degraded minutes objective was found to be a more stringent criterion and therefore more difficult to meet.

4.7   Diversity

Diversity is the simultaneous operation of two or more systems or parts of systems. It can be described as equipment redundancy or duplication. It is a means of achieving an improvement in the system reliability. A microwave path that has been expertly designed with respect to fade margin, path clearance, and elimination of reflections may still suffer from poor performance. Multipath fading can cause temporary failure in the best-designed paths. The system designer does not usually have the luxury of choosing the climate or terrain over which the microwave path must be designed. In regions where multipath fading conditions exist, it is necessary to incorporate diversity into the system design. The two types of DMR equipment diversity are

1. Space diversity
2. Frequency diversity

Note: Neither space diversity nor frequency diversity provides any improvement or protection against rain attenuation.

4.7.1 Space diversity

In this mode of operation (Fig. 4.34a), the receiver of the microwave radio accepts signals from two or more antennas that are vertically spaced apart by many wavelengths. The signal from each antenna is received, then simultaneously connected to a diversity combiner. Depending upon the design, the function of the combiner is to either select the best signal for its output or to add the signals.

For a space diversity protected system, the direct signal travels two different path distances from the transmitter to the two receiver antennas, as indicated in Fig. 4.35. In addition, there may be reflected paths, where the signal entering each antenna has also traveled dif-

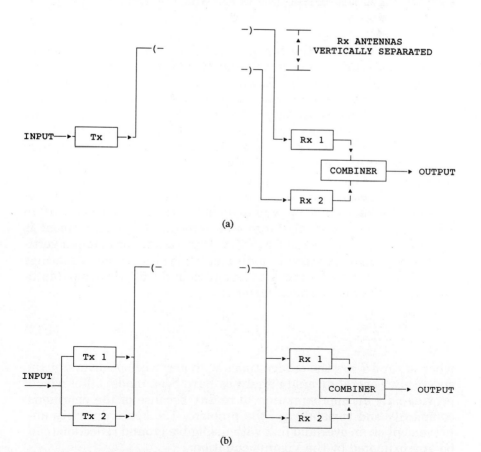

Figure 4.34 (a) Space diversity and (b) frequency diversity.

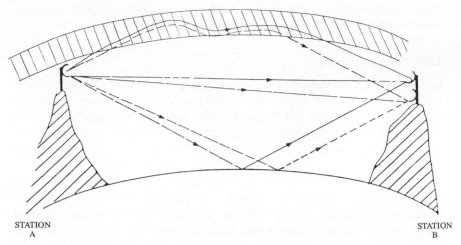

Figure 4.35 Signal paths for space diversity protection.

ferent distances from the transmitter. Experience has shown that
when the reflected path causes fading by interference with the direct
signal, the two received signals will not be simultaneously affected to
the same extent by the presence of multipath fading, because of the
different path lengths. Although the path from the transmitter to one
of the receiver antennas may cause phase cancellation of the direct
and reflected path waves, it is statistically unlikely that multiple
paths to the other antenna will cause phase cancellation at the same
time.

Statistical analysis has shown the improvement in reliability (or reduction of fading-caused outage) to be in the range of a factor of 10 to
200. This is a surprisingly large improvement. The improvement is
enhanced by an increase in frequency, fade margin, or antenna vertical spacing and a decrease in path length. Typical antenna spacings
are at least 200 wavelengths (at least 10 m in the 6-GHz band). Mathematically, the improvement factor is

$$I = \frac{T}{T_d} \qquad (4.13)$$

where T_d and T are the outage times with and without diversity, respectively. Several diversity analyses have been made, all of which
provide only an approximation to reality because of the enormous
complexity and variability of the problem. The space diversity improvement on an overland link with negligible ground reflections can
be approximated by the Vigants equation:

$$I_s \approx \frac{1.2 \times 10^{-3}\eta s^2 f \, 10^{(F-V)/10}}{d} \tag{4.14}$$

where η = effectiveness of the diversity switch
   s = vertical spacing of the antennas (m) center to center (5 m $\leq s \leq$ 15 m)
   f = frequency (GHz)
   F = fade depth (dB)
   V = difference in gain between two antennas
   d = hop length (km)

Recently, research has indicated that it can be beneficial to have the two antennas *horizontally* spaced (side by side) on the tower instead of vertically spaced. In this case, each antenna has a different angle of elevation, which is why this technique is usually called *angle diversity*. Although the difference in elevation angles from one antenna to the other may be only 1° or so, there is enough difference in the received signal strengths in a multipath fading environment to give a significant improvement in performance. Research work is still proceeding in this field. At present, space diversity is the first choice for system protection. For a single channel, it is cheaper than the frequency diversity described in the next section. Also, it does not use extra bandwidth, whereas the frequency diversity system does.

4.7.2 Frequency diversity

System protection is achieved with this type of diversity by effectively *operating two microwave radios* between the same transmit and receive antennas. The information to be transmitted is simultaneously transmitted by two transmitters operating at different frequencies. They are coupled to the waveguide, which runs to the antenna, and then transmitted by the same antenna (usually with opposite polarization). At the receiving end, the antenna collects the information and passes it through a waveguide to a filter which separates the two signals, and separate receivers extract the voice, video, or data information. As in space diversity, a combiner is used to provide the maximized output (Fig. 4.34b). If the separation in frequencies of the two transmitters is large, the frequency selective fading will have a low probability of affecting both paths to the same extent, thereby improving the system performance. A frequency separation of 2 percent is considered adequate and 5 percent is very good. This means a separation of at least 120 MHz in the 6-GHz band.

The major disadvantage of frequency diversity is the extra bandwidth that the system occupies. In uncongested regions this is no problem, but in large cities, the number of channels available for new ra-

dios is limited and the additional channels for system protection are even more difficult to acquire. The improvement gained by frequency diversity is considerably less than that of space diversity.

Calculations show at least a factor of 10 improvement for frequency diversity over the nondiversity system. Several elaborate mathematical models have recently been proposed, but the improvement factor as indicated in CCIR Report 338 provides satisfactory results. The improvement factor is approximated by

$$I_f \approx \frac{80 \ \Delta f \ 10^{F/10}}{f^2 \ d} \tag{4.15}$$

where Δf = frequency separation (GHz)
  F = fade depth (dB)
  f = carrier frequency (GHz) ($2 \le f \le 11$)
  d = hop length (km) ($30 \le d \le 70$)

As mentioned earlier, depending upon conditions, space diversity can provide an improvement of more than a factor of 100. A combination of both space and frequency diversity is used in cases of extremely high multipath fading problems.

Maintenance technicians and engineers like frequency diversity because they can proceed with repairs on one of the two radios without interrupting service. Of course, during the repair time there is no diversity protection. On nondiversity systems maintenance has to be performed at the lowest traffic periods, usually between 2 A.M. and 4 A.M.

To ensure that the transmission quality is not degraded by the switching process itself, a *hitless switching* technique has been devised. The received bit streams in the regenerated baseband are compared on a bit-by-bit basis. If an error is detected in one of the bit streams, by using an elastic store, a decision is made to switch to the bit stream devoid of errors. By this method switching can be error-free, or *hitless*.

4.8   Link Analysis

4.8.1   Hop calculations

A microwave link can span a distance of a few kilometers to several thousand kilometers. Each hop is surveyed for a line-of-sight antenna path having the necessary clearance, as mentioned above. The size of the antennas, transmitter output power, minimum acceptable receive power, and hop length are all interrelated. The minimum receivable power is the hop design starting point. This is determined by barriers created by the fundamental laws of physics and the state of the art of technology. The receive power level has a threshold value below which satisfactory communication is not possible. As Fig. 4.36 indicates, for

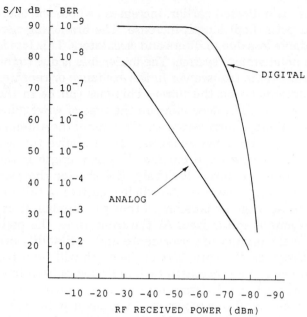

Figure 4.36 Comparison of how performance degrades for AMR and DMR systems.

an analog signal a gradual reduction of communication quality is experienced to the point where the noise is intolerably high. In practice, this effect is observed as a severe background hissing or crackling noise which eventually makes the talker's speech inaudible to the listener. In contrast, the threshold receive power level for digital systems is approached very abruptly. This received power level in dBm is

$$P_r = P_t - \alpha \tag{4.16}$$

where P_t = transmitted power level (dBm)
α = net path loss (dB)

The transmitted power level is limited primarily by cost and reliability considerations. The present day value of P_t is in the region of 1 W. The net path loss is

$$\alpha = \alpha_{fs} + \alpha_b + \alpha_f - G_t - G_r \tag{4.17}$$

where α_{fs} = free space loss
α_b = RF branching network loss
α_f = antenna feeder (waveguide) loss
G_t = transmitter antenna gain
G_r = receiver antenna gain

The free space loss, as indicated earlier, increases as both the operating frequency and path (hop) length increase. The branching network has an unavoidable loss due to filters and circulators. This loss is only about 2 dB in a nondiversity system. The feeder loss is due to the waveguide run from the radio transceiver in the exchange or repeater building up to the antenna(s) on the tower. This loss can be in the range 3 to 9 dB/100 m at 6 GHz, depending on the type of waveguide used. The gain of the antennas increases with diameter of the antenna and also as the operating frequency increases. To minimize the net loss, there are some conflicting requirements. The hop length is the most controversial parameter here. Obviously, the shorter the hop, the lower the free space loss. However, the whole objective of this exercise is to make hops as large as possible so that the number of expensive repeater stations is minimized. At the same time, the hops cannot be so long that the quality of performance and availability are compromised. In flat regions, the curvature of the earth will limit the hop length. The distance can be increased by building very high towers. Towers are expensive, but it is often cost effective to have towers even as high as 100 m so that fewer repeaters are necessary. In mountainous regions where long hops are possible, the free space loss is the limiting factor. The size of the antennas compensates to some extent, but there is a practical limit to the size of an antenna. The larger the antenna, the more robust and therefore expensive the tower must be to hold the antenna in position. Also, the beamwidth decreases with increasing antenna diameter, so alignment of the transmit and receive antennas can be a problem if the antennas are too large. In practice, 2- or 3-m-diameter antenna dishes are usually used, with 4 m being the exception. The receive power level P_r must be substantially higher than the minimum receivable or threshold power level P_{th}; otherwise during atmospheric fading conditions, the signal will be lost. This leads to the term *fade margin*, which is the difference between the threshold level and the operational level:

$$\text{Fade margin} = P_r - P_{th} \qquad (4.18)$$

The fade margin is a composite parameter, which is primarily associated with dispersive fading due to atmospheric effects but also contains the components of fading attributable to interference and thermal noise.

The fade margin of each hop in a link is determined by the required overall link availability. This is a statistical problem, due to the statistical nature of fading. Also P_{th} is fixed by the minimum acceptable BER, which is related to availability. P_{th} is usually considered to be the received power level at which the BER is 1×10^{-3}. The unavailability caused by fading obviously improves as the fade margin is increased. Mathematically, using Rayleigh-type single frequency fading, the probability that fading will exceed the fade margin (FM) in the worst month can be approximated by

$$P_F = 7 \times 10^{-7} c \, f^B \, d^C \, 10^{-FM/10} \tag{4.19}$$

where c = terrain and climate factor (4 for over the sea or coastal areas; 1 for medium rough terrain, temperate climate, noncoastal areas; 1/4 for mountainous terrain and dry climate areas)

f = carrier frequency (GHz) and $0.85 \le B \le 1.5$ (usually about 1)

d = hop length (km) and $2.0 \le C \le 3.5$ (usually about 3)

Note: The values of B and C depend on geographical location, and $p_F\% = 100\% \times P_F$.)

More rigorous statistical analyses have recently been applied to fade margin calculations, but the above equation still provides a reasonable approximation. An availability of 99.9 percent results in a fade margin designed to be approximately 30 dB. Ideally, in severe fading areas, hops are designed to have a comfortable fade margin of at least 40 dB. The microwave power gains and losses in a microwave link are shown graphically in Fig. 4.37. This figure gives the typical orders of magnitude of power levels at various points in a system. For example, a 30-dBm (1-W) transmitter output has a small waveguide run loss followed by an antenna gain. The free space loss between the antennas is the most significant loss, and the received signal power level has some uncertainty because of atmospheric effects. The receive

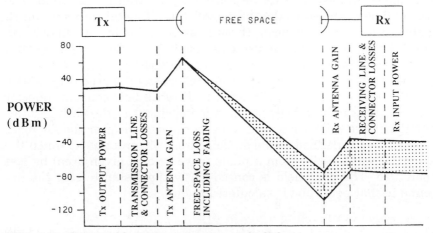

Figure 4.37 Gains and losses in a microwave radio link.

antenna has gain and the interconnecting waveguide and connectors have loss; this produces the resulting power level which must be well above the threshold level ready for processing.

4.8.2 Passive repeaters

When a microwave hop is required in a place which has some unavoidable physical obstacles, a passive repeater can sometimes solve the problem. For example, suppose a satellite earth station is built outside a major city, and the large satellite dish is situated in a hollow, surrounded by small hills about 150 m higher than the earth station. This is frequently a realistic situation because the earth station dish is then shielded from the city noise. A microwave link is required to access the earth station from the city 40 km away. It would be expensive to place a repeater station on the hilltop and the hop from the earth station to the hilltop might be less than 2 km (only a mile or so). A passive repeater is effectively a "mirror" placed on the hilltop to reflect the microwave beam down to (and up from) the earth station. There must be a clear line-of-sight path between the passive repeater and the other two end points. Another instance where a passive repeater is often used is when a mountain peak has to be surmounted. It may be so inaccessible that power cannot be provided for a usual active repeater. Even in sites where a solar power source is used, it may be too inaccessible for maintenance purposes. Whereas helicopter access for installation of a passive repeater may be practical, it is usually excessively expensive to use a helicopter for maintenance.

There are two possible solutions to these passive repeater problems. First, two parabolic antennas could be placed back to back with a length of waveguide from one feed horn to the other (Fig. 4.38a). Each antenna is aligned with its respective hop destination antenna. The second type is known as the "billboard" type of metal reflector, which deflects the microwave beam through an angle as in Fig. 4.38b. Provided the angle is less than about 130°, only one reflector is necessary. If the two paths are almost in line (i.e., less than about 50° between the two paths), a double passive reflector repeater is used (Fig. 4.38c). For the case of a single reflector, the distance between the reflector and the terminal parabolic antenna is important in the path loss calculations. The reflector is said to be in the *near field* or the *far field*. The far field is defined as the distance from the antenna at which the spherical wave varies from a plane wave (over a given area) by less than $\lambda/16$. The near field is consequently the distance from the antenna to that point and is calculated to be

$$r = \frac{2D^2}{\lambda} \tag{4.20}$$

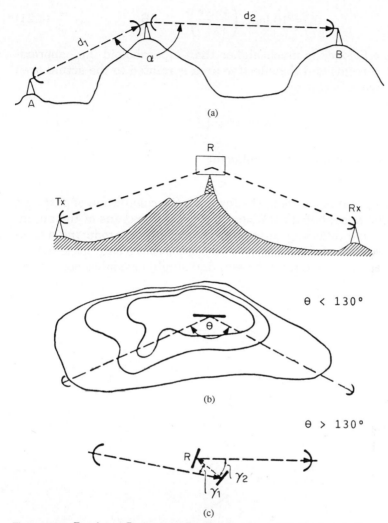

Figure 4.38 Passive reflectors. (*a*) Passive repeater with two parabolic reflectors. (*b*) Passive repeater with one plane reflector. (*c*) Passive repeater with two plane reflectors at one location.

where D = the diameter of a circular reflector or length of each side of
       a square reflector
      λ = the wavelength in the same units as D

As an example, the extremity of the near field at 6 GHz for D = 6 m is
r is about 1.4 km.

The gain of a passive reflector G_p is that of two back-to-back parabolic antennas:

$$G_p = 10 \log_{10} \left(\frac{4\pi a^2 e}{\lambda^2} \right)^2 \quad \text{dB} \tag{4.21}$$

where the efficiency is much higher than a paraboloid (i.e., approximately 95 percent) and the effective area is related to the actual area of the passive reflector by

$$a^2 = A^2 \sin\left(\frac{\theta}{2}\right)$$

where A^2 = actual area of the reflector
θ = deflection angle

If the passive reflector is in the far field, the calculation of path loss is straightforward, but for the near field case the gains of the antennas and passive reflectors interact with each other, producing a complex path loss calculation. Figure 4.39 compares the path losses for all combinations of near field, far field, and single or double passive re-

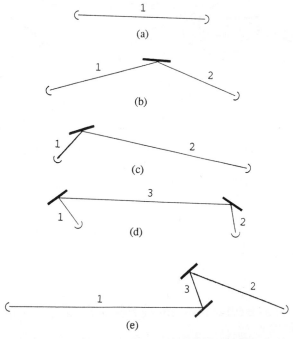

Figure 4.39 Path loss comparison for passive reflectors. (a) Path with no passive; $L = L_1 - G_t - G_r$. (b) Single passive, far field; $L = L_1 + L_2 - G_t - G_p - G_r$. ($c$) Single passive, near field; $L = L_2 - G_t - G_r - \alpha_n$. ($d$) Double passive, far field; $L = L_1 + L_2 + L_3 - G_t - G_{pa} - G_{pb} - G_r$. ($e$) Double passive, close coupled; $L = L_1 + L_2 - G_t - G_{pa} - \alpha_c - G_r$.

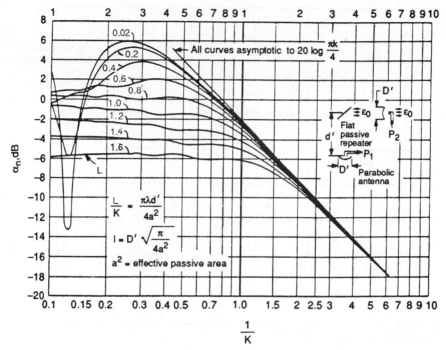

Figure 4.40 Antenna-reflector efficiency curves. (*Reproduced with permission from Ref. 95, © 1972 McGraw-Hill.*)

flectors. For the single passive, far field case (Fig. 4.39b), the path loss is simply the sum of the gains of the antennas and the passive minus the free space loss of each section. However, as Fig. 4.39c indicates, for the single passive, near field case, the free space loss of the short section is not included. Instead, a factor α_n is used, which can be a gain or loss depending upon the size of the passive, the distance between the passive and the antenna, and the frequency. Figure 4.40 is a graph which is used to evaluate α_n, which depends on the frequency and the physical dimensions of the system. Similarly, the path loss for the double passive in the far field of both antennas is an algebraic sum of the gains of the antennas, the passives, and the section free space losses. Whenever the two passives are closely coupled, the term α_c replaces the term $(G_p - L_3)$, where α_c is found from the graph in Fig. 4.41.

Passive reflectors are mainly used at 6 GHz and above since they are more efficient as the frequency increases, because the passive gain factor appears twice, whereas the free space loss appears only once. The gain of a passive reflector increases with its size. A maximum size of 12 m × 18 m is used up to 12 GHz, and this maximum size gradu-

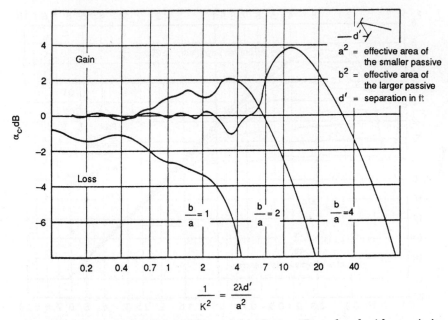

Figure 4.41 A closely coupled double passive reflector. (*Reproduced with permission from Ref. 95, © 1972 McGraw-Hill.*)

ally decreases as the operating frequency increases. This is because the 3-dB beamwidth decreases as the area of the passive increases, and it should not be allowed to fall below 1°. Even for a 1° beamwidth, twisting movements of the tower holding the passive must not be more than 1/4° since this angle appears as 1/2° for the deflecting angle.

The single passive, near field case can also be used in the "periscope antenna" mode, as shown in Fig. 4.42. Here, the antenna is located at ground level and is pointed directly upward toward the passive reflector mounted at the top of the tower. This system has the multiple benefits of reducing waveguide runs and, with appropriate choice of dimensions, has a net gain greater than that of the parabolic antenna alone. Numerous mathematical analyses have been made for all possible combinations of dimensions including various curvatures for the

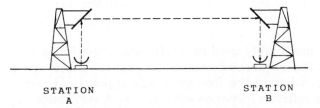

STATION A

STATION B

Figure 4.42 The periscope antenna arrangement.

reflecting surface. Despite the benefits, potential intermodulation noise problems have limited their widespread use. Also, their use is usually confined to the 6- to 11-GHz frequency range.

4.8.3 Noise

Interference.   Each receiver in a digital radio-relay network is exposed to a number of interference signals, which can degrade the transmission quality. The main sources of interference are:

- Intrasystem interference

  Noise
  Imperfections
  Echo

- Interchannel interference

  Adjacent channel
  Cochannel cross-polarization
  Transmitter and receiver
  Spurious emission

- Interhop interference

  Front to back
  Overreach

- Extra-system interference

  Satellite systems
  Radar
  Other radio systems

Intrasystem interference.   This type of interference is generated within a radio channel by thermal receiver noise, system imperfections, and echo distortions. Good system design ensures that imperfections do not introduce significant degradation. However, echo distortion caused by reflections from buildings or terrain and due to double reflections within the RF path (antenna, feeder) cannot be neglected in higher-order QAM systems. Echo delay causes interference, which increases as the delay increases. Ground reflections cause echo delays in the region of 0.1 to 1 ns. Reflections near the antennas or from distant buildings are usually greater than 1 ns and can cause severe echo interference. Transversal equalizers (as described later) help to improve the situation. Echoes caused by double reflection in the waveguide feeder cause very long delays of 100 ns or more. The only countermea sure against this problem is to ensure that it does not occur. In other words, good return loss (VSWR) is essential.

Interchannel Interference. Interchannel interference between microwave transmission frequency bands is described by the diagram in Fig. 4.43. Adjacent channel interference can be either:

1. Cross-polar
2. Copolar

Cochannel interference can only be cross-polar.

For adjacent channels, the copolar interference can be suppressed by filtering, and adjacent channel cross-polarization interference is not usually a problem with today's antennas.

Figure 4.44 is a graph of the cochannel interference for various S/N values. The noise introduced by one channel into the other is due to inadequate cross polarization discrimination of the antennas, and the S/N values are effectively the cross-polarization discrimination values for channels operating at the same output power level.

This type of interference can be more serious than adjacent channel interference, and the value of the S/N should be better than 25 dB and preferably at least 30 dB. For example, at a received power level of -70 dBm, a S/N $= \infty$ would give a BER equal to about 10^{-10}. At S/N $= 30$ dB the BER would degrade to about 2×10^{-8} and, if the S/N were only 20 dB, the BER would be worse than 10^{-5}.

Interhop Interference. This type of interference can occur because of front-to-back or nodal interference from adjacent hops and by *overreach* interference. The signal-to-interference ratio (S/I) is determined by the angular discrimination of antennas and can decrease during fading. Careful route and frequency planning is necessary to keep the degradation smaller than 1 dB. Some examples of the above intrasystem, interchannel, and interhop sources of interference are shown in Fig. 4.45.

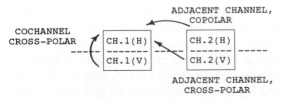

Figure 4.43 Interchannel interference.

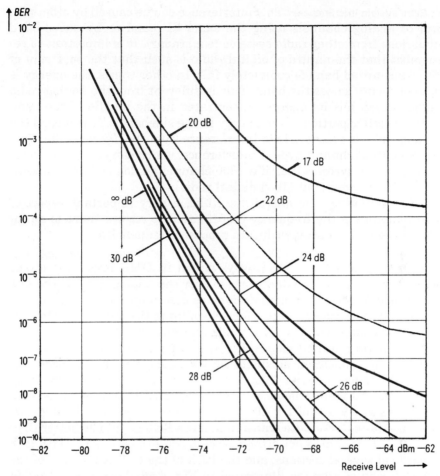

Figure 4.44 Cochannel interference. (*Reproduced with permission from Siemens Telcom Report: Special "Radio Communication," vol. 10, 1987, p. 95, Fig. 10.*)

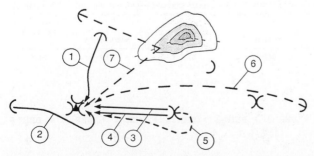

Figure 4.45 Various sources of interference. 1 = cochannel or adjacent channel signal from a different hop direction; 2 = opposite hop front-to-back reception; 3 = adjacent channel (same hop); 4 = cross-polarization (same hop); 5 = front-to-back radiation; 6 = overreach; 7 = terrain reflections.

Extra-system interference. This interference can be caused by other digital or analog channels using the same RF band or by out-of-band emissions from other radio systems (e.g., radar). It is important to remember that the nature of digital radio is such that the spectrum of the transmitted band is completely full. In other words, the energy is spread evenly across the band. This is different from the analog radio signal which has its energy concentrated in the middle of the band with a smaller portion of the energy in the sidebands. This means the *analog* radio is susceptible to interference from the *digital* radio. This is an adjacent channel type of interference, which can be a problem for high-capacity systems [i.e., if a 1800-channel analog radio is adjacent to a 1920-channel (140-Mb/s) digital radio].

When planning new microwave links, a very important aspect of this planning is the frequency coordination to ensure there is no interference with existing or future proposed frequencies.

DMR noise calculations. As always stated in DMR texts, one of the main advantages of the digital radio over the analog radio is the fact that thermal and intermodulation noise accumulation does not occur. The quantity of major interest for determining the quality for the digital link is the *BER*. The BER values recommended in CCITT G.821 are minimum required values. For speech communication, a value of 1×10^{-6} is considered adequate for excellent quality performance. When the value is worse than 1×10^{-6}, the link is considered to be *degraded,* and maintenance should be initiated to improve the BER. After 10 s at a value of 1×10^{-3}, the link is considered to be *unavailable* (i.e., failed). For data transmission the BER should be better than 1×10^{-7}. This goal is not always easy to achieve because a DMR link consists of several sections, and the BER of the link is the sum of the BER of each section, as illustrated in Fig. 4.46. This means that in

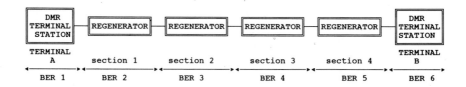

$$BER = \sum_{i=1}^{n} BER_i = BER\ 1 + BER\ 2 + BER\ 3 + BER\ 4 + BER\ 5 + BER\ 6$$

Figure 4.46 BER accumulation.

order to have a satisfactory link to support data traffic, each section should have a BER better than 10^{-9} or even 10^{-11}.

While BER is a practical, measured value, a more fundamental term is often used to describe the performance of a DMR—*the energy per bit per noise density ratio E_b/N_o:*

$$E_b = \frac{\text{carrier peak power}}{\text{bit rate}} = \frac{C}{B_r} \qquad (4.22)$$

$$N_o = \frac{\text{noise power in } B_{eq}}{\text{equivalent Rx 3-db noise bandwidth}} = \frac{N}{B_{eq}} \qquad (4.23)$$

$$= -204 \text{ dBW} + \text{NF}_{dB}$$

for a perfect receiver at room temperature, 290 K, where NF_{dB} is the receiver noise figure, and N_o is the noise in a 1-Hz bandwidth. So,

$$\frac{E_b}{N_o} = \left(\frac{C}{N}\right)\frac{B_{eq}}{B_r} \qquad (4.24)$$

Calculations have been made to relate BER to E_b/N_o, and graphs have been plotted for various modulation schemes. The graphs for several QAM schemes are plotted in Fig. 4.47, which shows that as the QAM level increases, the E_b/N_o required to achieve a particular BER increases.

So, for a required minimum BER, the E_b/N_o value can be evaluated, which can be related to parameters of the path by

$$\frac{E_b}{N_o} = P_r - (-204 + \text{NF}_{dB}) - 10 \log B_r \qquad (4.25)$$

when the modulated carrier level is considered to be the received signal level. Substituting for P_r [Eq. (4.16)],

$$\frac{E_b}{N_o} = P_t - \alpha + 204 - \text{NF}_{dB} - 10 \log B_r \qquad (4.26)$$

Using this equation, the hop can then be designed to provide an adequate fade margin.

Design example. Consider a 140-Mb/s DMR operating at 6 GHz with 16-QAM over a 64-km hop with 3-m antennas on 30-m towers. Establish the fade margin and therefore determine if any system specifications need to be modified.

The theoretical E_b/N_o for a BER of 1×10^{-3} is about 21 dB. The practical value could be up to 5 dB higher than this value. A 1-W (0-dBW) transmit power is relatively cheaply available with high

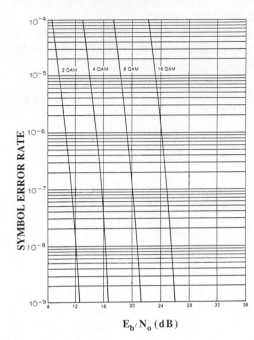

Figure 4.47 Error performance versus E_b/N_o. (*Reproduced with permission from Ref. 78, © 1987, J. Wiley and Sons.*)

linearity which requires no predistorter (see Chap. 5). The noise figure of a typical receiver in the 6-GHz band is 5 dB, so, from Eq. (4.26),

$$26 = 0 - \alpha + 204 - 5 - 81.5$$

$$\alpha = 91.5 \text{ dB}$$

This is the maximum acceptable total loss to provide the system with the 1×10^{-3} BER. Since the radio should be designed to operate in the region of BER $= 1 \times 10^{-9}$ for one hop and not the failure point (1×10^{-3}), the E_b/N_o would increase by about 5 dB from the 1×10^{-3} value to approximately 31 dB:

$$\frac{E_b}{N_o} = 26 \text{ dB} + 5 \text{ dB implementation} = 31 \text{ dB}$$

So,

$$\alpha = 86.5 \text{ dB}$$

since, from Eq. (4.17),

$$\alpha = \alpha_{fs} + \alpha_b + \alpha_f - G_t - G_r$$

Free space loss $= 144$ dB

Branching network loss $= 1.4 \times 2 = 2.8$ dB

Suppose for adequate obstacle clearances, two towers each 30 m high are needed. Then

$$\text{Waveguide feeder loss} = 1.5 \times 2 = 3 \text{ dB}$$

$$G_t = G_r = 44 \text{ dB}$$

$$\alpha = +144 + 2.8 + 3 - 44 - 44$$

Therefore,

$$\alpha = 61.8 \text{ dB}$$

Comparing this figure with 86.5 dB, there is a safety factor, or fade margin, of 24.7 dB. If the radio is in a region with no excessive rain or fading problems, this margin may be unnecessarily high. Costs could then be reduced by using smaller, lower-performance antennas (e.g., 1.8 m), having gain of about 39 dB, reducing the margin to 14.7 dB. Further savings could be made by using a transmitter with lower power output, say 500 mW, reducing the margin by 3 dB to only 11.7 dB.

Taking fading into account, the required nondiversity fade margin can be calculated from Eq. (4.19), that is,

$$P_F\% = 7 \times 10^{-5} \, cf^B d^C \, 10^{-FM/10}$$

Therefore,

$$FM = -10 \log\left(\frac{P_F}{7 \times 10^{-5} \, cf^B d^C}\right)$$

Suppose $f = 6$ GHz, $c = 1$, $B = 1$, $C = 3$, and $d = 64$ km. Also, consider the maximum outage time for the worst month of the year to be 0.01 percent (i.e., the BER of one of the transmission directions is worse than 1×10^{-3} for 0.01 percent of the month). Since the outage percentage has to be divided equally between the two transmission directions, for one channel $P_F = 0.005$ percent:

$$FM = -10 \log(0.0000454) = 43.4 \text{ dB}$$

The above calculated fade margin (24.7 dB) is now inadequate, under these fading conditions. Since the disparity is so large (18.7 dB), increasing transmitter power and antenna size to make up the difference is impractical. A diversity system would be the solution. Space diversity would decrease the required fade margin as calculated by Eq. (4.14):

$$I_s = \frac{1.2 \times 10^{-3} \, \eta \, s^2 f \, 10^{(F-V)/10}}{d}$$

If $s = 10$ m, $f = 6$ GHz, $V = 0$, $d = 64$ km, $\eta = 1$, and the outage with diversity is to be the same as without diversity (i.e., 0.005 percent), $I_s = 1$. Therefore $F = 19.5$ dB.

This decreases the *required* fade margin from the nondiversity value of 43.4 dB to 19.5 dB with space diversity. The initially calculated fade margin of 24.7 dB is 5.2 dB greater than the required 19.5. This could be consumed by reducing the transmitter power level or reducing the antenna size. Reducing the antenna diameter to 2.4 m would approximately consume the 5.2 dB, and this would be more cost effective than reducing the transmitter power level.

The inclusion of space diversity provides the necessary improvement in performance to overcome fading. Notice how the diversity antenna spacing affects the *required* fade margin:

If $s = 5$ m, $F = 25.5$ dB.

If $s = 15$ m, $F = 16.0$ dB.

A 10-m separation would be the required separation, since this is 200 wavelengths at 6 GHz. The 15-m separation could be used only if the lower antenna in these circumstances satisfies the path profile requirements of maintaining adequate line-of-sight path clearance of any obstacles. Incidentally, a 64-km hop is relatively long compared to a typical average hop length of about 40 km. In practice, a 64-km hop would have antennas placed on hill top vantage points, or if the hop is in a flat region, the antennas would need to be placed on towers higher than 30 m.

If this system is located in an area of heavy rainfall, during heavy rainfall periods there will be no multipath fading. In such circumstances the space diversity (or frequency diversity if it were included) would not provide any performance improvement. The entire 24.7-dB fade margin initially calculated would then be available to combat the rain attenuation. At 6 GHz a cloudburst (100 mm/h rainfall) would cause an attenuation of 1 dB/km. To consume all of the 24.7-dB fade margin, the cloudburst would have to be covering 24.7 km of the 64-km hop. The statistical likelihood of this happening is extremely remote. Rainfall of such intensity usually occurs over only a few (< 10) kilometers, although extreme weather conditions such as hurricanes or cyclones can exceed this figure.

In this example, space diversity is a sufficient system improvement, and frequency diversity would not be necessary as an additional countermeasure. Frequency diversity may be included as a telecommunications organization's policy on backbone routes to provide protection against equipment failure.

It must be emphasized that the above calculation is an over-simplification which does not take into account all of the countermeasure characteristics of the radio. However, it does provide a starting point to see how various parameters, including diversity, affect the availability. In early DMR designs, the unavailability caused by fading was so alarming that its future viability was in question. Since then, several countermeasures in addition to diversity have been designed and have proven to be reasonably effective. These include protection switching, IF adaptive equalizers, transversal equalizers, and FEC, all of which will be discussed in detail in Chap. 5.

Digital Microwave Radio Systems and Measurements

5.1 System Protection

As discussed in Chap. 4, space and frequency diversity methods are implemented to enhance the performance (increase the availability) of microwave radio systems. The manner in which diversity is incorporated into the system design is now considered. In addition, there are further measures which can be taken to provide an even higher availability, such as IF adaptive equalizers and baseband adaptive transversal equalizers. These will be described in detail.

5.1.1 Diversity protection switching

In a space diversity protected system, the received signals are usually combined at the IF. In a frequency diversity system, the received signals are either combined or the stronger of the two signals is used. This arrangement is called a 1-for-1 protection frequency diversity system. A switching technique is used to ensure that the stronger signal is accepted and the weaker signal is unused. This technique is often used in an M-for-N protection switching arrangement, where M protection channels are used to protect N information channels. For example, one protection channel may be used for protecting seven information channels. Obviously, 1-for-1 offers a much better level of protection than 1-for-7. Figure 5.1 shows the extent to which the availability of a typical single-channel DMR is improved by a 1-for-1 protected frequency diversity system. At a C/N value of 25 dB, the availability of the unprotected channel is not much better than 99.0 percent, whereas the frequency diversity protected system is improved by more than an order of magnitude to a very respectable value of bet-

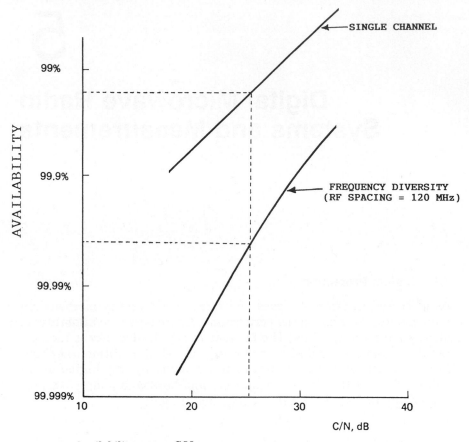

Figure 5.1   Availability versus C/N.

ter than 99.95 percent. This is the difference between an acceptable and unacceptable performance. Clearly, considerably extra cost is incurred in upgrading a system with frequency diversity protection.

5.1.2 Hot-standby protection

A hot-standby protection system is a fully redundant radio configuration operating in a *power ON* mode, ready for switching into operation in the event of a failure. The *frequency diversity system is one form of hot-standby protection.* The system illustrated in Fig. 5.2 is a hot-standby configuration used in a space diversity system. If a transmitter failure occurs, the RF switch disconnects the failed transmitter and connects the standby transmitter to the antenna. Unfortunately, this type of switching causes a brief disruption of the digital *bit stream,* resulting in an *error burst.* Since the equipment is built for a

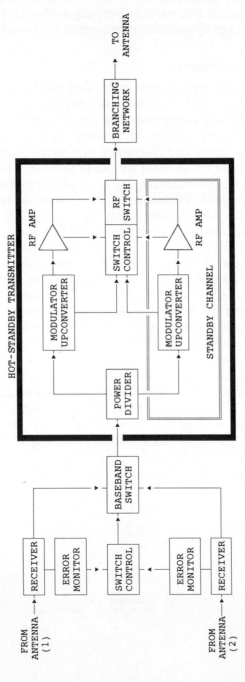

Figure 5.2 Hot-standby and space diversity DMR repeater.

high degree of reliability, this type of switchover occurs very seldom, and therefore the BER degradation due to the switchover is negligible. As in the frequency diversity system, the switching at the receiver end is done at the baseband by comparing the regenerated bit streams.

5.1.3 Combining techniques

The protection switching or signal combining can be done at the RF or the IF or in the baseband. Post-detection switching (at the baseband) provides the better reliability since the receive switch is the last component of the diversity system. In general, space diversity systems combine at the IF level, and frequency diversity systems combine in the baseband. Figure 5.3 shows a typical IF combined signal. Note the presence of the *adaptive equalizer*. This is a very important component which will be discussed in detail later. In this system, the two IF modulated carriers are dynamically delay equalized prior to combining. The combiner adds these IF carriers on a voltage basis to improve the C/N. If the C/N is the same on both channels, the combined C/N could be up to 3 dB higher than that of the individual channels. IF carriers must be combined in phase because a small phase error can cause a large BER. For example, Fig. 5.4 shows the C/N degradation as a function of phase error for 4-PSK systems. As the phase error becomes worse than 30° and approaches 40°, the C/N degrades dramatically. Higher levels of modulation (QAM, etc.) are even more susceptible to phase errors. An example of a baseband combiner is shown in Fig. 5.5. The signals from the *main* and *diversity* channels are down-converted and then demodulated to the baseband before combining. The advantage of this type of combiner is that it does not require such careful delay equalization as does the IF or RF combiner. In addition, this combiner can be set for hitless selective switching if the C/N for the

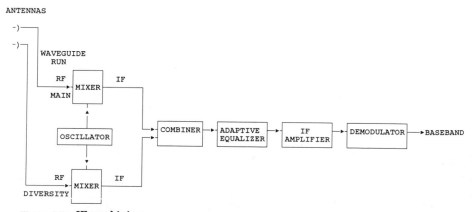

Figure 5.3 IF combining.

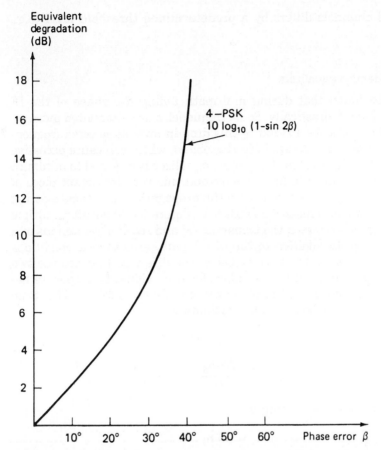

Figure 5.4 Degradation of C/N versus phase error for QPSK. β = phase error between the combined modulated carriers and recovered carrier phase. (*From K. Feher, Digital Communications: Microwave Applications, © 1981, p. 226; reprinted by permission of Prentice-Hall, Englewood Cliffs, N. J.*)

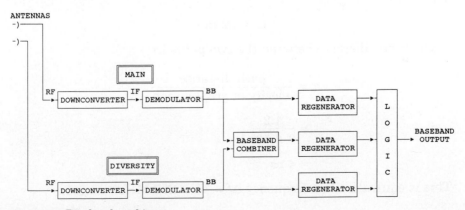

Figure 5.5 Baseband combiner.

individual channels differs by a predetermined threshold level (e.g., 10 dB).

5.1.4 IF adaptive equalizers

It has been found that during multipath fading the shape of the IF passband changes drastically. Some frequencies are attenuated more severely than others. As a result of this drop in level at selected frequencies, severe distortion occurs in the baseband, which can cause excessive errors and link failure. An IF adaptive equalizer is required to minimize this problem. In order to fully understand this very important piece of equipment, it is necessary to review the propagation conditions existing between the two antennas of a DMR hop. Figure 5.6 illustrates multiple path propagation between the transmitting and receiving antennas of a microwave hop. In addition to the direct path, atmospheric conditions can produce other paths above or below the direct path. Since the two paths differ in length, the time taken for each signal to propagate between the transmitter and receiver will clearly be different. This time difference can easily be calculated as follows:

Path difference:
$$d = D_2 - D_1$$

$$= \frac{2h_1h_2}{D}$$

For example, assume the values

$$h_1 = 60 \text{ m}$$

$$h_2 = 500 \text{ m}$$

$$D = 50 \text{ km}$$

Therefore,

$$d = 1.2 \text{ m}$$

The time difference between the two paths is

$$\tau = \frac{\text{path distance}}{\text{velocity of propagation}}$$

$$= \frac{1.2}{3 \times 10^8}$$

$$= 4 \text{ ns}$$

This is equivalent to a frequency $\Delta f = 1/\tau = 250$ MHz.

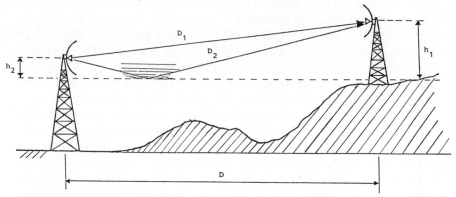

Figure 5.6 Multipath propagation.

A graph of the vector-added signal amplitudes from the two paths plotted against the inverse of the propagation time difference results in the *cycloid* curve. This curve (as shown in Fig. 5.7) has a maximum value when the two signals are in phase. This combined signal level has a 6-dB increase over the individual direct signal. However, when the two signals are in phase opposition, the resulting power level is theoretically zero. Fortunately, because of irregularities at the point of reflection, the combined level will never be zero for a significant length of time. Nevertheless, 40-dB fades are not uncommon when severe atmospheric conditions exist. In the above example, these in phase and phase opposition points occur every 250 MHz. All remaining

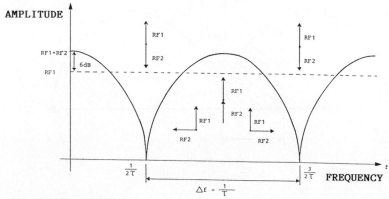

Figure 5.7 The cycloid curve.

points on the cycloid curve will have a phase difference between the two waves ranging between 0° and 180°. If the IF band to be demodulated is, for example, 36 MHz, only a portion of the cycloid curve is under consideration. This means that the IF passband will have a shape of part of the cycloid curve. Optimum demodulation conditions occur whenever the IF band to be demodulated falls in the middle of the cycloid curve (Fig. 5.8a). Conversely, demodulation is most adversely affected when the demodulation IF band falls on the notch (Fig. 5.8b). The notch can even be in the middle of the IF band. The situation is made more difficult to correct because of the *dynamic* nature of fading. This means that the IF band to be demodulated moves about the cycloid curve in an unpredictable manner. A *dynamic* or *adaptive* IF equalizer is consequently necessary to correct for the unflat IF passband response caused by multipath fading.

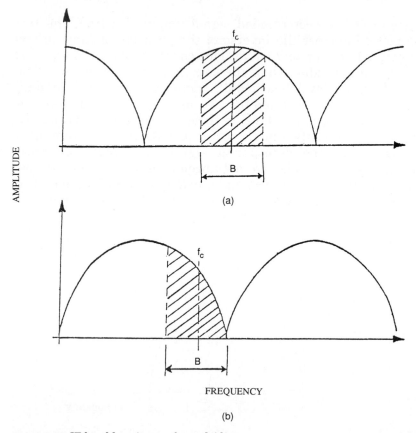

Figure 5.8   IF band location on the cycloid curve.

Figure 5.9 shows the actual IF passband for three cases of multipath fading. Figure 5.9*a* and *b* show the multipath fading causing a null at the IF band edges. Figure 5.9*a* shows the multipath distortion affecting the center of the IF passband. In each of these cases, the inclusion of the adaptive equalizer allows restoration of the IF passband to an almost flat condition. It also maintains a flat group delay. This eliminates (or minimizes) the existence of errors which would otherwise be caused by the multipath fading. *The word "adaptive" is used to indicate that the equalizer has the flexibility to change its character to suit the disturbance.* The equalizer operation is based upon the principle of creating an IF curve complementary to the curve of the channel affected by multipath fading. It is usually composed of two parts:

1. A slope equalizer

2. A "hump" (or inverted notch) equalizer

A bandwidth meter detects the amplitude distortions present in the spectrum by measuring the signal power in the proximity of three selected frequencies (e.g., 57, 70, and 83 MHz for a 70-MHz IF or 127, 140, and 153 MHz in the case of a 140-MHz IF). The process illustrated in Fig. 5.10 allows the introduction of slope or hump equalization, as necessary, to account for the multipath fading. The diagrams in Fig. 5.9*a* and *b* show how the slope equalizer restores the attenua-

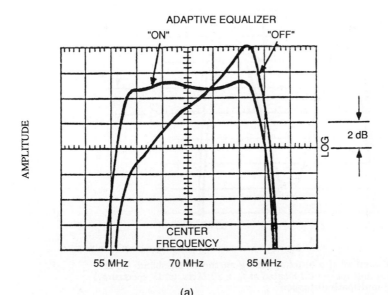

(a)

Figure 5.9 The effect of multipath fading on the IF passband. (*a*) IF spectrum with multipath null located at f_0 − 20 MHz.

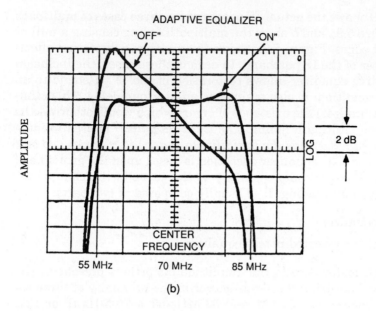

(b)

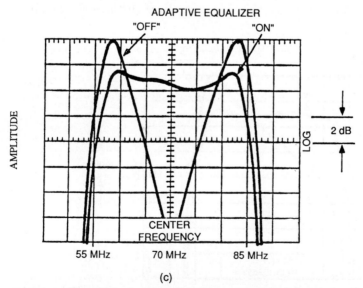

(c)

Figure 5.9 The effect of multipath fading on the IF passband. (*b*) IF spectrum with multipath null located at $f_0 + 20$ MHz. (*c*) IF spectrum with in-band multipath distortion.

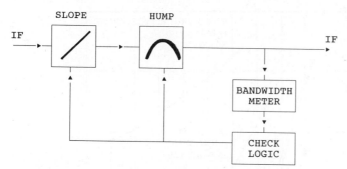

Figure 5.10 Slope and hump IF adaptive equalization.

tion at the edge of the passband to a relatively flat response. Figure 5.9c shows how the hump equalizer removes the notch in the passband. The improvement in the BER due to IF adaptive equalization is illustrated in Fig. 5.11 for a 6-GHz 90-Mb/s 8-PSK DMR. The graph shows two values of fade depth (i.e., 23.8 and 25.5 dB). The fade is due to 6-GHz multipath signals destructively interfering with each other at the worst possible phase of 180°. For the 23.8-dB fade, without the equalizer, a −70-dBm received signal is degraded to a BER of 1×10^{-3}. This is a failure condition. With the equalizer, the BER improves to 2×10^{-4}. At higher received signal levels the improvement is even more noticeable. The 25.5-dB fade at a received signal level of −66-dBm improves the BER from worse than 1×10^{-3} to better than 1×10^{-7}. In each example, the presence of the equalizer almost completely restores the BER to the value obtained with no multipath fade present.

Multipath fading causes dispersive delay distortion, which can cause crosstalk between the I and Q signals in a QAM radio system. The IF adaptive slope equalizer can improve the situation to some extent, but additional equalization is often required.

5.1.5 Baseband adaptive transversal equalizers

The operation of the transversal equalizer can best be explained by looking at the impulse response of a distorted channel. Figure 5.12 compares an ideal channel with a distorted one. The impulse response of the ideal channel has *equal zero crossing at the symbol interval*. However, the impulse response of the distorted channel has some positive or negative amplitude at the points where there should be zero crossings. In other words, the *ringing effect* has some delay distortion included. The transversal equalizer effectively forces the zero crossings to occur at the points where they should occur.

Although this is a simple problem to describe, the circuit required to

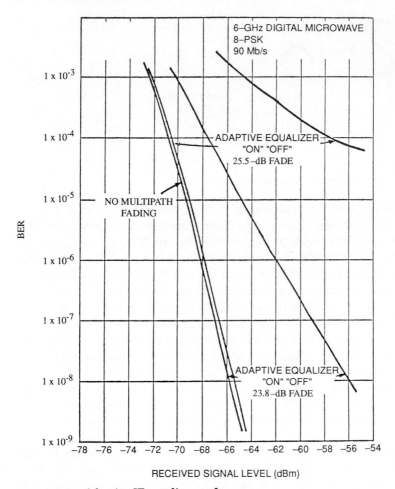

Figure 5.11 Adaptive IF equalizer performance.

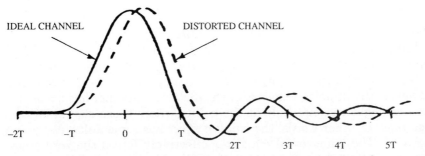

Figure 5.12 Impulse response of an ideal channel and a distorted channel.

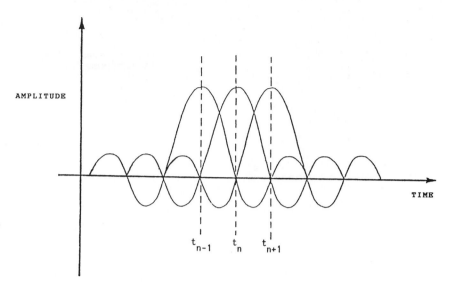

Figure 5.13   Optimal conditions for providing zero ISI.

solve this problem is rather complicated. Since this is a very important aspect of DMR, it will be discussed in more detail and simplified
as much as possible. It is probably fair to say that the success of DMR
to a large extent can be attributed to the design and incorporation of
these equalizers to counteract the potentially devastating effects of
multipath fading. The objective of the baseband equalizer is to minimize the intersymbol interference caused by adverse propagation conditions. As shown in Fig. 5.13, under optimal conditions the zero crossing instants of the "tails" of a PAM signal coincide perfectly with the
sampling instants of adjacent PAM information. In this situation, it is
easy to see that a cancellation of the tails produces zero intersymbol
interference. During multipath fading conditions, the adjacent re-
sponses do not have tails that cross the time axis at exactly equal intervals (see Fig. 5.14). The summation of the tails, called precursors
and postcursors, can result in intersymbol interference if their amplitudes are sufficiently large, as in the case of severe multipath fading.
These pre- and postcursors proceed to infinity with decreasing amplitude. For a simplified analysis, it is sufficient to consider only the first
postcursor and the first precursor. Figure 5.15 shows how the
postcursor of the preceding PAM signal (X_{k-1}) and the precursor of
the proceeding PAM signal (X_{k+1}) increase the value of the transmitted PAM signal (X_k). Mathematically,

$$X_k = \hat{X}_k + V_a + V_b$$

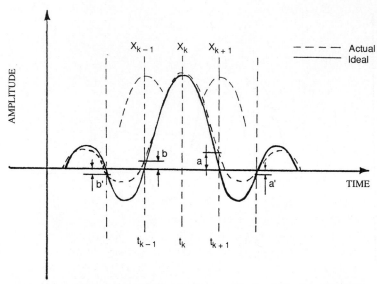

Figure 5.14 Distorted PAM X_k interferes with the following PAM X_{k+1} and previous PAM X_{k-1}. *Note:* Voltages (a, a') are postcursor interferences and (b, b') are precursor interferences.

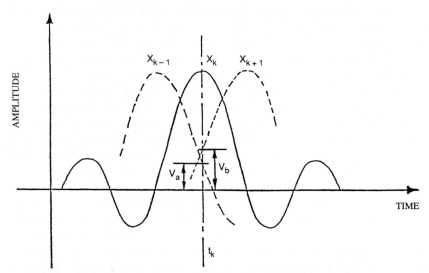

Figure 5.15 At sampling instant τ_k, V_a (postcursor interference by PAM X_{k-1}) and V_b (precursor interference by PAM X_{k+1}) are to be added to the value of the transmitted PAM X_k.

where $\hat{X}_k$ = the amplitude of the transmitted PAM signal
　　V_a = the postcursor interference amplitude
　　V_b = the precursor interference amplitude

that is,

$$V_a = aX_{k-1}$$

$$V_b = bX_{k+1}$$

where a and b are coefficients whose values lie between -1 and $+1$ and directly indicate the interference present. Therefore,

$$X_k = \hat{X}_k + aX_{k-1} + bX_{k+1} \tag{5.1}$$

The simplified circuit block diagram in Fig. 5.16 allows the interfering components to be subtracted from the X_k signal, resulting in X_{0k}. This equalized signal is theoretically the same as the transmitted signal. X_{0k} is passed through a decision circuit to restore the binary information and then a D/A converter produces the original transmitted signal $\hat{X}_k$. That is,

$$X_{0k} = X_k - a\hat{X}_{k-1} - bX_{k+1}$$

as

$$|a| \ll 1 \qquad \text{then} \qquad a\hat{X}_{k-1} \approx aX_{k-1}$$

so

$$X_{0k} \approx X_k - aX_{k-1} - bX_{k+1}$$

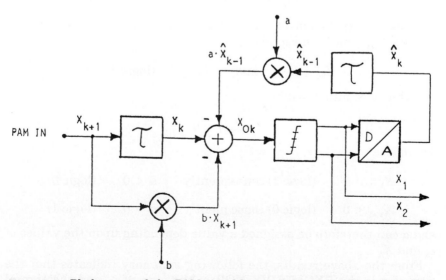

Figure 5.16 Blocks τ retard the PAM signal by one decision period to make PAM X_{k-1} available at the same time instant as PAM X_k.

Substituting for $X_k \to X_{0k} \approx \hat{X}_k$. The coefficients a and b must next be established.

Postcursor equalization (to determine a). The difference between the equalized PAM signal and the transmitted PAM signal can be defined as the error ϵ_k, where

$$\epsilon_k = X_{0k} - \hat{X}_k \qquad (5.2)$$

Considering an unequalized signal which for simplicity has only postcursor interference, Eq. (5.1) becomes

$$X_{0k} = X_k - a\hat{X}_{k-1}$$

It can be seen that if X_k increases in amplitude because of postcursor interference, X_{0k} will correspondingly increase. From Eq. (5.2), if X_{0k} is greater than X_k,

$$\epsilon_k > 0 \to \text{logic 1}$$

In order to counteract the increase in X_k it is necessary for

$$a\hat{X}_{k-1} > 0$$

from which it follows that

If $\hat{X}_{k-1} > 0$ (logic 1)   then $a > 0$ (logic 1)

If $\hat{X}_{k-1} < 0$ (logic 0)   then $a < 0$ (logic 0)

Similarly, a reduction in X_k caused by postcursor interference decreases X_{0k}, and from Eq. (5.2)

$$X_{0k} < \hat{X}_k \quad \text{and} \quad \epsilon_{k < 0} \quad \text{(logic 0)}$$

and this is counteracted by

$$a\hat{X}_{k-1} < 0$$

Therefore,

$\hat{X}_{k-1} > 0$ (logic 1) consequently $a < 0$ (logic 0)

$\hat{X}_{k-1} < 0$ (logic 0) consequently $a > 0$ (logic 1)

and a can therefore be assigned a value depending upon the values of ϵ_k and $\hat{X}_{k-1}$.

From the above results, the following summary indicates that the sign of a is the EXCLUSIVE NOR of the sign of ϵ_k and the sign of $\hat{X}_{k-1}$:

Sign of ϵ_k	Sign of $\hat{X}_{k-1}$	Sign of a
1	1	1
1	0	0
0	1	0
0	0	1

The analog values of a are found simply by integrating the digital values of the sign of a. Figure 5.17 shows a simplified diagram of the circuit which equalizes postcursor interference.

Precursor equalization (to determine b). An analysis similar to the above precursor equalization would result in b being evaluated by passing the sign of ϵ_k and the sign of X_{k+1} through an EXCLUSIVE NOR gate. However, X_{k+1} is not available since it has not yet passed through the decision circuit. This problem is solved by shifting the time reference by one decision instant (period). Therefore, as in Fig. 5.18, the input is now X_{k+1}, the sign of ϵ_{k-1} is used instead of the sign of ϵ_k, and the sign of $\hat{X}_k$ is used instead of the sign of X_{k+1}. This is acceptable because the signal equalization takes place over a time

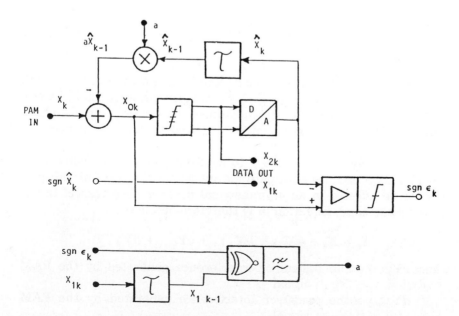

Figure 5.17   Simplified block diagram for postcursor equalization only.

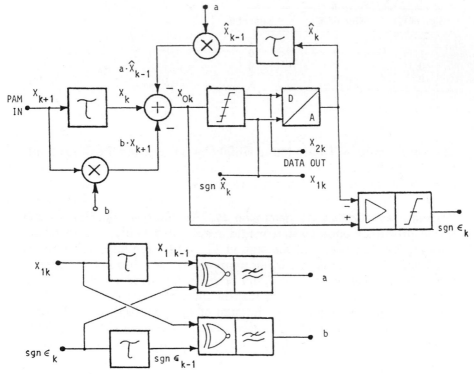

Figure 5.18   Block diagram for postcursor and precursor equalization.

interval corresponding to thousands of bits (depending upon the RC time constant of the integrator), so the precursor distortion of X_{k+1} on X_k is approximately the same as X_k on X_{k-1}.

So far, equalization has been considered only for the PAM X signal. For 16-QAM, there is also a PAM Y component, such that the two components are subsequently modulated by carrier signals phase shifted by 90°. The Y component is equalized by a circuit identical to that of Fig. 5.18.

Unfortunately, it is necessary to consider interactions between X and Y. This is because multipath fading causes the transmitted spectrum to be affected in an asymmetrical manner with respect to the center of the band. Equation (5.1) therefore becomes

$$X_k = \hat{X}_k + aX_{k-1} + bX_{k+1} + cY_{k-1} + dY_{k+1}$$

where cY_{k-1} = the postcursor interference generated by the PAM Y_{k-1} signal

dY_{k+1} = the precursor interference generated by the PAM Y_{k+1} signal.

It is sufficient to perform cross-correlations between ϵX_k and Y_{k-1} and between ϵX_{k-1} and Y_k as indicated in Fig. 5.19.

As one can appreciate, this simplified circuit is starting to become complex and present-day baseband equalizers are very sophisticated. The baseband transversal equalizer together with the IF slope equalizer provides most links with the necessary protection to withstand

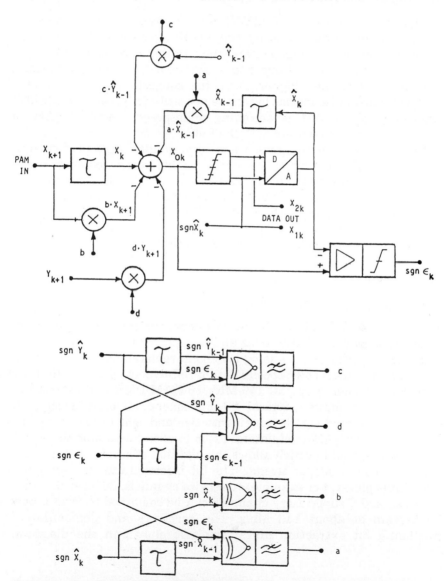

Figure 5.19 Complete equalizer block diagram.

very deep multipath fades. Usually, the only links that still have a problem even after incorporating both types of equalizer are the long hops over *water,* in hot, humid climates.

5.2 Digital Microwave Radio Systems

Perhaps the most widely used DMR in recent years has been the long-haul, medium- to high-capacity radio (140 Mb/s), used for backbone links between major metropolitan areas. In the more developed countries the ever-increasing hunger for more capacity has led to transmission at 565 Mb/s and higher. Also, the congestion within the prescribed frequency bands has led to the necessity for greater bandwidth efficiency, so 64-QAM is becoming widespread, and 256-QAM is emerging from the laboratory to the field.

As technology has progressed and high-quality, high-performance microwave components have become available in the frequency range above 10 GHz, a significant interest in using DMRs in the local network has evolved. This has been in the form of rural area spur hops from a backbone link or short hops within a city between company premises or from a large company to the exchange. In each case the transmission distance is short, usually in the range of 2 to 15 km.

5.2.1 140-Mb/s DMR with 16-QAM

The 140-Mb/s 16-QAM is still the workhorse system which is used extensively in developed countries and is rapidly being adopted by developing countries as an analog backbone replacement system. Each manufacturer has its own special design refinements but all follow the same basic theme. A typical 140-Mb/s 16-QAM DMR is illustrated in Fig. 5.20. This shows a typical 1 + 1 frequency diversity system. A 140-Mb/s baseband signal is split into two and sent to two transmitters via two 16-QAM modulators. Prior to each modulator there is a bit insertion module which allows the addition of several extra channels to the 140-Mb/s bit stream. The bit insertion unit is simply another multiplexer. For example, 10 service channels, 30 local channels (2 Mb/s), and 2 supervisory channels can be combined to form a new bit stream at about 143 Mb/s. The receivers and demodulators (including bit extraction units) are also shown in the diagram.

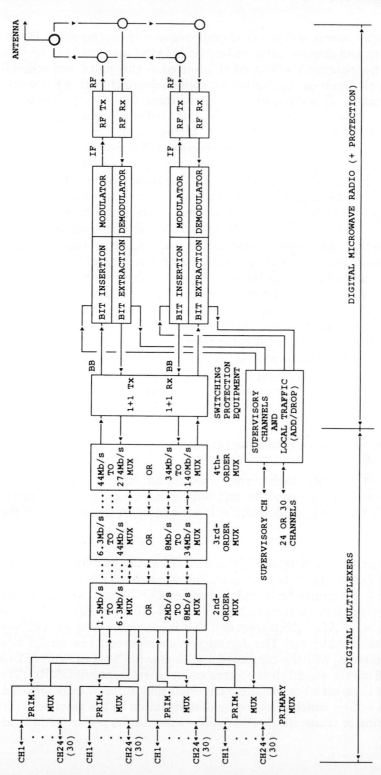

Figure 5.20   A typical 16-QAM DMR system.

221

Figure 5.21 shows a typical digital regenerative repeater. Down-conversion and demodulation to baseband is the usual procedure. The service channels can be dropped or inserted at this point, but individual voice channels cannot unless a complete demultiplexing procedure is undertaken. This diagram also shows space diversity in one direction, indicating that two repeaters and a combiner are now necessary for each half of the space diversity system, which means that two transmitters and four receivers are required. Figure 5.22 illustrates a frequency diversity system where there is only one protection channel available for up to 12 bearers (12 + 1). This is clearly more economical than 1 + 1 protection, but in the unlikely event of two bearers simultaneously falling below a BER of 1×10^{-3}, one of the bearers is lost. For the 12 + 1 case, the protection switch splits the signal of each bearer so that the signal is sent to the transmitter and also to a microprocessor controlled switch. In the event of a failure, the microprocessor connects the faulty bearer to the standby channel and at the receiver redirects the standby to the appropriate faulty channel while disconnecting the faulty signal. The switching decision is made based on the BER of each channel. If there is not an instantaneous failure of a channel, as the BER rapidly degrades (over a period of a few microseconds) at a predetermined BER of say 1×10^{-6}, the microprocessor ensures a hitless switching operation. An instantaneous failure (in the nanosecond time frame) will not allow switching without an error burst. Even dropped calls may result.

5.2.2 Digital microwave radio transceiver components

The higher-capacity, higher-order modulation scheme DMRs tend to have the most complex system and component design. For this reason, the 140-Mb/s DMR components will be studied in more detail in the following.

Transmit path. The diagram of Fig. 5.23 shows the building blocks of a typical 140-Mb/s DMR transmitter. The signal inputs to the transmitter include the multiplexed voice, data, and/or TV channels, together with service channel information. The 140-Mb/s bit stream has the CMI line code. After clock recovery, this is converted into an NRZ signal in the transmitter. The digital service channels are mixed into this bit stream, and the combined signal is scrambled to remove line spectra from the signal (line spectra cause interference in DMRs). The signal is then serial-to-parallel converted and differentially encoded. This is a coding which allows easy demodulation in the receiver without needing to transmit absolute phase information. After QAM mod-

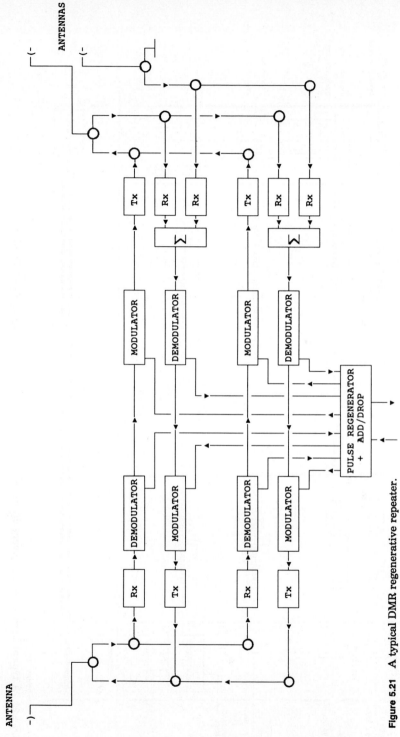

Figure 5.21 A typical DMR regenerative repeater.

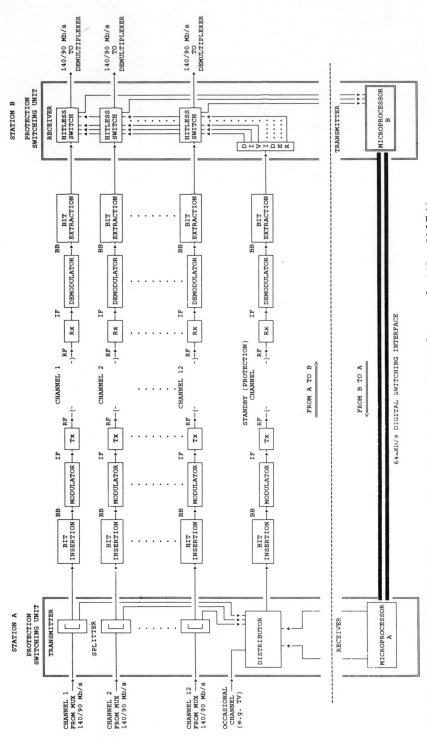

Figure 5.22 A typical DMR 12 + 1 frequency diversity system (operating at, for example, 140 or 90 Mb/s).

224

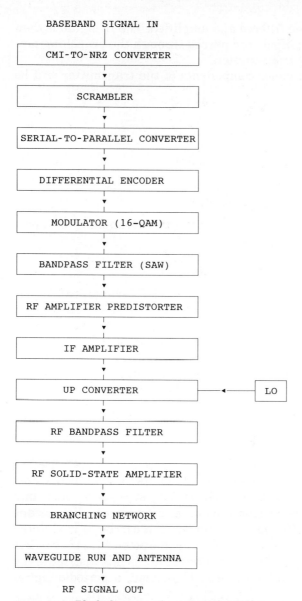

Figure 5.23 Block diagram of a 140-Mb/s DMR transmitter.

ulating, the signal is then filtered and amplified. The IF signal is then up-converted to RF, amplified, and passed through the branching network to the antenna for transmission.

Each of the following major components of the transmitter will be discussed in detail:

1. CMI to NRZ converter

2. Scrambler

3. Serial-to-parallel converter in the modulator

4. Differential encoder

5. RF amplifier predistorter/IF amplifier

6. RF section mixer, local oscillator, amplifier, branching network, waveguide run, and antenna

The modulator is treated in detail in Chap. 3.

CMI to NRZ decoding. Information from the MUX equipment is presented to the 140-Mb/s radio in the CMI form. The preferred code for dealing with information in the radio is the NRZ code. The simple circuit in Fig. 5.24 converts the incoming CMI signal to NRZ. The CMI signal is inverted and then passed to an EXCLUSIVE OR gate together with the ½-bit delayed CMI signal. The output of the EXCLUSIVE OR gate is the input to a D flip-flop. The output of the D flip-flop is the NRZ required signal. The timing diagram in Fig. 5.25 shows the progress of the signal through the circuit. The final NRZ signal is 1/2 bit delayed from the original CMI input.

Scrambling. In recent years, the scrambler circuit has been used in DMRs even in small-capacity systems (2 Mb/s). Its function is to transform repeated sequences of bits, such as long strings of zeros, into *pseudorandom* sequences. Pseudorandom, as discussed earlier, means almost random but not 100 percent random. If a carrier is modulated with a random digital signal, the spectrum characteristics in the first lobe would be of the type $(\sin x)/x$. This would uniformly distribute the energy over the transmitted spectrum. If, instead, a periodic digital

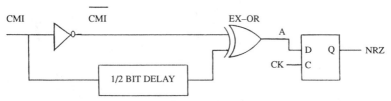

Figure 5.24 A typical CMI-to-NRZ converter.

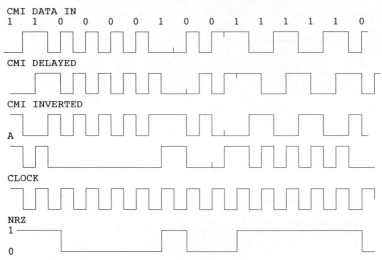

Figure 5.25   Timing diagram for the CMI-to-NRZ converter.

signal modulates the carrier, there will be concentrations of energy in the spectrum known as line spectra. The lines appear at discrete frequencies, depending upon the digital words, which keep repeating. This situation must be avoided because:

1. RF energy concentrations will increase the possibility of producing interferences in the adjacent RF channels.

2. Permanent lines in the spectrum can create serious problems in reception. The receiver VCO can lock onto a line in the spectrum instead of the incoming carrier, causing a loss of signal information.

3. In high-capacity DMRs an adaptive equalizer circuit is incorporated at the IF stage to restore the signal which has been impaired by selective fading. This circuit was developed on the basis that a pseudorandom signal would be transmitted.

The scrambler/descrambler pair does have a disadvantage. Any errors which occur because of noise or interference during transmission (between the scrambler and descrambler) will be *multiplied* by the descrambler, thereby slightly worsening the S/N (by approximately 0.5 dB).

The principle of the scrambler can be understood by observing Fig. 5.26. The truth table shows that the output of the second EXCLUSIVE OR gate, at y, is exactly the same as the input to the first EXCLUSIVE OR gate at a. This means that the output y is totally *independent* of the input c. So, provided c is the same in both halves of the

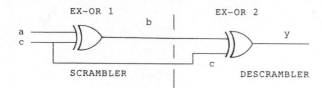

$$b = \overline{a}c + a\overline{c}$$
$$y = a$$

	a	c	b	y
TRUITH TABLE	0	0	0	0
	0	1	1	0
	1	0	1	1
	1	1	0	1

Figure 5.26   The principle of the scrambler.

circuit, the circuit can be split so that the *scrambler* is the left half of the circuit and the *descrambler* is the right half of the circuit. The input c is made pseudorandom by using shift registers. A simple three-flip-flop shift register pseudorandom circuit in Fig. 5.27 shows how the shift registers together with the EXCLUSIVE OR gate create an output which repeats itself after $2^n - 1$ words. This circuit when used in the configuration of Fig. 5.26 results in a 3-bit scrambler circuit. If the periodicity of the a input is 2 (i.e., 01010101...), the repetition occurs after

$$(2^n - 1) \times \text{input periodicity} = (2^3 - 1) \times 2$$

$$= 14 \text{ words}$$

Figure 5.28 illustrates the circuit and truth table.

Scramblers used in present-day DMR equipment have more than three flip-flop shift registers. To ensure equipment compatibility from different manufacturers, the CCITT has specified a scrambler circuit in Annex B of Recommendation G.954. A circuit similar to the CCITT scrambler, shown in Fig. 5.29a, is made up of a shift register which uses 10 D flip-flops with feedback via EXCLUSIVE OR gates. It can be seen from the chart in Fig. 5.29b that the configuration of the output bits (*data out*) is not dependent only on the incoming bits (*data in*) but also on the Q-outputs of the flip-flops. Therefore, when switching

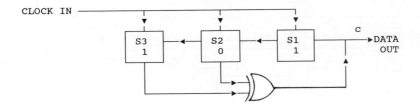

	S1	S2	S3	c	WORD
INPUT STATE	1	0	1	1	1
	1	1	0	1	2
	1	1	0	0	3
	0	1	1	0	4
	0	0	1	1	5
	1	0	0	0	6
	0	1	0	1	7
	1	0	1	1	
	1	1	0		

$2^3 - 1 = 7$
⟵ REPETITION HERE

Figure 5.27 Formation of the pseudorandom signal.

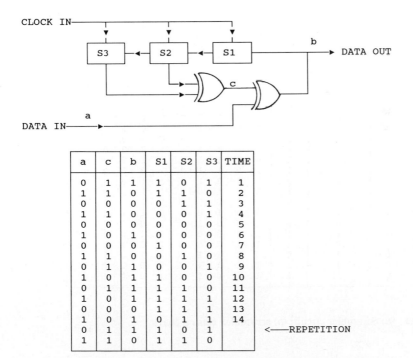

a	c	b	S1	S2	S3	TIME
0	1	1	1	0	1	1
1	1	0	1	1	0	2
0	0	0	0	1	1	3
1	1	0	0	0	1	4
0	0	0	0	0	0	5
1	0	1	0	0	0	6
0	0	0	1	0	0	7
1	1	0	0	1	0	8
0	1	1	0	0	1	9
1	0	1	1	0	0	10
0	1	1	1	1	0	11
1	0	1	1	1	1	12
0	0	0	1	1	1	13
1	0	1	0	1	1	14
0	1	1	1	0	1	
1	1	0	1	1	0	

⟵REPETITION

Figure 5.28 A 3-bit scrambler circuit and truth table. *a* has a periodicity of 2.

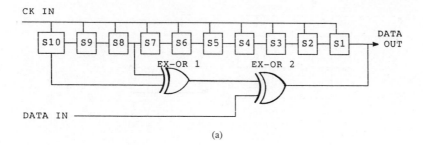

(a)

TIME	DATA IN	BIT OUT EX-OR 1	DATA OUT	S1	S2	S3	S4	S5	S6	S7	S8	S9	S10
T0				1	1	0	0	1	0	0	0	1	1
T1	1	1	0	0	1	1	0	0	1	0	0	0	1
T2	1	1	0	0	0	1	1	0	0	1	0	0	0
T3	1	1	0	0	0	0	1	1	0	0	1	0	0
T4	0	0	0	0	0	0	0	1	1	0	0	1	0
T5	0	0	0	0	0	0	0	0	1	1	0	0	1
T6	1	0	1	1	0	0	0	0	0	1	1	0	0
T7	1	1	0	0	1	0	0	0	0	0	1	1	0
T8	0	0	0	0	0	1	0	0	0	0	0	1	1
T9	1	1	0	0	0	0	1	0	0	0	0	0	1
T10	0	1	1	1	0	0	0	1	0	0	0	0	0
T11	1	0	1	1	1	0	0	0	1	0	0	0	0
T12	1	0	1	1	1	1	0	0	0	1	0	0	0
T13	0	1	1	1	1	1	1	0	0	0	1	0	0
T14	0	0	0	0	1	1	1	1	0	0	0	1	0
T15	1	0	1	1	0	1	1	1	1	0	0	0	1
T16	0	1	1	1	1	0	1	1	1	1	0	0	0
T17	1	1	0	0	1	1	0	1	1	1	1	0	0
T18	1	1	0	0	0	1	1	0	1	1	1	1	0
T19	1	1	0	0	0	0	1	1	0	1	1	1	1
T20	1	0	1	1	0	0	0	1	1	0	1	1	1
T21	1	1	0	0	1	0	0	0	1	1	0	1	1
T22	1	0	1	1	0	1	0	0	0	1	1	0	1
T23	1	0	1	1	1	0	1	0	0	0	1	1	0
T24	1	0	1	1	1	1	0	1	0	0	0	1	1
T25	0	1	1	1	1	1	1	0	1	0	0	0	1
T26	1	1	0	0	1	1	1	1	0	1	0	0	0
T27	0	1	1	1	0	1	1	1	1	0	1	0	0
T28	0	0	0	0	1	0	1	1	1	1	0	1	0
T29	0	1	1	1	0	1	0	1	1	1	1	0	1
T30	0	0	0	0	1	0	1	0	1	1	1	1	0
T31	1	1	0	0	0	1	0	1	0	1	1	1	1
T32	1	0	1	1	0	0	1	0	1	0	1	1	1
T33	0	1	1	1	1	0	0	1	0	1	0	1	1
T34	1	0	1	1	1	1	0	0	1	0	1	0	1
T35	1	1	0	0	1	1	1	0	0	1	0	1	0
T36	0	1	1	1	0	1	1	1	0	0	1	0	1
T37	0	1	1	1	1	0	1	1	1	0	0	1	0

(b)

Figure 5.29 (*a*) Scrambler circuit; (*b*) scrambler timing chart.

on the equipment, the Q-outputs from the 10 stages of the shift register can assume any of the 2^{10} (1024) possible bit combinations. So, before any transmission, the shift register must be filled with 10 *real* bits. In other words, the first 10 bits cannot be recovered.

Serial-to-parallel conversion. The data information from the multiplexer is always transmitted in serial form. During modulation, the information must be processed in the parallel form. A serial-to-parallel conversion is therefore necessary. The serial-to-parallel converter transforms a serial bit stream into two separate bit streams as shown in the example of Fig. 5.30a. Figure 5.30b shows a typical logic circuit for serial to four parallel streams conversion together with its timing diagram in Fig. 5.30c. The $\overline{Q}$-output of FF1 is used as the input to FF2. This means that the incoming NRZ signal is delayed, inverted, and then presented to FF2. As the $\overline{Q}$-output from FF2 is passed on to FF6, the signal input to the D flip-flop FF6 is the same as the incoming NRZ signal except that it is now delayed by *two* clock periods. By a similar process, the output from FF3, which is the input to FF7, is the same as the incoming NRZ signal except that it is delayed by *three* clock periods. Also, the $\overline{Q}$-output from FF4, which is the input to FF8, is the same as the incoming NRZ signal except that it is delayed by *four* clock periods. FF5, FF6, FF7, and FF8 are all clocked by a signal that is one-quarter of the rate of the clock input to FF1, FF2, FF3, and FF4. This enables the output of FF5 to be constructed from every fourth bit in the original NRZ bit stream. Similarly, FF6 is composed of every adjacent fourth bit; it is the same for FF7 and FF8. The resulting outputs from the buffer amplifiers are parallel signals, thereby fulfilling the initial objective of separating the continuous input bit stream into four parallel output streams.

Differential encoding. It was previously stated that PSK and QAM types of modulation have the double sideband suppressed carrier (DSB-SC) characteristic. In other words, the carrier is not present in the received signal. A difficulty therefore arises. At the receiver, how does one establish the carrier phase as sent from the transmitter? This is further complicated because the phase delay introduced by the transmission medium is not constant because of atmospheric effects. To overcome this problem, the carrier is modulated with information by changing the phase of one state, depending on the previous state. At the receiving end, the phase in the present interval is compared to the phase in the previous interval. The signal received in the previous interval is delayed for one signal interval and is used as a reference to demodulate the signal in the next interval. This is differential encoding and decoding. *It allows detection without needing to know the absolute phase of the signal.* The differential encoder forms a part of the

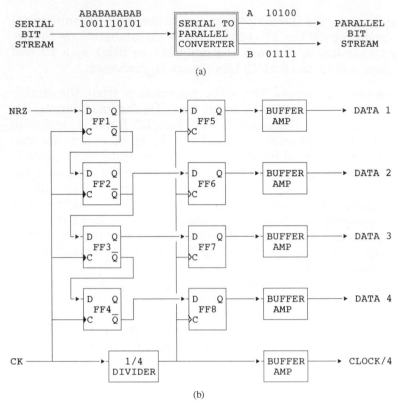

Figure 5.30   (a) Serial-to-parallel converter block diagram; (b) a typical serial-to-parallel converter circuit.

modulator. The simplified block diagram of the QAM modulator shown in Fig. 5.31 is an extension of the modulators described in Chap. 3. An incoming 140-Mb/s bit stream is split into *four* bit streams by a serial-to-parallel converter. Two of the bit streams are passed through one low-pass filter, and the other two bit streams are passed through another low-pass filter. The output of each filter is a two-level PAM signal. An LO then mixes with each PAM signal. Because of the 90° phase shift in one side of the circuit, the resulting output is a modulated signal that has 16 states (i.e., 16-QAM).

The differential encoder is placed between the serial-to-parallel converter and the low-pass filters. The differential encoder incorporates a feedback mechanism which allows the phase associated with each pair of bits to be fixed depending upon the previous pair of bits (Fig. 5.32a). The combination of all possible binary word outputs, X1,X2,Y1,Y2, will represent the 16 possible states as shown in Fig. 5.32b.

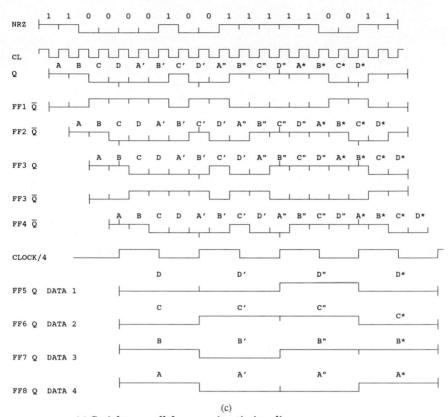

(c)

Figure 5.30 (c) Serial-to-parallel conversion timing diagram.

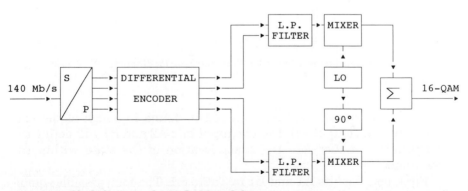

Figure 5.31 Block diagram of a 16-QAM modulator.

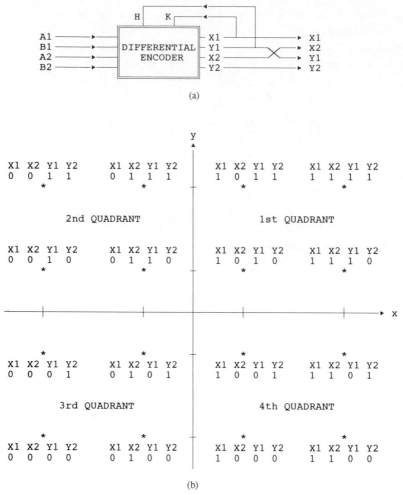

Figure 5.32 (a) The differential encoder block diagram; (b) differentially encoded states.

The input bits A1 and B1 are used to define in which of the four quadrants the output will lie. The input bits A2 and B2 will define the amplitude and therefore the exact location of the state within the quadrant.

First, observe bits A1 and B1 in Table 5.1. The four possible combinations of these bits represent the phase with respect to the previous state. The outputs X1 and Y1 are related to the input A1 and B1 by the following equation, which includes the feedback bits H and K:

$$\phi\,X1Y1 = \phi\,HK + \phi\,A1B1$$

TABLE 5.1

A1	B1	$\Delta\phi$
1	1	0°
0	1	+90°
1	0	−90°
0	0	180°
H,X1	K,Y1	

TABLE 5.2

X1	Y1	Quad.	X2	Y2
0	0	3d	$\overline{A2}$	$\overline{B2}$
1	0	4th	A2	$\overline{B2}$
0	1	2d	$\overline{A2}$	B2
1	1	1st	A2	B2

NOTE: X1 and Y1 provide *quadrant* information and X2 and Y2 represent *level* (position in the quadrant) information.

The outputs X2 and Y2 are related to the inputs A2 and B2 as in Table 5.2.

As an example, suppose the following bit sequence is to be transmitted:

A1 B1 A2 B2

0 1 1 1   0101 0100 1011 0000 1101 0101 0100 0001 1001

The inputs and outputs are summarized in Fig. 5.33.

Notice that H and K are the feedback bits, which are the same as the previous X1 and Y2 bits. The inputs and outputs follow the equation

$$\phi \, X1Y1 = \phi \, HK + \phi \, A1B1$$

For example, consider the first 4 input bits 0111. In this case, A1 = 0 and B1 = 1, so from Table 5.1 the phase change with respect to the previous state is +90°. Similarly, A2 = 1 and B2 = 1, so the phase is 0°. Since the previous X1 and Y1 bits were both H = 0 and K = 0, the phase is 180°.

Substituting this information into the equation

$$X1Y1 = 180° + 90°$$
$$= 270° \quad \text{or} \quad -90°$$

Tx SIDE: φX1Y1 = φHK + φA1B1									
BIT TO BE TRANSMITTED				PREVIOUS X1,Y1 BIT		DIFFERENTIALLY ENCODED BIT			
QUADRANT		LEVEL				QUADRANT		LEVEL	
A1	B1	A2	B2	H	K	X1	Y1	X2	Y2
0	1	1	1	0	0	1	0	1	0
0	1	0	1	1	0	1	1	0	1
0	1	0	0	1	1	0	1	1	0
1	0	1	1	0	1	1	1	1	1
0	0	0	0	1	1	0	0	1	1
1	1	0	1	0	0	0	0	1	0
0	1	0	1	0	0	1	0	0	0
0	1	0	0	1	0	1	1	0	0
0	0	0	1	1	1	0	0	1	0
1	0	0	1	0	0	0	1	1	1

Figure 5.33 Transmission of differentially encoded bits.

From Fig. 5.33, $X1 = 1$ and $Y1 = 0$, which corresponds to a phase of $-90°$.

The output bits X1, X2, Y1, and Y2 are then filtered. Each pair (X1,X2 and Y1,Y2) represents a four-level PAM signal, having analog voltages $+3$, -3, $+1$, -1. This signal is subsequently sent to the 4-PSK section of the modulator for completion of the 16-QAM process.

RF amplifier predistorter and IF amplifier. The RF predistorter is an important component in the DMR system. Its function is to purposely insert a nonlinearity into the IF signal which is the complement of the nonlinearity caused by the RF power amplifier. Ideally, the RF power amplifier should have an output power versus input power curve that is a perfect straight line (i.e., linear). Unfortunately, the harmonic content of the RF amplifier output power produces a graph that is slightly curved. This can result in serious degradation of the system performance if it is not corrected. Since nonlinearity compensation is rather difficult to accomplish at the microwave RF, it is performed at the lower IF, which is considerably simpler to achieve. One method is to incorporate a circuit which enhances the third harmonic and introduces it at a phase and level which can be varied by a technician. On alignment of the system, the technician observes the RF output spectrum and adjusts the predistortor to produce an output spectrum having the required performance (see the subsection "Transmitter Distortion Level").

The transmitter IF amplifier is usually very simple compared to the receiver amplifier. Its function is to supply the mixer with the modulated baseband signal at a frequency of 140 or 240 MHz.

RF components. The RF section is typically composed of the following components:

1. Mixer

2. Local oscillator

3. RF amplifier

4. Branching network

5. Waveguide run

6. Antenna(s)

These components are fabricated using a variety of "media." For example, in order to minimize losses and group delay distortion, filters are still made using waveguide cavities. The miniaturized stripline or microstrip integrated circuit form is inadequate because of higher losses. Similarly, the transmission line from the output amplifier to the antenna must be as low loss as possible to ensure a low noise figure and to maximize system gain. Waveguide is therefore necessary for this purpose. However, the mixer, local oscillator, and amplifiers are all now made in the microwave integrated circuit (MIC) form. In fact, the RF amplifiers are now at the peak of microwave technology since they have evolved from the hybrid technology to the *monolithic* technology. Monolithic MICs are smaller, more reliable, cheaper, and have better performance than the previous hybrid or discrete component structures.

Mixers.   The mixer is part of the upconverter. It is usually made using hybrid MIC technology. This means the circuit is etched on a ceramic (Al_2O_3) substrate, and the diode devices are very small chips which are bonded into place on the substrate. Because of the miniaturization of these circuits, it is not possible to repair them in the field. They must be returned to the manufacturer if a fault occurs.

Local oscillators.   The local oscillator, which is the other main component of the upconverter, is also fabricated in the miniaturized MIC construction. Local oscillators are usually designed using a GaAs FET transistor oscillator that is frequency stabilized by a dielectric resonator. The dielectric resonator makes the output frequency relatively insensitive to ambient temperature changes. The complete circuit is constructed in the microstrip circuit form and is small, cheap, and reliable.

RF amplifiers.   The two main requirements for the RF power amplifier in a high-capacity DMR system are

1. High output power

2. Good linearity

Based on a typical fade margin of 35 to 45 dB and the system gain requirements, a typical output power requirement is now in the region of 100 mW to 1 W (20 to 30 dBm). Until relatively recently, traveling wave tube (TWT) amplifiers and Klystrons were used for RF power amplification. These "tube" devices suffer from a short life and therefore contribute to system failure. Considerable effort has gone into increasing the lifetime of TWTs, and they are still used in some modern systems. However, the rapid rate of progress of the *solid-state* (particularly GaAs FET type) amplifiers leads one to believe that the days of the TWT are numbered. The solid-state amplifiers have longer life. It is this improvement in reliability which makes the solid-state amplifier so attractive. At present, the output power from a GaAs FET amplifier is not quite as high as the TWT, but as research progresses, the gap is narrowing rapidly. Eventually, solid-state amplifiers will surpass the TWT in both output power and linearity. Already, GaAs FET amplifiers are being made with output power values of approximately 10 W at the 1-dB compression point. This is comparable to the TWT. However, the main advantage of the TWT at present is its superior linearity performance. In order to operate the GaAs FET amplifier at an adequate linearity, an output power *backoff* of about several decibels is necessary. To compensate for nonlinearity, adaptive *predistorters* have been introduced, as already mentioned. GaAs FET power amplifiers with a 1-dB compression point of 10 W can reliably produce output power levels for 64-QAM in the 1-W range without predistortion and in the 2-W range with predistortion. Finally, high output power is less important than it used to be because recent enhancement in semiconductor device performance has improved the receiver sensitivity values.

Branching networks. A branching network is necessary to allow several transmitters and receivers to operate on a single common antenna. An example of a branching network is shown in Fig. 5.34. Channel branching filters are used to combine the various signal paths in the transmission section without mutual interference and to separate them again in the receiver. In modern DMR systems multiple-resonator bandpass filters are interconnected via circulators. Normally, the orthogonally polarized channel groups (1,3,5,7 and 2,4,6,8 in Fig. 5.34) are mutually displaced in frequency by one-half of the channel branching filter spacing in order to achieve adequate adjacent channel displacement (interleaved pattern). In digital systems, the orthogonally polarized channel groups can, however, also be operated on the same radio frequencies (cochannel pattern), thereby improving bandwidth utilization. The attenuation of the filters must be kept to a minimum. The metal cavity filters typically have a center band loss of 0.8 dB and the stripline circulators have a loss of 0.1 dB.

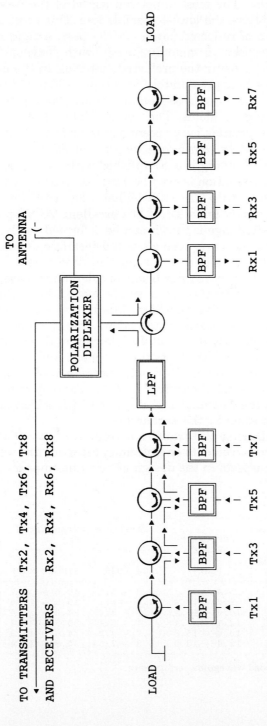

Figure 5.34 Schematic diagram of channel branching filters.

Waveguide run. The most important aspect of the waveguide run is that it should have the lowest possible loss. This must be achieved with a minimum of reflected power. In the past, a rigid rectangular waveguide has been commonly used, with oxygen-free, high-conductivity copper being the preferred material. In the 6-GHz band, WR137 has a loss of approximately 6.7 dB per 100 m. Although a circular waveguide has lower loss than a rectangular waveguide, it has significant *moding* problems. The semiflexible elliptical waveguide has become very popular over recent years. It is very attractive because it is supplied in continuous lengths on a drum. This enables easy installation. Bending is accomplished without the need for transitions, and its attenuation is even better than that of a standard rectangular waveguide. At 6 GHz, EW-59 has a loss of about 5.75 dB per 100 m. When correctly installed, it has excellent VSWR performance. However, a disadvantage is that it can be deformed easily, and if it is not treated with great care, even a small deformation (dent) can cause a mismatch that results in severe echo distortion noise.

Recently, a new approach has been considered for waveguide runs. Instead of operating in the usual single mode, the waveguides are operated in the *overmoded* condition. Traditional waveguides are designed to have a cross-section which allows only the fundamental frequency to propagate. By deliberately allowing multiple modes (over-modes) to exist, a substantial loss improvement can be achieved. The loss of the overmoded waveguide run is approximately *one-third* of the single mode loss, as tabulated in Fig. 5.35. The key element in this new approach is the design of a *mode filter* which selects the required mode for transmission by the antenna(s).

Antenna(s). During the transition from AMR to DMR systems, more stringent requirements on antenna properties are required. The most important aspects in the design of new antennas are:

TYPE OF WAVEGUIDE FEEDER		FREQUENCY RANGE	WAVEGUIDE ATTENUATION (dB/100m)	
LOW LOSS	SINGLE MODE	(GHz)	LOW LOSS	SINGLE MODE
S40	S 70	5.4 - 10.7	2.8 - 3.1	10.0 - 6.7
S58	S100	8.2 - 12.4	4.9 - 5.1	19.9 - 13.6
S58	S120	10.0 - 15.4	4.9 - 5.4	24.0 - 16.4
S58	S140	12.4 - 15.4	5.1 - 5.4	30.9 - 24.2
S70	S120	10.0 - 15.8	6.7 - 7.2	24.0 - 16.3
S70	S140	12.4 - 18.0	6.8 - 7.5	30.9 - 22.4

Figure 5.35 Overmoded waveguide performance.

- Higher sidelobe attenuation in all areas of the azimuth radiation pattern outside the main lobe
- Substantially improved cross-polarization properties
- Considerably increased bandwidth, allowing dual-band antennas which can be operated simultaneously in two frequency bands with two planes of polarization in each band

The construction of a typical present-day antenna is shown in Fig. 5.36. The electrical data for the parabolic-reflector type of antenna is shown in Fig. 5.37. When the multiband use of antennas is required, the shell type of construction is necessary. This is illustrated in Fig. 5.38.

Receive path.   The diagram of Fig. 5.39 shows the building blocks of a typical 140-Mb/s DMR receiver. First, the incoming RF signal from the antenna and waveguide run is downconverted to the IF. After preamplification, the signal level is maintained as flat as possible across the IF band by the adaptive equalizer. It is then filtered, amplified, and delay equalized. The AGC circuit is incorporated at this point. Next is the demodulation process. This is followed by regeneration and timing extraction, after which the signal is differentially decoded. Service channels are then extracted via a parallel-to-serial converter, after which the signal is descrambled. Finally, the NRZ signal is converted into CMI for transmission to the demultiplexer for recov-

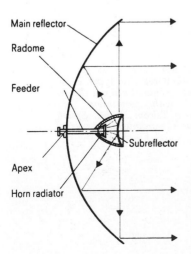

Figure 5.36 A symmetrical parabolic-reflector antenna. (*Reproduced with permission from Siemens Telcom Report: Special "Radio Communication," vol. 10, 1987, p. 139, Fig. 5a.*)

ANTENNA DIAMETER (m) AND TYPE	3.0 DIRECTLY FED WITH BLINDERS	3.0 CASSEGRAIN WITH BLINDERS	3.0 CASSEGRAIN WITH SHROUDED RADOME	3.0 CASSEGRAIN	0.6 CASSEGRAIN WITH SHROUDED RADOME
FREQUENCY RANGE (GHz)	1.7-2.1	3.4-4.2	6.425-7.125	10.7-11.7	17.7-19.7
GAIN (dBi)	32.8	38.5	43.6	48.5	39.0
3dB BEAMWIDTH (DEG)	3.8	1.5	0.8	0.7	1.6
SIDELOBE ATTENUATION (x dB) FROM BORESIGHT (DEG)	84(52 dB)	90(55 dB)	65(60 dB)	90(60 dB)	90(60dB)
FRONT/BACK RATIO (dB)	52	52	70	70	60
XPD (dB)	>30	>32*	>40	>35*	>37*
REFLECTION COEFFICIENT (%)	6	4.5	3	3.5	8.5

Figure 5.37 Electrical data for several parabolic-reflector antennas. * = over entire azimuth range.

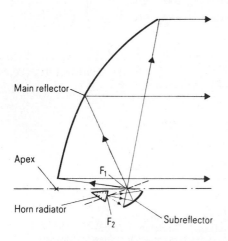

Figure 5.38 The shell type of antenna (Gregorian type). F_1, F_2 = focal points of ellipse. (*Reproduced with permission from Siemens Telcom Report: Special "Radio Communication," vol. 10, 1987, p. 139, Fig. 5b.*)

Main reflector

Apex

Horn radiator

F_1

F_2

Subreflector

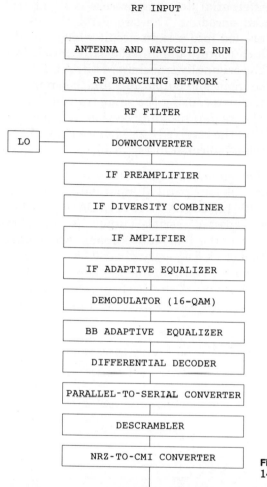

RF INPUT

| ANTENNA AND WAVEGUIDE RUN |
| RF BRANCHING NETWORK |
| RF FILTER |
| DOWNCONVERTER |
| IF PREAMPLIFIER |
| IF DIVERSITY COMBINER |
| IF AMPLIFIER |
| IF ADAPTIVE EQUALIZER |
| DEMODULATOR (16-QAM) |
| BB ADAPTIVE EQUALIZER |
| DIFFERENTIAL DECODER |
| PARALLEL-TO-SERIAL CONVERTER |
| DESCRAMBLER |
| NRZ-TO-CMI CONVERTER |

LO

BASEBAND OUT

Figure 5.39 Block diagram of a 140-Mb/s DMR receiver.

ering the voice, data, or TV channels. The RF section has already been discussed, which leaves the following processes to be described:

1. Differential decoding
2. Parallel-to-serial conversion
3. Descrambling
4. NRZ to CMI encoding
5. IF in-phase combiner (for space diversity).

These circuit functions are basically the inverse of those discussed in the transmit path.

Differential decoding. The differential decoding process is simply the reverse process of differential encoding. The two PAM signals obtained in the demodulator are fed into a logic circuit which extracts bits X1, X2, Y1, and Y2, which contain the differentially encoded information. From these bits, the originally transmitted bits A1, B1, A2, and B2 can be recovered. The input and output bits are shown in Fig. 5.40 for the receive-side demodulator, using the same transmitted bit sequence example as in the *differential encoding* section. It is a useful exercise to convince oneself that the received bit sequence is correct by applying the equation and table indicated in Fig. 5.40.

Parallel-to-serial conversion. The circuit of Fig. 5.41*a* is an example of a parallel-to-serial converter. The four parallel input data streams in this example are DATA-1, 10011; DATA-2, 10111; DATA-3, 0110; and DATA-4, 0010. The expected serial output is therefore 1100,0010,0111,1100,11.... First, the four data streams are clocked through four *D* flip-flops with a clock frequency divided by 4 (in accordance with the parallel data speed). The data bits streams 3 and 4 are delayed by ½ bit prior to entering the selector to ensure that when the last address is set by the counter, the data will be read in the center

X1	Y1	QUAD	A2	B2
0	0	3rd	$\overline{X2}$	$\overline{Y2}$
1	0	4th	X2	$\overline{Y2}$
0	1	2nd	$\overline{X2}$	Y2
1	1	1st	X2	Y2

Rx SIDE:	$\phi A1B1$	$=$	$\phi X1Y1$	$-$	ϕHK			
BITS RECEIVED				PREVIOUS X1,Y1		RECEIVED BIT		
X1	X2	Y1	Y2	H	K	A1	B1	A2	B2
1	1	0	0	0	0	0	1	1	1
1	0	1	1	1	0	0	1	0	1
0	1	1	0	1	1	0	1	0	0
1	1	1	1	0	1	1	0	1	1
0	1	0	1	1	1	0	0	0	0
0	1	0	0	0	0	1	1	0	1
1	0	0	0	0	0	0	1	0	1
1	0	1	0	1	0	0	1	0	0
0	1	0	0	1	1	0	0	0	1
0	1	1	1	0	0	1	0	0	1

SEQUENCE TRANSMITTED:
A1 B1 A2 B2
0 1 1 1   0101   0100 1011 0000 1101 0101 0100 0001 1001

Figure 5.40 Differential decoding.

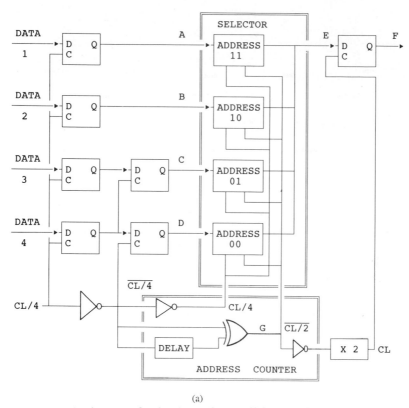

Figure 5.41   (*a*) An example of a circuit for parallel-to-serial conversion.

position. Otherwise some bits could be lost in the event of small changes in clock or data speed. The four address words are generated in the address counter as indicated in the timing diagram of Fig. 5.41*b*. The $\overline{\text{CL/4}}$ is inverted to provide a CL/4 input to each address unit in the selector. Also, a $\overline{\text{CL/2}}$ signal is derived in the address counter by feeding $\overline{\text{CL/4}}$ and $\overline{\text{CL/4}}$ delayed by one input bit time clock period (one-quarter of a $\overline{\text{CL/4}}$ period) into an EXCLUSIVE OR gate. The resulting $\overline{\text{CL/2}}$ signal at G is fed into each address unit. One can see that the combination of CL/4 and $\overline{\text{CL/2}}$ creates the address words 11, 10, 01, and 00. This allows the input data streams to be sequentially selected to produce the serial output, E.

The serial data clock is generated by multiplying the output from the inverted EXCLUSIVE OR gate (EXCLUSIVE NOR) by 2. This clock signal is then used to read the data into the final flip-flop, producing the serial bit stream at the output, F. The parallel-to-serial conversion is now complete.

Descrambling.   The circuit of Fig. 5.42 is an example of a descrambler.

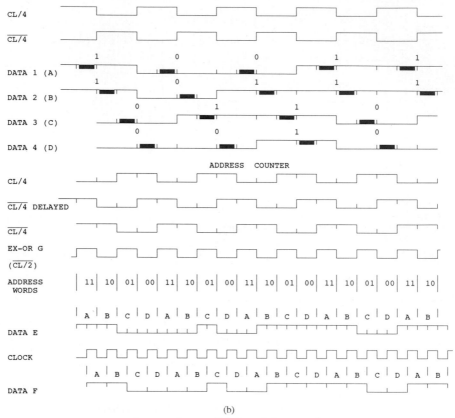

Figure 5.41   (b) Timing diagram for parallel-to-serial conversion.

The operation of the descrambler can be observed by analyzing the timing sequence as shown in the timing chart (Fig. 5.42b). In order to be able to recover data from time T1, it is necessary to have the same bit sequence in the descrambler in T0 as when starting the scrambling process. A comparison of the descrambler timing chart with that of the scrambler timing chart shows that the only difference is that the *data in* and *data out* bits are reversed. In other words the output from the descrambler is exactly the same as the input to the scrambler. This fulfills the initial objective.

NRZ to CMI encoding.   The NRZ waveform is suitable for manipulating data within the radio equipment, but it cannot be transmitted in this form to or from the multiplexer. The reason is that *the NRZ signal spectrum contains a dc component.* When long sequences of 1s occur, dc wander produces errors caused by 1s being detected as 0s. Figure 5.43 illustrates how dc wander causes uncertainty in the position of

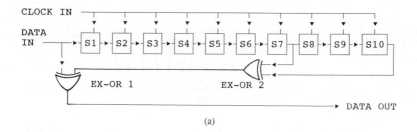

(a)

TIME	DATA OUT	BIT OUT EX-OR 2	DATA IN	REGISTER OUTPUTS									
				S1	S2	S3	S4	S5	S6	S7	S8	S9	S10
T0				1	1	0	0	1	0	0	0	1	1
T1	1	1	0	0	1	1	0	0	1	0	0	0	1
T2	1	1	0	0	0	1	1	0	0	1	0	0	0
T3	1	1	0	0	0	0	1	1	0	0	1	0	0
T4	0	0	0	0	0	0	0	1	1	0	0	1	0
T5	0	0	0	0	0	0	0	0	1	1	0	0	1
T6	1	0	1	1	0	0	0	0	0	1	1	0	0
T7	1	1	0	0	1	0	0	0	0	0	1	1	0
T8	0	0	0	0	0	1	0	0	0	0	0	1	1
T9	1	1	0	0	0	0	1	0	0	0	0	0	1
T10	0	1	1	1	0	0	0	1	0	0	0	0	0
T11	1	0	1	1	1	0	0	0	1	0	0	0	0
T12	1	0	1	1	1	1	0	0	0	1	0	0	0
T13	0	1	1	1	1	1	1	0	0	0	1	0	0
T14	0	0	0	0	1	1	1	1	0	0	0	1	0
T15	1	0	1	1	0	1	1	1	1	0	0	0	1
T16	0	1	1	1	1	0	1	1	1	1	0	0	0
T17	1	1	0	0	1	1	0	1	1	1	1	0	0
T18	1	1	0	0	0	1	1	0	1	1	1	1	0
T19	1	1	0	0	0	0	1	1	0	1	1	1	1
T20	1	0	1	1	0	0	0	1	1	0	1	1	1
T21	1	1	0	0	1	0	0	0	1	1	0	1	1
T22	1	0	1	1	0	1	0	0	0	1	1	0	1
T23	1	0	1	1	1	0	1	0	0	0	1	1	0
T24	1	0	1	1	1	1	0	1	0	0	0	1	1
T25	0	1	1	1	1	1	1	0	1	0	0	0	1
T26	1	1	0	0	1	1	1	1	0	1	0	0	0
T27	0	1	1	1	0	1	1	1	1	0	1	0	0
T28	0	0	0	0	1	0	1	1	1	1	0	1	0
T29	0	1	1	1	0	1	0	1	1	1	1	0	1
T30	0	0	0	0	1	0	1	0	1	1	1	1	0
T31	1	1	0	0	0	1	0	1	0	1	1	1	1
T32	1	0	1	1	0	0	1	0	1	0	1	1	1
T33	0	1	1	1	1	0	0	1	0	1	0	1	1
T34	1	0	1	1	1	1	0	0	1	0	1	0	1
T35	1	1	0	0	1	1	1	0	0	1	0	1	0
T36	0	1	1	1	0	1	1	1	0	0	1	0	1
T37	0	1	1	1	1	0	1	1	1	0	0	1	0

(b)

Figure 5.42 (a) An example of a descrambler circuit; (b) the descrambler timing chart.

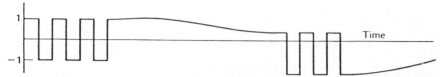

Figure 5.43 DC wander of an NRZ signal. (*Reproduced with permission from Bellamy, J., Digital Telephony, Fig. 4.6, © 1982, J. Wiley and Sons.*)

the pulses relative to the dc level. This problem exists not only for long strings of 1s or 0s but also whenever there is an imbalance in the number of 1s and 0s. Consequently, the final step before transmitting the pulses to the multiplexer for demultiplexing is to convert the NRZ signal back to CMI in the case of a 140-Mb/s signal or to HDB3 in the case of 34 Mb/s or less. The circuit in Fig. 5.44*a* converts NRZ to CMI. The timing diagram of Fig. 5.44*b* shows the steps in the conversion process. FF1 delays the input data stream by one clock period. The stream is passed on to FF2 via an EXCLUSIVE OR gate whose second input is derived from a feedback loop around FF2. The inverted output of FF2 is fed to FF4, which delays the signal by one clock period before entering the upper NOR gate. The second input to the upper NOR gate is the inverted output of FF1 delayed one clock period by FF3. The output of the NOR gate at point A is the CMI 1s, which are part of the required CMI output. The CMI 0s are formed at point B by feeding into the lower NOR gate (1) the clock signal together with (2) the original data stream, delayed two clock periods by the double inversion of FF1 and FF3. Finally, the CMI 1s and 0s are summed by the OR gate to produce the complete CMI output at C.

IF in-phase combiner (for space diversity). For a space diversity DMR, combining at the IF is usually the preferred technique. The purpose of the IF combiner is to combine the IF signals sent by the main mixer and the diversity mixer to offset any signal-level degradation caused by multipath fading. In-phase combining for the best case of equal levels and noise out of phase produces a 6-dB power level improvement and a 3-dB S/N improvement. Figure 5.45 illustrates one possible design for an in-phase IF combiner.

Two PIN diode attenuators, one on the main IF signal path and one on the diversity IF signal path, are driven by the AGC comparator circuit. The AGC comparator circuit continuously compares the RF signals received by the two mixers and drives the PIN diode attenuators as follows:

While the difference between the RF signals received by the two mixers is less than or equal to a preestablished value (e.g., 10 dB), the IF signal corresponding to the lower received RF signal is atten-

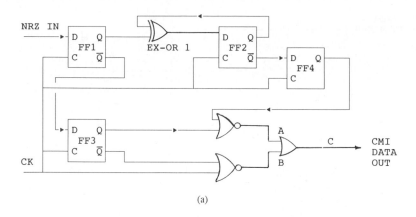

(a)

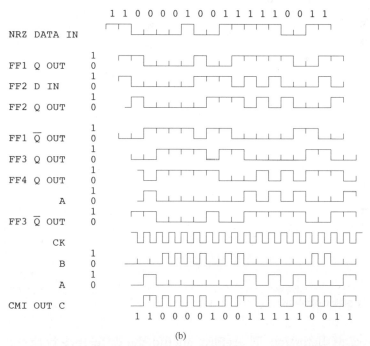

(b)

Figure 5.44 (a) CMI encoder circuit; (b) timing diagram for NRZ-to-CMI converter.

uated. This is to restore, at IF, the same decibel difference present at RF. Note that because of the AGC circuits, both IF signals are present at the combiner input with the same level.

There is no benefit in combining two signals when one of them has

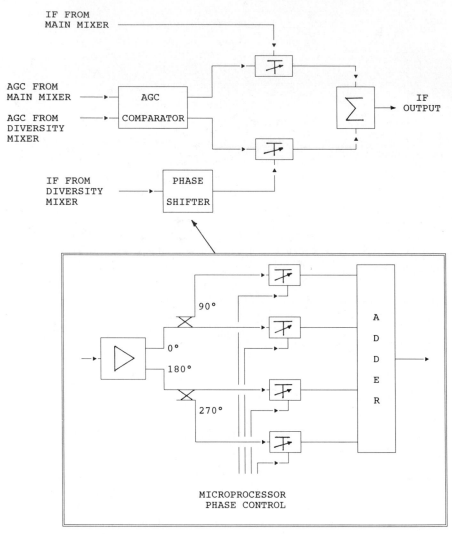

Figure 5.45 IF in-phase combiner.

a high level of distortion. Therefore, should the difference between the received RF signals exceed, say, 10 dB, the attenuation introduced into the lower level IF signal is proportionately increased to ensure the combined signal is weighted in favor of the higher-quality signal. For example, a 14-dB difference between the IF signals may be designed to cause a 20-dB difference between the IF levels at the combiner input.

The phase shifter is the key component necessary for successful operation of the in-phase IF combiner. This exists only in the diversity mixer path of the space diversity receiver. It has an IF input amplifier that has two outputs, phase shifted by 180°. Each output feeds a coupler, so the four coupler outputs supply IF signals with relative phases of 0°, 90° and 180°, 270° which are sent to four variable attenuators driven by a microprocessor, according to the following criteria:

At every instant, two of the four attenuators insert a large attenuation into the IF signal so that its contribution to the combiner is almost zero.

The two remaining attenuators are regulated to produce two output signals with amplitudes such that the resultant (vector sum) has the required phase.

Therefore, by selecting the two signals to be combined and by modifying the amplitudes of the selected signals, a resulting *constant amplitude* IF signal is obtained, and the phase can continuously shift between 0° and 360°. This allows the incoming main and diversity signals which have traveled different path lengths over the microwave hop to be combined on an in-phase basis. In addition, if one incoming signal is too low and therefore is excessively noisy, the combined output signal is weighted in favor of the better *noise quality* RF input signal.

Service channel methods.   Service channel transmission is required for the efficient maintenance, performance monitoring, and control of digital microwave systems. This includes control, supervisory, and order-wire signal transmission. In order to have conversations between maintenance personnel at different locations, it is necessary to have a system in which it is easy to combine the main subscriber traffic with the service channel signal.

Furthermore, it is essential to continue maintenance conversations in the event of a failure of the main traffic signal. There are several techniques for transmitting the service channel information together with subscriber traffic. The main methods are as follows:

1. Digital TDM
2. FM of the PSK carrier frequency
3. AM of the PSK carrier frequency
4. Analog service channel below the digital band

Each of these methods is illustrated in Fig. 5.46. In Fig. 5.46a, the service channel information is A/D converted into a bit stream and then time division multiplexed with the subscriber bit stream. Clearly, an input bit stream of 140 Mb/s will be increased by the addition of the

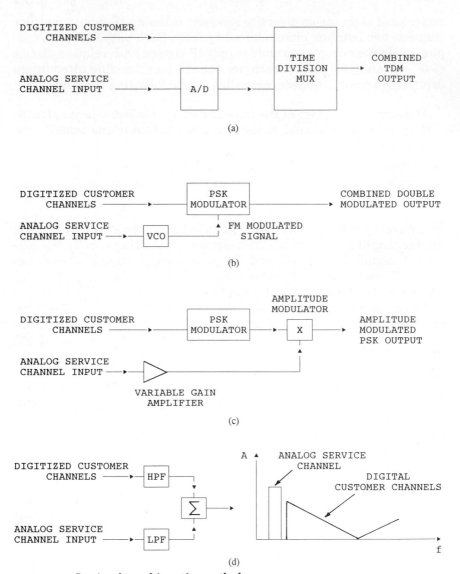

Figure 5.46 Service channel insertion methods.

service channel information. A practical system would have an actual transmitted bit rate of about 143 Mb/s, depending on the quantity of service channels. In Fig. 5.46*b* the analog service channel signal frequency modulates the subscriber signal using a VCO. Since the subscriber signal has already been modulated with PSK (or QAM, etc.), the resulting signal is a double modulated output. Similarly, in Fig. 5.46*c* the PSK modulated subscriber signal is amplitude modulated by the analog service channel. In Fig. 5.46d the analog service channel is inserted into the frequency band below the digital signal band (*subbaseband*). The digital TDM method tends to be the main choice for service channel transmission, although other methods may be simultaneously incorporated together with the TDM.

5.2.3 140-Mb/s DMR with higher modulation levels (64-QAM or 256-QAM)

The 64-QAM and 256-QAM DMRs have many similarities to the 16-QAM system. Theory dictates that the S/N for the 64-QAM system must be about 7 dB better than the 16-QAM system for a BER of 10^{-8} This translates into requiring the components of the 64-QAM system to have a higher linearity.

The FEC technique is a very satisfactory way of improving the C/N without having to place very tight tolerances on the linearity of the components. In other words, the designer accepts that "dribble" errors will occur because of the C/N value being inadequate, then an error correcting device is incorporated into the system to reduce these errors, thereby effectively improving the C/N. In a 64-QAM transmitter, the FEC encoder is placed between the differential encoder and the modulator, as in Fig. 5.47a. At the receive side (Fig. 5.47b), the FEC decoder is placed between the demodulator and the differential decoder.

With the exception of the FEC codec, one can see that this system is very similar to the 16-QAM system. The 64-QAM system transmitter has a series-to-six-path parallel converter whereas the 16-QAM system has a series-to-four-path parallel converter. The 64-QAM modulator primarily differs from the 16-QAM modulator in that the two inputs are three-path parallel inputs instead of two-path inputs. The reverse conversions (parallel to serial) occur in the receiver.

64-QAM system. Figure 5.47a and b shows simplified block diagrams of a typical 140-Mb/s 64-QAM DMR transmitter and receiver, respectively. As stated previously, the higher-level QAM systems should be designed for high linearity even though the FEC codec accounts for some degree of nonlinearity. The nonlinearities in the microwave portions of QAM radios are primarily the AM to AM and AM to phase modulation (PM) conversions. AM to AM conversion can be defined as a device nonlinearity in which the gain of the device is dependent upon the input power. AM to PM conversion can be defined as a device nonlinearity in which the *transfer phase* of the device is dependent upon the input power.

Upconverters must have minimal AM to AM and AM to PM distor-

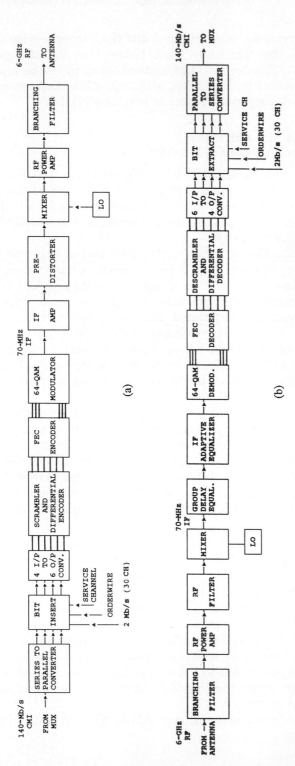

Figure 5.47 (*a*) Simplified block diagram of a typical 140-Mb/s, 64-QAM DMR transmitter; (*b*) simplified block diagram of a typical 140-Mb/s, 64-QAM DMR receiver.

tion. The mixer must therefore be operated at output levels well below the 1-dB compression point. The LO leakage is consequently large relative to the desired sideband signal. Additional filtering following the up-converter is therefore necessary.

The RF power amplifier is the main culprit in causing nonlinearity problems. The predistorter helps to correct nonlinearity created by the amplifier. In addition, power reduction (back-off) for 64-QAM radios is of the order 7 or 8 dB, giving an output power in the region of 20 to 30 dBm (about 100 mW to 1 W). The power reduction increases to 10 to 13 dB if no predistorter is used. Power amplifiers operating at 7-dB reduction have a poor dc-to-microwave conversion efficiency of less than 5 percent.

The LOs used in the up- and the downconverters can be almost identical designs. For space diversity receivers it is convenient to use the same LO for each downconverter, with the power being split and fed to the respective mixers. The frequency variation and phase noise must be minimized for the LO. This can be done by temperature controlling a dielectric resonator FET oscillator (DRO) with a thermoelectric heater/cooler. This type of oscillator approaches the stability of a crystal oscillator with the advantages of lower cost and smaller size.

If cochannel operation is used, cross-polarization interference cancelers (XPIC) are necessary for the 64-QAM systems. The XPIC circuits closely resemble those of the adaptive transversal equalizers used for ISI reduction. Using custom-built integrated circuit technology, these components are now inexpensive.

Figure 5.48 displays the RF channel arrangements for the 6-GHz band. The adjacent channel spacing for the 140-Mb/s DMR is 29.65 MHz as designated by CCIR Rec. 383. The radio can be operated in the interleaved pattern with the adjacent channels orthogonally polarized or in a cochannel arrangement with two polarizations of each channel frequency being used independently to transmit separate 140-Mb/s bit streams. The cochannel arrangement obviously doubles the number of RF channels transmitted. Within the 5925- to 6425-GHz band either 8-interleaved-channel or 16-cochannel-bidirectional 140-Mb/s bit streams can be transmitted. AMR systems also have an adjacent channel spacing of 29.65 MHz, so during the transition period of changing

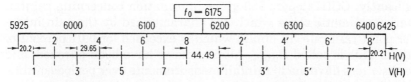

Figure 5.48 RF channel arrangement for the 6-GHz DMR band (CCIR Recommendation 383-4).

from analog to digital technology, *it is quite acceptable to operate an 1800-voice-channel FM system on the same route as a 64-QAM, 140-Mb/s (1920 channels) system using adjacent channels with different polarizations.* If the unfaded cross-polarization discrimination is at least 28 dB, there should be no problem. In areas of excessive fading, it may be necessary to incorporate extra filtering to make sure there is no interference between the two adjacent channels.

5.2.4 Low-capacity DMR

There is a large demand for low-capacity radio relay systems spanning short distances in the range of 2 to 20 km. The applications are mainly for rural area spur routes from a backbone link or intercity hops between large company premises and exchanges. There are two different approaches that can be used to satisfy this objective: (1) low-microwave-frequency bands in the 1.5- to 2.5-GHz range or (2) high-microwave-frequency bands in the 14- to 24-GHz range.

In the past, the low-frequency approach has been the most widely used method. Recent technological advancements at the higher-frequency microwave bands have increased the interest in this style of link. Despite the rain attenuation problems, the size and cost of these systems is becoming very attractive to network planners.

Since the gain of an antenna is proportional to $\log f^2$, in the Ku-band or Ka-band (about 12 to 26 GHz) the size of antennas becomes manageably small. For example, at 18 GHz a 1-m-diameter antenna has a gain of about 42 dB. Technology has recently improved to the point where high-quality microwave components are now readily available at 18 GHz. When these factors are combined with the fact that many large cities have heavily congested 2-, 4-, and 6-GHz frequency bands, the obvious next step is to move to higher operating frequencies.

As stated previously, the major drawback of the higher-frequency bands is the high rain attenuation. A rainfall analysis must be done to find out the margin available under the worst possible rainfall conditions. Rain is placed in the category of flat fading because, although the rain attenuation does increase with frequency, the variation over a relatively narrow radio band is very small. When the raindrop size approaches the wavelength of operation, the attenuation increases significantly. CCIR Report 563 gives information concerning rainfall for various climatic zones which are characterized by the rain intensity or point rain rate (in mm/h) which is exceeded for 0.01 percent of the time. This is called $R_{0.01\%}$. Many studies have been pursued, and some observers have made rainfall measurements over periods of time up to 10 years or more. There is clearly merit in using their zone information when contemplating a new design, but the most valuable information is the rainfall information at the specific location being considered. Although guidelines can be followed and some characteristics apply to all locations, no two paths are the same. For example, raindrops become flattened as they fall, making the size of the horizontal dimension greater than the vertical dimension. This favors a vertically polarized propagation to minimize rainfall attenuation irrespective of location. However, the size and direction of a rain cell is very significant in determining rainfall attenuation. A tropical thunder-cell will release rain well in excess of 100 mm/h over a distance of only 1 or 2 km for half an hour or less, whereas a temperate climate cloud overcast may release rain at about 20 mm/h but over many kilometers for several hours. Hurricanes and cyclones are perhaps the least predictable because they release enormous amounts of rainfall for 24 h or more, but they may pass over a particular tropical location only once every 10 years. The prevailing wind can also play an important role in rain attenuation because if the rain cells tend to pass *across* a path, they will cause less disruption than if they proceed *along* a path. A nearby lake or mountain range can affect the rainfall pattern significantly.

Because of the enormous number of variables concerned, it is evident that the subject of how rain affects a radio link is not a very exact science. When all the data have been taken and manipulated to predict the maximum path length, the following rules of thumb seem to apply to the 18-GHz band. For an unavailability between 0.001 and 0.01 percent per year, radio hop lengths up to 20 km for an $R_{0.01\%}$ of 30 mm/h are allowed and up to 7 km for an $R_{0.01\%}$ of 60 mm/h. The picture looks rather bleak for tropical regions. Depending upon specific local conditions, the maximum hop length may be as low as 5 km or less in the 18-GHz band. This is rather unsatisfactory from a planner's point of view because some tropical regions have periods of several months where the rainfall is very low or even zero, during which time a 20-km hop would work very well. Unfortunately, during the rainy season such a link would be unavailable for several minutes almost every day.

RF channel arrangement. CCIR Recommendation 595 indicates the channel arrangements for 34- and 140-Mb/s systems operating in the 17.7- to 19.7-GHz band. Figure 5.49 summarizes this Recommendation for a 4-PSK modulation scheme. There are a total of 16 channels (8 pairs) for 140 Mb/s with a channel spacing of 110 MHz. For 34-Mb/s transmission, the band is subdivided to allow 70 channels (35 pairs) with a channel spacing of 27.5 MHz. This is similar to the 13- and 15-GHz bands which have a channel spacing of 28 MHz for 34-Mb/s trans-

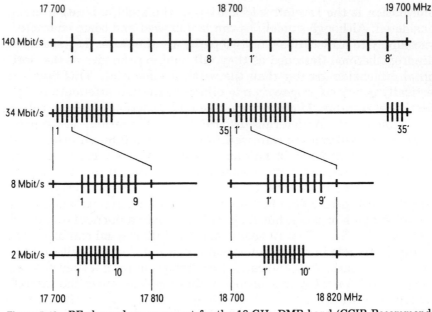

Figure 5.49 RF channel arrangement for the 18-GHz DMR band (CCIR Recommendation 595-2). (*Reproduced with permission from Siemens Telcom Report: Special "Radio Communication," vol. 10, 1987, p. 53, Fig. 2.*)

mission. The channel arrangements for 2 and 8 Mb/s can be decided at the discretion of each individual administration. A typical acceptable subdivision would provide a channel spacing of between 7 and 15 MHz for 8-Mb/s transmission and between 2.5 and 5.5 MHz for 2-Mb/s transmission. Figure 5.49 shows 7.5- and 5-MHz channel spacings.

Modulation. 4-PSK is usually the preferred modulation technique for low-capacity systems up to 34 Mb/s. In order to prevent interchannel interference by a 4-PSK modulator, spectrum limiting must be provided by RF or channel branching filters. This is achieved at the expense of increased attenuation and complexity.

A significant cost saving and efficiency improvement can be attained by using the technique of directly modulating the RF band. The benefit is particularly appreciated at the higher RF frequencies, for example in the 18-GHz band. For 2- and 8-Mb/s systems (low capacity), the filter requirements are too stringent to allow 4-PSK direct modulation in the 18-GHz band. Direct FSK of the RF oscillator is a good solution for producing a low-cost band-limited signal spectrum having acceptable efficiency. It is not even necessary to use a baseband filter with 4-FSK, since the modulated signal occupies a narrow enough bandwidth without filtering. Furthermore, the 4-FSK modulation has a relatively high immunity to gain nonlinearities. In addition, incoherent demodulation can be used in the receiver. This involves the use of a frequency discriminator, which is simpler and cheaper than the PSK demodulator since the oscillator specifications are more relaxed for 4-FSK. The S/N for incoherent demodulation is a little higher than for coherent modulation, but this is compensated by the elimination of the modulator attenuation.

Transmitter/receiver. Figure 5.50 shows the block diagram of a typical 18-GHz band transmitter and receiver. This system employs a transmitter which has 4-FSK direct modulation of the RF oscillator and incoherent demodulation in the receiver.

The HDB3 baseband signal is equalized, regenerated, and then converted into two parallel NRZ bit streams at half the incoming bit rate. The signals are then dejitterized, scrambled, and converted to four amplitude levels. The low-pass filter then limits the baseband spectrum, which subsequently modulates an RF oscillator to produce the 4-FSK output. The RF oscillator is voltage controllable and can be fabricated using a Gunn or FET device, giving an output in the region of 100 mW (20 dBm). A PLL can be included for improved frequency stability.

At the receiving end, the RF signal is filtered and double downconverted via 790 MHz to the 70-MHz IF. The double down-conversion allows the use of low-loss, relatively wideband RF filters. A low-noise amplifier is used between the two down-converters. The 70-MHz local oscillator can be a surface-acoustic-wave, resonator stabilized, transistor oscillator or a traditional crystal multiplied oscillator. The IF signal then passes through a bandpass filter for image frequency rejection and onto a limiter to provide a constant level signal to the frequency discriminator. The output from the discriminator is the baseband signal, which passes through a low-pass filter prior to pulse recovery in the timing recovery and regenerator circuit. Also in the regenerator, the four-level, band-limited baseband signal is converted into two parallel signals, which are then descrambled. Finally, the two streams are combined and an NRZ-to-HDB3 conversion completes the receiver process.

Perhaps the most attractive feature of this system is the compactness of the design. The 0.6-m parabolic antenna can be mounted on a pole, together with the transmitter and receiver, which are placed in a weatherproof housing (Fig. 5.51). This arrangement has the advantage of a very short waveguide run. The cables into the weatherproof housing carry the baseband, service channel, power supply, and lightning protection. No expensive towers are necessary, and these systems can be installed very rapidly.

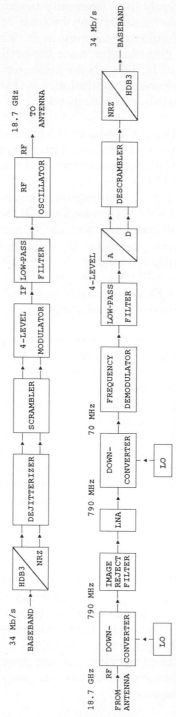

Figure 5.50 Schematic diagram of a typical 18-GHz band DMR transceiver

Figure 5.51 Weatherproof case and antenna installation. (*Reproduced with permission from Siemens Telcom Report: Special "Radio Communication," vol. 10, 1987, p. 56, Fig. 7.*)

5.3   Performance and Measurements

DMR testing involves performance measurements in the RF, IF, and baseband sections of the system. The usual additional features such as protection switching, service channel operation, supply voltages, and alarm conditions are also checked. The measurements in this chapter refer to a typical 6-GHz band 140-Mb/s 16-QAM DMR. In the RF section, the following tests are usually made:

1. Transmitter RF output power level
2. Transmitter RF output frequency

3. Transmitter local oscillator frequency

4. Transmitter local oscillator power level

5. Receiver local oscillator frequency

6. Receiver local oscillator power level

7. Receiver RF input power (and AGC curve)

8. Transmitter distortion level

9. Waveguide return loss and pressurization

In the IF section, the following tests are usually made:

1. IF input power level (to the mixer)

2. IF frequency and bandwidth

3. IF input power level (to the demodulator)

4. IF output spectrum

5. IF-IF frequency response and group delay

6. Demodulator eye pattern

7. Absolute delay equalization (on space diversity receivers)

At the baseband, the following measurements are usually made on the received bit stream before demultiplexing:

1. Error analysis (BER versus S/N)

2. Jitter analysis

5.3.1 RF section tests

Power and frequency measurements. RF power level and frequency measurements are made on the transmitter and receiver simply by using an RF power meter and frequency counter at the Tx output and Rx input. Usually, an RF coupler is provided for this purpose, with a coupling factor of typically 30 dB. The transmitter output power level is mostly in the range of 0.1 to 10 W (20 to 40 dBm), so, in addition to the coupler, an attenuator is sometimes required to reduce the level to ensure that no damage occurs to the power meter or frequency counter.

The LO in the up- and down-converters is usually less than 0 dBm, so excessive power is no problem here. For the LO frequency measurements, very accurate values are required, since the frequencies are fixed to a very high tolerance. For example, a frequency may have a tolerance of ±30 ppm which at 6 GHz is only ±0.18 MHz (or ±180 kHz).

In the receiver, the so-called AGC curve is plotted. This requires measurement of the IF output level as the input RF power level is varied from about -25 dBm down to the threshold level. The IF output level should remain constant until about -70 dBm, after which an abrupt drop in level is observed as the threshold level is approached.

Transmitter distortion level. The RF sections of *analog* and *digital* radios primarily differ in the required linearity. As already stated, the digital radio transmitter must be very linear compared to an analog radio. This is necessary because, for example, the 16-QAM has both amplitude and phase components. If the transmitter amplifiers are not extremely linear, AM to AM and AM to PM conversion can cause serious problems. The result would be intersymbol interference and subsequent errors. The higher-order QAM systems such as 64-QAM and eventually 256-QAM require even higher levels of linearity. There are two tests that can be used to check the transmitter distortion level (see below).

Using a digital signal. The pattern generator is used to send a CMI signal with a pseudorandom bit sequence (PRBS) $2^{23} - 1$ periodicity to the modulator input. A spectrum analyzer then monitors the RF output from the transmitter amplifier. Figure 5.52 shows the test setup and expected response. The first sidelobe should be attenuated by a certain minimum value depending on the specifications of individual administrations. As a guideline, for good performance of a 140-Mb/s

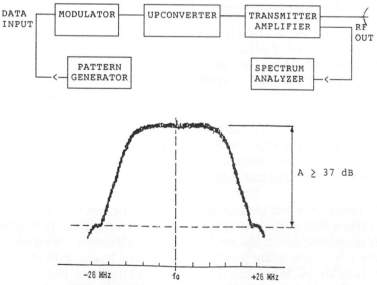

Figure 5.52 Tx distortion using a digital signal.

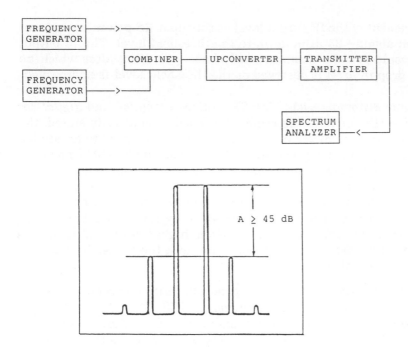

Figure 5.53   Tx distortion using two tones.

DMR, the first sidelobe should be at least 37 dB down at ±25 MHz from f_0. If the transmitter is out of specification, a linearization adjustment would be necessary to correct the problem.

Using the two-tone method.   Two frequency generators are used to send IF signals at 130 and 150 MHz to the transmitter upconverter (as in the test setup of Fig. 5.53). The spectrum analyzer is connected to the output of the transmitter RF amplifier. The spectrum analyzer displays the two tones, together with the intermodulation products. The level of the intermodulation products is a direct indication of the linearity of the transmitter. Typically, these intermods should be at least 45 dB down from the two input tones. As in the previous method, if the transmitter is out of specification, adjustments can be made while observing the spectrum analyzer display.

Waveguide return loss and pressurization.   On completion of installation, the return loss of the waveguide and antenna system is a very important measurement. If the return loss is inadequate, *echo distortion* results in a group delay problem. The return loss should ideally be better than 25 dB. When the waveguide and antenna section has been commissioned, the waveguide must be maintained in a pressurized condition. This is necessary because otherwise moisture would condense in the waveguide and degrade the return loss. It is therefore an important part of the maintenance process to closely monitor the pressure of the gas in the waveguide. Usually, the pressurization mechanism is connected to an alarm so that a drop in pressure below the acceptable value alerts the technician to provide the necessary maintenance.

5.3.2 IF section tests

Power and frequency measurements. These measurements simply involve the use of a power meter and frequency counter, as in the RF power and frequency measurements. In this case, the power levels are relatively low, so there is no danger of damaging the instruments with excessive input power.

IF output spectrum. A PRBS pattern $(2^{23} - 1)$ is sent to the input to the modulator in the NRZ plus clock format at a level of 1 $V_{pp}/75$ Ω. The spectrum analyzer is connected to the modulator output to display the output spectrum. It is recommended that the output should comply with the mask shown in Fig. 5.54.

IF-IF frequency response and group delay. The IF amplitude should be as flat as possible over the bandwidth of operation (e.g., <0.5 dB at 140 ±20 MHz). The IF amplitude versus frequency response can be measured using the microwave link analyzer (MLA) test setup in Fig.

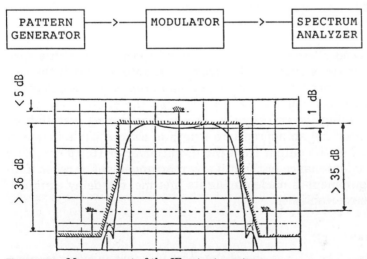

Figure 5.54 Measurement of the IF output spectrum.

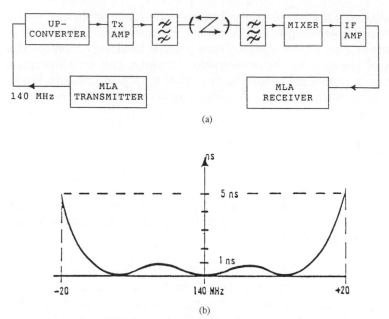

Figure 5.55 (*a*) IF-IF frequency response and group delay test setup; (*b*) group delay characteristics.

5.55*a*. With the same setup, the group delay can be displayed on the MLA receiver. The group delay should typically be within the specifications suggested in Fig. 5.55*b* (i.e., <1 ns at 140 ±18 MHz and <4 ns at 140 ±20 MHz).

A group delay that is out of specification by 2 or 3 ns can be corrected by adjusting the IF amplifier module. If it is out by more than 3 ns, a microwave filter may need retuning.

Absolute delay equalization.   In a space diversity system, it is necessary to ensure that the signals coming from each antenna reach the two receivers simultaneously. Otherwise, *in-phase combining* is not possible. The path lengths of the two signals must therefore be equalized prior to combining the two signals. This is done during installation, usually by inserting an appropriate length of coaxial cable in one of the two paths at the output from the downconverter (IF) point. This test is not a maintenance test, but if any rearrangement or repositioning of the equipment is made during its lifetime, the delay equalization must be reestablished.

5.3.3 Baseband tests

Error analysis.   The BER is one of the most frequent measurements a technician or engineer has to perform on a DMR system. This is the measurement which defines transmission quality and acceptability. In other words, if the BER is too high, the system is declared failed or unavailable. As stated previously, a link will be affected by several types of distortion (e.g., thermal noise, RF interference, group delay, phase jitter, etc.). The result of these combined distortions is observed as the reception of "false bits," or errors, at the receiving end. It is important to know the rate at which the errors occur, because if they are excessive, the communication link is disrupted. If the errors exceed a critical value, all calls are dropped. This situation is more severe compared to the analog system, where excessive noise impairs the audibility of the speaker but does not drop the call. Fortunately, good digital equipment design ensures that the link disruption is a rare occurrence.

The BER performance objectives for international ISDN connections are stated in CCITT Recommendation G.821. This Recommendation is summarized in Fig. 5.56 and also elaborated in Chap. 4. Over a long period of time (e.g., 1 year) a link is designed to provide availability as high as possible; 100 percent availability is not possible. The term *errored seconds* is an important parameter in performance evaluation in order to make the distinction between the occurrence of random errors and error bursts. BER measurements are made in the DMR station equipment. Depending upon the manufacturer, the measurements are used to provide an alarm if the error ratio becomes worse than 1×10^{-5} or 1×10^{-6}. This is a degraded state. Another alarm is activated if the error ratio drops below 1×10^{-3}. Some manufacturers provide the facility for observing the actual error ratio in addition to the alarm requirements. This is very useful because complete digital links or portions of a link can then be evaluated by using the loopback mode. If the error ratio cannot be read on an LED display at the DMR station, the measurement can be made by using a test set

PERFORMANCE CLASSIFICATION	OBJECTIVES
DEGRADED MINUTES	Fewer than 10% of one-minute intervals to have a BER worse than 1×10^{-6}
SEVERELY ERRORED SECONDS	Fewer than 0.2% of one-second intervals to have BER worse than 1×10^{-3}
ERRORED SECONDS	Fewer than 8% of one-second intervals to have any errors (equivalent to 92% error-free seconds)

Figure 5.56 CCITT Recommendation G.821 summary. (*Reproduced by permission from ITU, CCITT Red Book, vol. III, Fascicle III.3, Digital Networks Transmission Systems and Multiplexing Equipment, VIIIth Plenary Assembly, Malaga-Torremolinos, Oct. 1984, Geneva 1985, p. 12, Table 1/G.821.*)

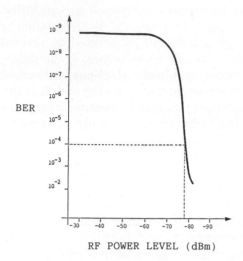

Figure 5.57 BER versus RF received level for a typical DMR.

RF POWER LEVEL (dBm)

to introduce a pseudorandom digital bit stream at one terminal station, looping back at the desired receiving terminal station, and feeding the returned bit stream into the test set to evaluate the BER.

For performance evaluation, it is very useful to relate the BER to the S/N or the C/N. DMRs have a BER which increases rapidly as the receiver threshold is approached (Fig. 5.57). In fact, the receiver threshold for a DMR is often defined as the RF input power that degrades the BER to 1×10^{-3}. In QAM systems, because of the amplitude and phase characteristics, it is not possible to measure the carrier peak power C. However, the mean carrier (signal) power S can easily be measured at the demodulator input. For 16-QAM, S is approximately 2.5 dB lower than C. The received power P_r is related to the normalized S/N as follows:

$$P_r = KTB + \text{NF} + (\text{S/N})_n   \text{dBm}$$

where K = Boltzmann's constant = 1.38×10^{-23}
T = absolute ambient temperature (about 298 K)
B = equivalent noise bandwidth
NF = Rx system noise figure

So, for a 1-MHz bandwidth,

$$KTB = -114 \text{ dBm}$$

$$P_r = -114 + 10 \log B_s + \text{NF} + (\text{S/N})_n$$

where B_s = the symbol rate (MHz)
= bit rate/number of bits forming a symbol
= B_r/n

For example, for a 140-Mb/s system $n = 4$, so $B_s = 35$ MHz. Also, if NF = 6,

$$P_r = -114 + 10 \log 35 + 6 + (S/N)_n$$

$$P_r = -92.6 + (S/N)_n$$

From this equation, the received power can be calculated if the normalized S/N is known, or vice versa. (*Note:* The normalized S/N $[(S/N)_n] = (S/N) \cdot (\text{Rx 3-dB noise bandwidth})/B_s$.)

BER measurement. The method in Fig. 5.58, which uses an RF variable attenuator inserted at the input to the receiver, is the most suitable for training purposes. However, in the field, it is probably more convenient to use a noise generator to degrade the S/N. The attenuator is adjusted to provide the required BER value, after which the received input power to the radio is measured. Following a sequence of such measurements, a graph can be plotted of P_r against BER. The curve is then compared with the manufacturer's installation curve or factory recommended curve to see if any degradation has occurred. Corrective maintenance is necessary if the system has deteriorated significantly.

Demodulator eye pattern. The eye pattern check is a useful, quick, *in-service, qualitative* method of observing the digital signal transmission quality. In the absence of a BER instrument, an experienced engineer or technician can quickly evaluate the eye pattern and decide whether the quality is adequate or not. A constellation pattern check is another qualitative check which provides considerably more information regarding the source of equipment faults. Nevertheless, the eye pattern has some benefit. It is displayed on an oscilloscope by synchronizing the oscilloscope with the demodulator clock and connecting the X and Y axes of the oscilloscope to the two eye pattern test points (I and Q) provided in the demodulator. The eye patterns of Fig. 5.59 show the deterioration of the eye as the BER degrades. Figure 5.59a is the *normal* condition (BER $< 10^{-10}$). When the BER $= 10^{-6}$, the eye opening

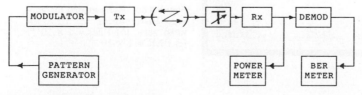

Figure 5.58 BER measurement.

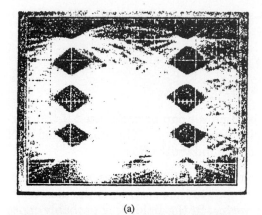

(a)

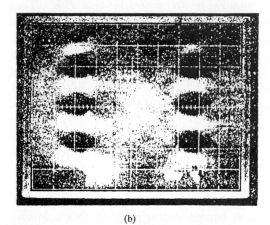

(b)

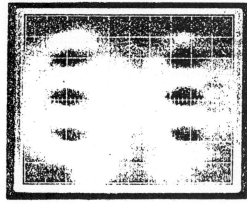

(c)

Figure 5.59 Eye pattern for various BER values. (a) BER almost zero; (b) BER = 1×10^{-6}; (c) BER = 1×10^{-3}.

starts to become less clearly defined. By the time the BER has increased to 10^{-3} (the failure condition), the eye is very badly distorted. Incidentally, the number of eye openings depends on the level of modulation. The example shows three pairs of openings for 16-QAM. A 64-QAM signal would have seven pairs of openings.

Jitter analysis.   Jitter is defined as the short-term variations of the significant instants (meaning the pulse rise or fall points) of a digital signal from their ideal positions in time. Figure 5.60 illustrates the definition of jitter more clearly. The main jitter-creating culprits are regenerators and multiplexers. In the regenerators, the timing extraction circuits are the sources of jitter, and since a transmission link has many regenerators, jitter accumulates as the signal progresses along the link. The main type of multiplexer presently in operation is the asynchronous type which uses the justification technique described in detail in Chap. 2. Jitter is generated in the justification process and also because of the waiting time between the justification decision threshold crossing and the justification time slot. Furthermore, the new synchronous type of multiplexers are not immune to jitter generation. The pointer processing technique used in these multiplexers introduces jitter into the payloads that they carry.

The extent of jitter acceptance depends on the hierarchical inter-

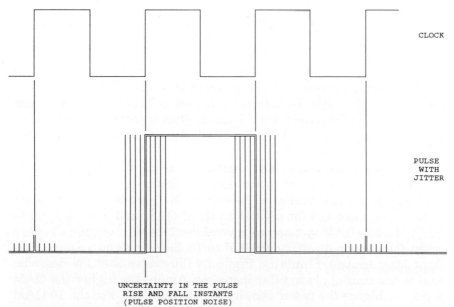

Figure 5.60   An example of jitter.

face. That simply means the output bit rate. The CCITT has defined maximum permissible jitter values in Recommendations G.823 and G.824. Figure 5.61*a* shows the values for the 2.048-, 8.448-, 34.368-, and 139.264-Mb/s European interfaces and Fig. 5.61*b* shows the 1.544-, 6.312-, 32.064-, 44.737-, and 97.727-Mb/s North American interface values. There are three categories of jitter measurement:

- *Maximum tolerable input jitter.* This is tested by superimposing jitter onto an input bit stream and determining the point at which errors start to occur.

- *Maximum output jitter.* This is the value observed when an input bit stream of known jitter content is applied.

- *Jitter transfer function.* This is a measure of how the jitter is attenuated (or amplified) by passing through the system. This measurement is necessary to assess any jitter accumulation within a link.

Jitter is usually measured with the same transmission analyzer test setup that is used for the BER measurements.

CCITT Recommendation G.824 (1988 blue book) has been established to control jitter and "wander" within digital networks. It also provides equipment designers with guidelines to ensure that equipment performance reaches certain minimum standards of jitter and wander. Figure 5.62*a* and *b* gives the Recommendation G.824 jitter and wander tolerance of equipment input ports for European and North American hierarchies, respectively.

The North American equivalent jitter specifications are similar to CCITT and found in Bell Technical References 43501 and 43806. In addition, further jitter information can be found in CCITT Recommendations G.735/6/7/8/9, G.742/3/4/5/6/7, and G.751/2/3/4/5 (1988 blue book), which address the jitter tolerance, jitter transfer, and jitter generation requirements for both the North American and European hierarchies.

Higher-level QAM signals are extremely susceptible to degradation in the C/N that is caused by carrier and timing phase jitter. Figure 5.63 shows the calculation of how the C/N degrades as the carrier phase jitter increases for several levels of QAM and a BER equal to 10^{-6}. The 256-QAM system has several decibels of power penalty even when the carrier jitter is only 0.5° to 1°. Such systems require very tight jitter control. Similarly, Fig. 5.64 illustrates the C/N degradation as the timing phase jitter increases. Again, the higher the QAM level, the higher the power penalty for a given BER. For the 16-QAM case, the curves are drawn for several values of filter roll-off factor.

PARAMETER	NETWORK	LIMIT	MEASUREMENT	FILTER	BANDWIDTH
	B_1	B_2	Bandpass filter having lower cutoff frequency f_1 and f_3 and an upper cutoff frequency f_4		
Bit rate (kb/s)	Unit interval peak-peak	Unit interval peak-peak	f_1 (Hz)	f_3 (kHz)	f_4 (kHz)
64 (Note 1)	0.25	0.05	20	3	20
2048	1.5	0.2	20	18 (0.700)	100
8448	1.5	0.2	20	3 (80)	400
34368	1.5	0.15	100	10	800
139264	1.5	0.075	200	10	3500

(a)

PARAMETER	NETWORK	LIMIT	MEASUREMENT	FILTER	BANDWIDTH
	B_1	B_2	Bandpass filter having lower cutoff frequency f_1 and f_3 and an upper cutoff frequency f_4		
Bit rate (kb/s)	Unit interval peak-peak	Unit interval peak-peak	f_1 (Hz)	f_3 (kHz)	f_4 (kHz)
1544	5.0	0.1a	10	8	40
6312	3.0	0.1a	10	3	60
32064	2.0	0.1a	10	8	400
44736	5.0	0.1	10	30	400
97728	1.0	0.05	10	240	1000

(b)

Figure 5.61   (a) Maximum permissible output jitter (European hierarchies). *Note 1:* For the codirectional interface only. *Note 2:* The frequency values shown in parentheses apply only to certain national interfaces. *Note 3:* UI = unit interval. For 64 kb/s, 1 UI = 15.6 μs; for 2048 kb/s, 1 UI=488 ns; for 8448 kb/s, 1 UI = 118 ns; for 34368 kb/s, 1 UI = 29.1 ns; for 139264 kb/s, 1 UI = 7.8 ns. (*Reproduced by permission from ITU, CCITT Red Book, Vol. III, Fascicle III.3, Digital Networks Transmission Systems and Multiplexing Equipment, VIIIth Plenary Assembly, Malaga-Torremolinos, Oct. 1984, Geneva 1985, p. 322, Table 1/ G.823.*) (b) Maximum permissible output jitter (North American and Japanese hierarchies). *Note (a):* This value requires further study. For systems in which the output signal is controlled by an autonomous clock (e.g., quartz oscillator), more stringent output jitter values may be defined in the relevant equipment specifications (e.g., for the muldex in Recommendation G.743, output jitter should not exceed 0.01 UI rms). (*Reproduced with permission from Trischitta, P. R., et al., Jitter in Digital Transmission Systems, © 1989 Artech House, Inc., Norwood, MA, p. 217.*)

BIT RATE (kb/s)	JITTER AMPLITUDE PEAK-TO-PEAK			FREQUENCY					TEST SIGNAL
	A_0 (UI)	A_1 (UI)	A_2 (UI)	f_0 (Hz)	f_1 (Hz)	f_2 (Hz)	f_3 (kHz)	f_4 (kHz)	
64 (Note 1)	1.15 (18 μs)	0.25	0.05	1.2×10^{-5}	20	600	3	20	$2^{11} - 1$ (Rec.0.152)
2048	36.9 (18 μs)	1.5	0.2	1.2×10^{-5}	20	2400 (93)	18 (0.7)	100	$2^{15} - 1$ (Rec.0.151)
8448	152 (18 μs)	1.5	0.2	1.2×10^{-5}	20	400 (10700)	3 (80)	400	$2^{15} - 1$ (Rec.0.151)
34368	*	1.5	0.15	*	100	1000	10	800	$2^{23} - 1$ (Rec.0.151)
139264	*	1.5	0.075	*	200	500	10	3500	$2^{23} - 1$ (Rec.0.151)

(a)

Figure 5.62 (a) Input port jitter and wander tolerance (European hierarchies). * = values under study. *Note 1:* For the codirectional interface only. *Note 2:* For interfaces within national networks the frequency values (f2 and f3) shown in parentheses may be used. *Note 3:* UI = unit interval. For 64 kb/s, 1 UI = 15.6 μs; for 2048 kb/s, 1 UI = 488 ns; for 8448 kb/s, 1 UI = 118 ns; for 34368 kb/s, 1 UI = 29.1 ns; for 139264 kb/s, 1 UI = 7.8 ns. *Note 4:* The value for A_0 (18 μs) represents a relative phase deviation between the incoming signal and the internal local timing signal derived from the reference clock. This value A_0 corresponds to an absolute value of 21 μs at the input to a node (i.e., equipment input port) and assumes a maximum wander of the transmission link between the two nodes of 11 μs. The difference of 3 μs corresponds to the 3 μ allowed for the long-term phase deviation in the national reference clock [Recommendation G.811, Sec. 3(c)]. (*Reproduced by permission from ITU, CCITT Red Book, Vol. III, Fascicle III.3, Digital Networks Transmission Systems and Multiplexing Equipment, VIIIth Plenary Assembly, Malaga-Torremolinos, Oct. 1984, Geneva 1985, p. 325, Table 2/G.823.*)

It is interesting to see that the less steep roll-off value of $\alpha = 0.7$ has a better jitter characteristic than the more expensive, steeper $\alpha = 0.3$ roll-off filter. Remember, if α is too large, the spectral emission mask specification may be exceeded.

Constellation analysis. The constellation measurement is becoming an indispensable technique for *in-service* evaluation of a DMR system. It is a simple measurement to perform, requiring the connection of the I, Q, and symbol-timing clock monitor points in the demodulator to a constellation display unit. Whereas the eye diagram indicates a problem only by eye closure, the constellation can often identify the source of a problem by the type of irregularity observed in the pattern. Figure 5.65 illustrates the constellation pattern and the eye diagram for a 16-QAM radio under *normal operating conditions*. The 16 states of the constellation are correctly placed on the rectangular grid, and the size of each dot representing each state indicates low thermal noise

BIT RATE (kb/s)	JITTER AMPLITUDE PEAK-TO-PEAK			FREQUENCY					TEST SIGNAL
	A_0 (μs)	A_1 (UI)	A_2 (UI)	f_0 (Hz)	f_1 (Hz)	f_2 (Hz)	f_3 (kHz)	f_4 (kHz)	
1544	18 (Note 2)	5.0	0.1 (Note 2)	1.2×10^{-5}	10	120	6	40	$2^{20} - 1$ (Note 3)
6312	18 (Note 2)	5.0	0.1	1.2×10^{-5}	10	50	2.5	60	$2^{20} - 1$ (Note 2)
32064	18 (Note 2)	2.0	0.1	1.2×10^{-5}	10	400	8	400	$2^{20} - 1$ (Note 3)
44736	18 (Note 2)	5.0	0.1 (Note 2)	1.2×10^{-5}	10	600	30	400	$2^{20} - 1$ (Note 2)
97728	18 (Note 2)	2.0	0.1	1.2×10^{-5}	10	12000	240	1000	$2^{23} - 1$ (Note 2)

(b)

Figure 5.62 (b) Input port jitter and wander tolerance (North American and Japanese hierarchies). *Notes:* (1) Reference to individual equipment specifications should always be made to check if supplementary input jitter tolerance requirements are necessary. (2) This value requires further study. (3) It is necessary to suppress long zero strings in the test sequence in networks not supporting 64-kb/s transparency. (4) The value A_0 (18 μs) represents a relative phase deviation between the incoming signal and the internal local timing signal derived from the reference clock. (*Reproduced with permission from Trischitta, P. R., et al., Jitter in Digital Transmission Systems,* © *1989 Artech House, Inc., Norwood, MA, p. 219.*)

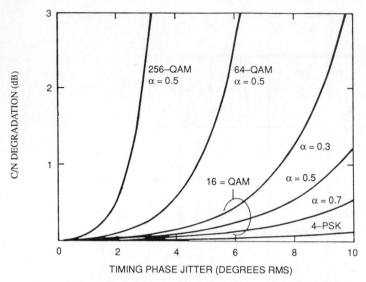

Figure 5.63 Graph of C/N degradation against carrier phase jitter. (*Reproduced with permission from Ref. 59, © 1986 IEEE.*)

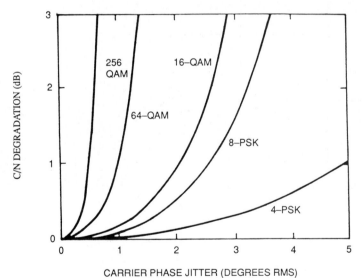

Figure 5.64 Graph of C/N degradation against timing phase jitter. (*Reproduced with permission from Ref. 59, © 1986 IEEE.*)

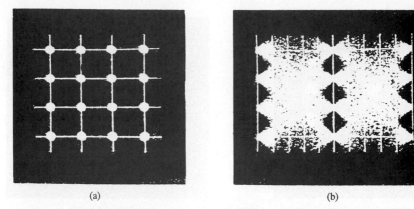

Figure 5.65 Constellation plot for a 16-QAM radio (normal operation). (*a*) Constellation plot; (*b*) eye diagram.

and low intersymbol interference. The corresponding eye diagram has a closure of approximately 10 percent with negligible angular displacement.

If the 16-QAM radio develops a fault, the constellation display, in many instances, can quickly identify the source of the problem. The following faults will be discussed:

- Degradation of C/N
- No input signal
- Sinusoidal interference
- Modem impairments

  I and Q carriers not exactly 90°
  Phase-lock error in the demodulator
  Amplitude imbalance
  Demodulator out of lock

- Phase jitter
- Transmitter linearity
- Amplitude slope

Figure 5.66 shows how the displays are modified when a radio has a *degraded C/N* (20 dB). The constellation states are enlarged and the eye closure is more pronounced (about 20 percent). The measurement of a higher BER would also confirm the degradation. In this case, the constellation pattern and the eye diagram provide the same information, or diagnostic result. For the case of no input signal at all, the constellation display produces a large cluster of random points (Fig.

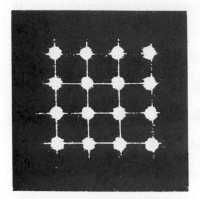 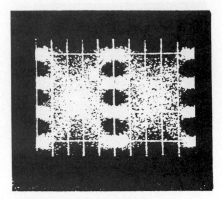

Figure 5.66 Constellation plot for a 16-QAM radio (degraded C/N = 20 dB).

5.67). The eye diagram would show complete eye closure for this condition without indicating the reason for the radio failure. Figure 5.68 shows a radical difference between the two displays. The small circles, or "doughnuts," generated for the constellation states are characteristic of a radio affected by a *sinusoidal interfering tone* such as the strong carrier component of an FM radio. Note that the deterioration of the eye closure provides degraded BER information only, without indicating the reason for the degradation.

Several modem impairments can be observed on the constellation diagram. First, Fig. 5.69 shows the display that would be observed when the I and Q carriers are not exactly 90° to each other. The states form the outline of a parallelogram. Figure 5.70 shows the result of a *phase-lock error* in the demodulator carrier-recovery loop. The constellation states are rotated about their normal positions, which is characteristic of the phase-lock problem. Furthermore, the clockwise rota-

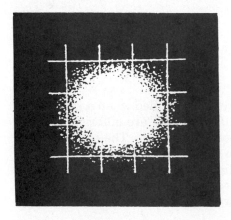

Figure 5.67 Constellation plot for a 16-QAM radio (no signal).

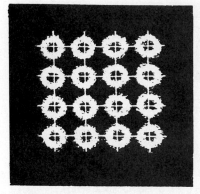 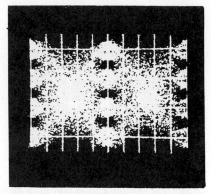

Figure 5.68 Constellation plot for a 16-QAM radio (sinusoidal interference). C/I = 15 dB.

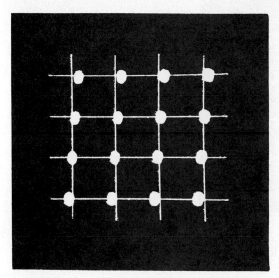

Figure 5.69 Constellation plot for a 16-QAM radio (quad without lock).

tion indicates a negative angle of lock error (in this case, −5.6°). Once again, the eye closure only indicates BER degradation (noise) and does not specify the source of the problem. Another modem impairment observable on the constellation display is the *amplitude imbalance* shown in Fig. 5.71. Here the amplitude levels of the quadrature carrier are not set correctly, which might be the result of a nonlinear modulator or an inaccurate D/A converter. When the demodulator is out of lock, circles, as illustrated in Fig. 5.72, indicate the total lack of phase coherence, culminating in total radio failure. Although the eye

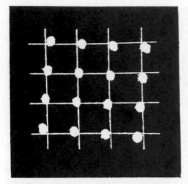 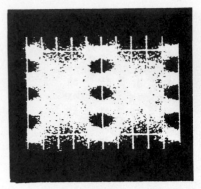

Figure 5.70   Constellation plot for a 16-QAM radio (– 5.6° carrier-lock error).

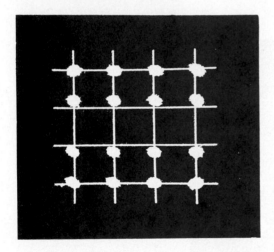

Figure 5.71 Constellation plot for a 16-QAM radio (amplitude imbalance).

diagram would show eye closure, it would not indicate the location of the fault in the equipment.

Phase jitter on a carrier, local oscillator, or recovered carrier causes a circular spreading of the constellation states as in Fig. 5.73. Again, note that the eye diagram does not indicate the specific source of this degradation. Transmitter power amplifier nonlinearities can be detected on the constellation plot. In this respect, Fig. 5.74 shows phase distortion (AM to PM) by the twisted and rotated states, and amplitude distortion (AM to AM) is indicated by compression of states at the corners of the plot. As one can appreciate, it is very useful to observe this constellation diagram when adjusting predistorters. Since the transmitter linearity gradually improves during the adjustment, the constellation diagram clearly indicates when a satisfactory performance is achieved. Finally, the *IF amplitude slope* conditions that oc-

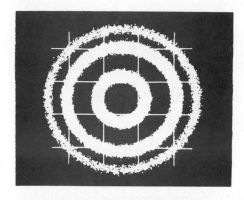

Figure 5.72 Constellation plot for a 16-QAM radio (carrier-recovery loop out of lock).

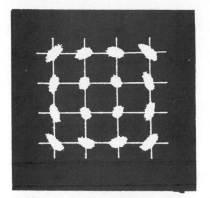
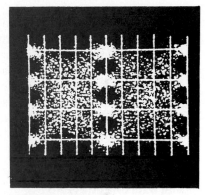

Figure 5.73 Constellation plot for a 16-QAM radio (phase jitter).

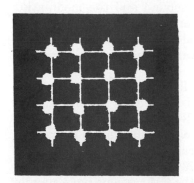
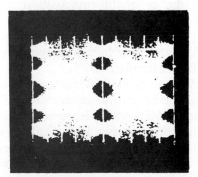

Figure 5.74 Constellation plot for a 16-QAM radio (amplitude nonlinearities).

 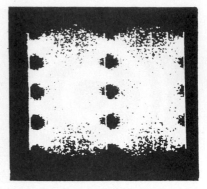

Figure 5.75   Constellation plot for a 16-QAM radio (amplitude slope across the channel).

cur during multipath fading can be viewed. The 45° oval shape of the states in Fig. 5.75 indicates crosstalk between the I and Q signals causing an asymmetrical distortion. Here is one instance where the eye diagram does indicate the fault since the eye opening in this case is shifted to the right of the position for normal operation. From the above constellation diagram observations, it is clear that pattern aberrations provide very useful information for in-service diagnosis of QAM DMR faults.

5.4   Comparison between Analog and Digital Microwave Radio

As stated in Chap. 1, the major differences between AMR and DMR systems lie in the composition of the baseband and the modulation techniques. Also, the repeaters for DMR are regenerative (they recreate, then retransmit the original baseband pulse stream), whereas the AMR repeaters do not demodulate the carrier but merely frequency shift and amplify it.

Most of the AMR systems use FM. Recent advances in single-sideband amplitude modulation (SSB-AM) technology promise to give future AMRs a new lease on life, especially since they are very bandwidth conservative. In fact, about 6000 voice channels can operate on one RF carrier. This compares very favorably with the 1800 channels per RF carrier for an FM system or the 2700 channels for the FM overlapping sideband scheme. Nevertheless, the FM systems have simpler designs and have a more mature technology. Furthermore, since networks are rapidly becoming digitalized, if an SSB-AM system is used, transmultiplexers are necessary to interface these analog radios with the rest of the digital network. This is not desirable. Also, in the near future, the bandwidth efficiency of a 256-QAM DMR employing frequency reuse (dual polarization) will exceed the value obtainable by SSB-AM. Since these two systems have a similar technical complexity, the 256-QAM would be preferable because of its better integratability with the digital network. Countries that have not yet taken the route to digitalization may benefit from SSB-AM, but it is probably fair to say that if their transmission capacity warrants SSB-AM, they should be considering network digitalization anyway.

5.4.1 Composition of the baseband

The AMR baseband is a FDM signal which contains voice channels and/or a video channel. It is formed by a sequence of heterodyning (up-converting) the voice channels as shown in Fig. 5.76. Initially, 12 voice channels are multiplexed to form a *group* occupying a frequency band of 60 to 108 kHz. Five groups can be multiplexed to form a 60-channel *supergroup* occupying 312 to 552 kHz. Next, five supergroups can be multiplexed to form a 300-channel *mastergroup* occupying 812 to 2044 kHz. Next, three mastergroups can be multiplexed to form a 900-channel *supermastergroup* occupying 8516 to 12,388 kHz. The baseband can be constructed by using a combination of the above levels of multiplexing. For example, 16 supergroups can be multiplexed to form a 960-channel baseband occupying 60 to 4028 kHz.

The baseband input to a typical AMR is shown in Fig. 5.77 for a message (voice) and video baseband. In the message baseband, the omnibus order-wire occupies the speech frequency range of 300 Hz to 4 kHz. Next, 12 subbaseband service channels occupy 12 to 60 kHz. For the case illustrated, the 960 FDM subscriber channels occupy the frequency range from 60 kHz to 4028 MHz. Although 960 channels is perhaps the most commonly used AMR transmission, channel capacities up to 2700 on a single carrier are in use. Figure 5.78 indicates the CCIR Recommendations for channel capacity and carrier channel spacing in various analog radio frequency bands.

Notice the radio continuity pilot tone in Fig. 5.77. Here it is considered part of the baseband, but it is really an independent signal which is inserted above the message baseband prior to modulation. This is an important feature of an AMR link. The pilot tone can have several functions. It is primarily used to detect a link disruption, hence the name continuity pilot. Its presence at the correct level in the receiver denotes satisfactory operation of the complete link from modulator through to demodulator. It is newly introduced at the input to each modulator in a multihop link. A missing pilot denotes link failure, and as a consequence, an alarm is given. The pilot tone level is also used in the AGC loop for gain regulation. Furthermore, the pilot level is often displayed as a fading monitor. Since the pilot is a direct mea-

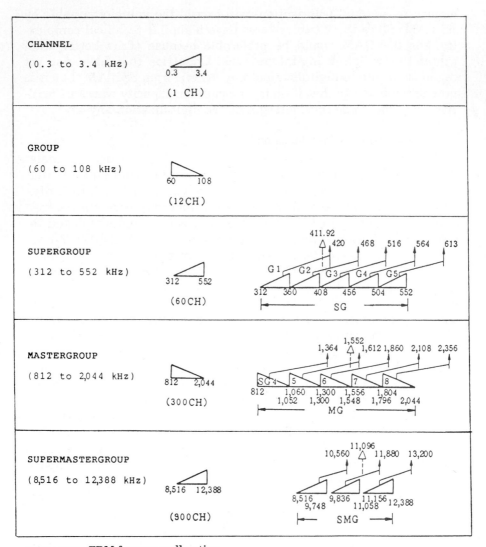

Figure 5.76 FDM frequency allocation.

sure of the "well being" of a link, it can also be used in a space or frequency diversity receiver to indicate to the combiner which of the signals is better, and therefore can dictate which signal is to be selected.

5.4.2 FM analog microwave radio

Figure 5.79 illustrates the major components of an FM microwave radio. The FDM signal enters the transmit terminal equipment containing either a video channel or up to 2700 voice channels. In some de-

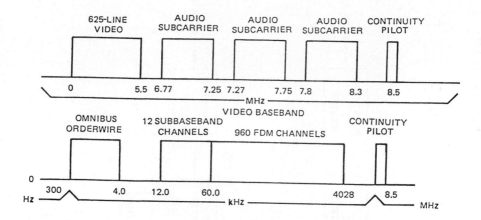

Figure 5.77 AMR baseband composition.

FREQUENCY BAND (GHz)	FREQUENCY RANGE (MHz)	CARRIER SPACING (MHz)	FDM CHANNEL CAPACITY	CCIR REC.
2	1700–1900	14	60,120,300	2
	1900–2100			
	2100–2300			
	2500–2700		960	
	1700–2100	29	600–1800	382–5
	or			
	1900–2300			
4	3700–4200	29	600–1800	382–5
		40	1260	382–5
6	5925–6425	29.65	1800	383–4
	6430–7100	40	2700	384–5
		20	1260	384–5
7	7425–7725	7	60,120,300	385–4
8	8200–8500	11.66	960	386–3
	7725–8275	29.65	1800	386–3
11	10700–11700	40	1800	387–5
13	12750–13250	28	960	497–3
		14	300	497–3
		35	960	497–3
15	14400–15350	28	MEDIUM	636–1
		7	SMALL	
21	21200–23600	3.5		637 AND
		2.5		REPORT 936–2

Figure 5.78 CCIR frequency allocations for analog microwave radio.

signs both can be accommodated simultaneously. At this first stage the service channel, incorporating the order-wire and alarms, is combined with the baseband signal, which is then shaped by a pre-emphasis circuit.

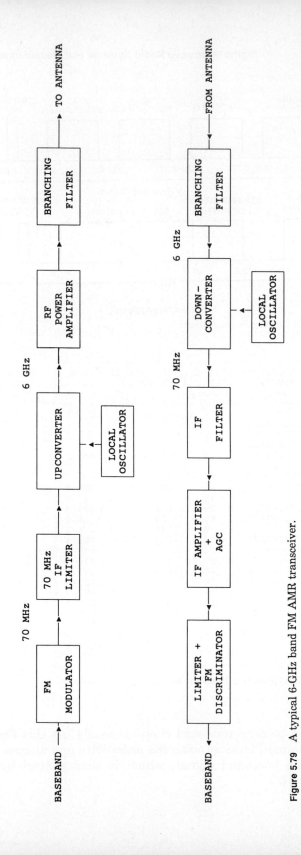

Figure 5.79 A typical 6-GHz band FM AMR transceiver.

Preemphasis and deemphasis. Preemphasis is a noise equalization process. Following FM demodulation, the thermal noise power increases with increasing frequency throughout the baseband, in a so-called triangular distribution, at a rate of 6 dB/octave. To equalize this noise over the baseband, a preemphasis circuit is included prior to the modulator which has an attenuation characteristic opposite to the noise characteristic (attenuation decreases with increasing frequency). Note, the total baseband energy input to the modulator is almost the same as without preemphasis. At the receiving end, a deemphasis circuit is incorporated after the modulator, which introduces attenuation complementary to the preemphasis circuit (i.e., attenuation increases with increasing frequency). The result is a *relatively flat S/N across the baseband,* while maintaining a flat power level across the baseband. The preemphasis and deemphasis circuits are detailed in CCIR Recommendations 275-2 (Ref.1) and 405-1 (Ref.1). These circuits are not required for the DMR systems.

FM modulator. The FM modulator superimposes the baseband signal onto a 70- or 140-MHz IF carrier. The IF bandwidth which must accommodate the RF bandwidth is established using Carson's rule:

$$B_{\text{IF}} \approx 2(\Delta f_p + f_m)$$

where Δf_p is the peak frequency deviation and f_m is the highest modulating frequency. The peak deviation depends on the number of voice channels (or video) which load the microwave radio channel. CCIR Rec. 276-2 indicates that a video system without preemphasis should have a peak deviation of ± 4 MHz. CCIR Rec. 404-2 indicates the peak deviation depending upon the number of channels loading the radio. The older designs used a Klystron oscillator as the oscillator to be modulated by the baseband signal. Varactor diode, all-solid-state oscillators have superseded those designs.

The IF signal is then filtered and amplified using components similar to the DMR system. The major difference here is that the IF filter of a DMR is a very critical item because it must satisfy the Nyquist criteria for pulse transmission. Its bandwidth determines the amount of RF spectrum occupied. The AMR IF filter does not have this excessive bandwidth problem. Its bandwidth is simply dictated by the peak deviation per channel (Carson's rule). The RF bandwidth for AMR is established from the knowledge of the IF bandwidth. Since the analog systems were established first, the digital systems must now compete with the analog systems in terms of bandwidth.

Transmit RF stage. For the analog systems the upconverter to RF is almost identical to the digital system. It is simply a mixer fed by a local oscillator. The older designs consisted of a rat-race hybrid incorporating Schottky diodes in the appropriate arms. The local oscillators used to be crystal oscillators multiplied up to the RF. Also, the upconverted RF was not the transmit frequency. If the transmit frequency was 6 GHz, the upconverted signal would be about 2 GHz. This was filtered, amplified at 2 GHz by a bipolar transistor amplifier, and then multiplied up to the 6-GHz output with a × 3 multiplier. Alternatively, the amplification would take place at the output frequency (e.g., 6 GHz) by a TWT or Klystron amplifier. Systems with this design are in operation in many parts of the world and will continue to operate until they are replaced by digital systems. The more recent upconverter designs include a microstrip double balanced mixer with a dielectrically stabilized microstrip oscillator operating directly at the RF output frequency. Solid-state GaAs FET amplifiers now replace the outdated and relatively short-lifetime Klystron or TWT amplifiers. For an FM system, operating the power amplifier at or even above the 1-dB compression point is not serious and is actually desirable from the power efficiency point of view. The AM to PM distortion of the amplifier causes intermodulation distortion, but this can be reduced by an equalizer or limiter. Further RF filtering is done, and the signal is passed to the branching network, which is very similar in design for both digital and analog systems. The bandwidths of the filters are different from one system to another, but they are usually all constructed in the waveguide medium as are the circulators which are designed to cover a full transmission band (e.g., 3.7 to 4.2 GHz or 5.925 to 6.425 GHz).

Receiver.   The AMR receiver is very similar to the DMR receiver at the RF and IF portions of the system. The incoming RF signal is filtered and mixed with a local oscillator to downconvert to the 70-MHz (or 140-MHz) IF. The signal is then filtered and equalized to be ready for the discriminator. The discriminator demodulates and deemphasizes the 70-MHz signal. At this point, the baseband is ready for demultiplexing. As previously stated, an important component of the baseband is the pilot continuity signal, which is filtered off and monitored. An alarm is given if the pilot level drops below a predetermined value.

Waveguide run and antennas.   The waveguide run and antennas used for AMRs are often interchangeable with those used for DMRs. In many cases, when an analog system is replaced by a digital system, the waveguide run and antenna are adequate for the digital system. Sometimes, in areas of problematic multipath fading, a second antenna and accompanying waveguide run must be installed for the digital system to allow operation in the space diversity mode. In other cases, the system gain of the new digital system may dictate a larger (higher gain), improved-performance (better front-to-back ratio) antenna.

5.4.3 Measurements

Many of the measurements made on an AMR are the same as for a DMR. The important differences are highlighted as follows.

Modem. For an AMR the transmitter deviation sensitivity must be set correctly and checked from time to time. This is usually done indirectly by observing the amplitude of the carrier, which gradually decreases with increasing deviation until it is eventually nulled at a specific value of deviation. A microwave link analyzer is used for this purpose.

Noise. Noise is perhaps the most significant parameter that differentiates AMR from DMR systems. It has already been said that noise accumulation occurs in analog systems, whereas the regenerative process in digital systems recreates the original signal at each repeater. Jitter is introduced in the digital system and is the ultimate distance-limiting factor. The maximum link distance for the analog system is limited by the accumulation of thermal and intermodulation noise. The sources of thermal noise, sometimes called idle noise, are, as usual, the RF transmitter and receiver, the modulator and demodulator, etc. The intermodulation noise sources are due to the modem and RF transmitter/receiver nonlinearities, but this noise depends on the channel loading and also on multipath propagation effects. Of course, the usual interference noise from other RF channels (cochannel and adjacent channel) also contributes to the overall noise.

Thermal noise limits the performance of a receiver for the case of *low* signal levels which occur during deep fading, whereas intermodulation noise limits the performance for *high* signal levels. The total noise accumulation in an FDM voice channel caused by the radio portion of a 2500-km hypothetical reference circuit is given by CCIR Rec. 395-2 as follows. The psophometrically weighted noise power at a point of zero relative level in the telephone channels of FDM radio relay systems of length L (where L is 280 to 2500 km) should not exceed:

1. $3L$-pW 1-min mean power for more than 20 percent of any month

2. 47,500-pW 1-min mean power for more than $(L/2500) \times 0.1$ percent of any month

Rec. 592-2 also gives noise requirements for links which differ from the hypothetical reference circuit. Also on the hypothetical reference circuit, the CCITT recommends a 2500-pW mean value of noise in any hour for the FDM equipment where 1 pWp = 0.56 × pW = − 90 dBm, and the North American noise parameter is dBrnc0 = 10 log pWp0 + 0.8 dB.

Noise power ratio (NPR) is the most meaningful noise measurement for AMRs. By definition, NPR (in decibels) is the ratio of the noise level in a channel being measured, with the baseband fully loaded, to the level in that channel with all of the baseband noise loaded except the channel being measured. This measurement is achieved by using a white noise generator to simulate the noise spectrum produced by the FDM equipment. The white noise signal is applied to the FM transmitter input at a level equal to the typical baseband level (Fig. 5.80). A bandstop filter is then switched in to clear a narrow window in the noise signal spectrum, which is usually equal to the bandwidth of an FDM voice channel. A selective level meter (or noise analyzer) is then connected to the baseband output at the receiving end of the hop or link, which is used to measure the noise power in the cleared window. This is equivalent to the total thermal and intermodulation noise present in the window bandwidth. The ratio of the input power to the measured noise power in the cleared bandwidth is the NPR. For a link containing heterodyne repeaters, a minimum NPR of 59 dB should be established. When diversity combining is included, the NPR is improved by approximately 3 dB.

A typical NPR test would be done at three points in the baseband at low, medium, and high frequencies. Figure 5.81 illustrates how the level (and deviation) affects the noise. As the per-channel input level increases, the deviation and therefore bandwidth also have to increase. The thermal noise therefore decreases at the expense of increased bandwidth. This improved overall noise reaches a minimum, after which the intermodulation noise increases rapidly and increases the overall noise. The balance between thermal noise and intermodulation noise is an important aspect of aligning an AMR. The best performance is obtained when operating at the optimum input level which produces the minimum overall noise.

Output spectrum.   The AMR spectrum differs from the DMR spectrum as shown in Fig. 5.82. Notice how the DMR spectrum almost completely fills the band, whereas the AMR spectrum has the peak at the carrier frequency and sidebands gradually decreasing in amplitude at frequencies away from the carrier. It is this characteristic which

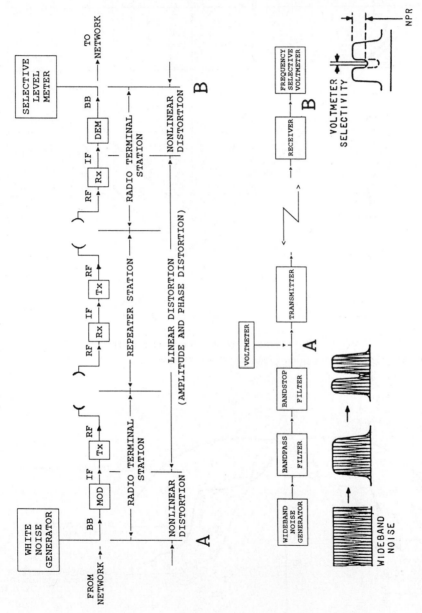

Figure 5.80 Noise power ratio measurement.

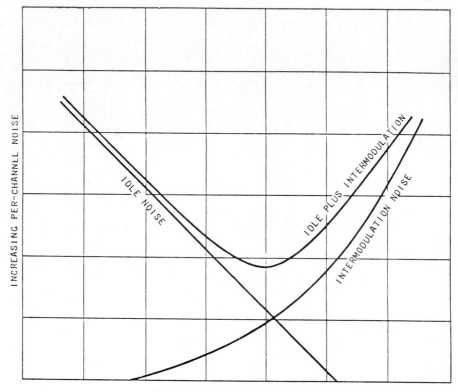

Figure 5.81 Frequency deviation versus per-channel noise.

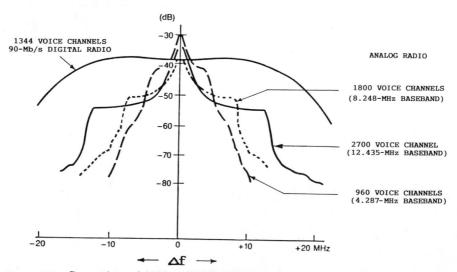

Figure 5.82 Comparison of AMR and DMR RF spectra.

	ANALOG	DIGITAL
Baseband	FDM-group, supergroup; mastergroup etc. 12,60,.... Emphasis/pre-emphasis	TDM - PCM (30/24ch) 2,34,140,565 Mb/s
Modulation	FM	FSK, PSK QAM (16,64,256)
Transmitter	Klystron or TWT (old) Solid state (new)	Solid state (all) Predistorter Differential encoder Scrambler
Receiver	Discriminator	Decrambler Differential decoder
Repeater	Baseband (drop/insert) Otherwise IF	Baseband (always)
Multipath fading countermeasure	Space diversity (sometimes) Frequency diversity (rarely)	Adaptive IF equalizer Transversal equalizer Space diversity (often) Frequency diversity (often)

Figure 5.83   Comparison of AMR and DMR systems.

makes AMR vulnerable to interference by DMR. If the frequency band of a DMR operates too close to that of an AMR, the large amplitude at the edges of the band of the DMR can swamp the low-amplitude sidebands of the AMR, causing interference so severe that the AMR fails. This point stresses the need for adequate IF and RF filtering of microwave radio signals whose bands are operating very close to each other on the same route.

5.4.4 Summary

Based on the above discussion, the differences between the AMR and DMR systems are summarized as shown in Fig. 5.83. A comparison of the two types of system initially highlights a considerable increase in the complexity of digital systems. Although this is true, a substantial portion of the construction is done using integrated circuits, which are relatively cheap and very reliable.

6

Introduction to
Fiber Optics

6.1 Introduction

Lightwave communication was first considered more than 100 years ago. The implementation of optical communication using light waveguides was restricted to very short distances prior to 1970. Corning Glass company achieved a breakthrough in 1970 by producing a fused silica (SiO_2) fiber with a loss of approximately 20 dB/km. The development of semiconductor light sources also started to mature at about that time, allowing the feasibility of transmission over a few kilometers to be demonstrated. Since 1970, the rate of technological progress has been phenomenal. Optical fibers are now used in transatlantic service. Besides the long distance routes, fibers are used in the interexchange (junction) routes, and the subscriber loop is the final link in what will eventually be the global interconnection chain. Optical fibers are associated with high-capacity communications. A lot of attention is presently being given to optical fibers to provide a very extensive ISDN. First, the characteristics of the fibers and components will be discussed.

6.2 Characteristics of Optical Fibers

The evolution of optical fibers has been extremely rapid over the past 15 years. Research and development has been directed toward reducing the attenuation of fibers and also increasing the digital transmission rate through the fibers. The attenuation defines the distance between repeaters, which directly affects the cost of a communication link. Initial costs increase as the number of repeaters increases, so the repeater spacing must be maximized. Also maintenance costs increase as the number of repeaters increases. The transmission rate (bits per second) directly determines the number of channels that the link can carry, so research has been aimed at maximizing this parameter. The bit rate is dependent upon the linewidth of the light source and the size and dispersion characteristics of the fiber. Dispersion causes transmitted pulses to spread and overlap as they travel along a fiber, limiting the maximum transmission rate. A figure of merit is often used to describe the optical fiber performance: the bit rate (B) times the unrepeatered length (L) in gigabits per kilometer.

Ease of coupling light into the fiber is also an important factor. This is related to the diameter of the light-carrying portion of the fiber and the characteristics of the glass waveguide. A parameter called the *numerical aperture* is associated with coupling. In summary, the three important aspects of optical fibers which must be discussed in detail are:

1. Numerical aperture

2. Attenuation

3. Dispersion

6.2.1   Numerical aperture

Figure 6.1 shows the nature of light entering a fiber waveguide. From elementary physics it is known that there is refraction of light at an air-glass interface. Similarly, there is refraction at the interface of two glass materials having different refractive indices, unless the critical angle is exceeded, in which case the light is totally internally reflected. The light waveguide is established by a glass fiber core whose refractive index is slightly higher than the glass cladding. Light propagates along the fiber by a series of "bounces" caused by internal reflection at the core-cladding interface. This is clearly illustrated in Fig. 6.2. However, if the light enters the fiber at an angle which is greater than the "cone of acceptance angle" (ray 2 in Fig. 6.1a), instead of being internally reflected at the core-cladding interface, it is refracted and lost, so the light does not propagate along the fiber. The numerical aperture is defined as:

$$\text{NA} = \sin \theta_c = \sqrt{n_1^2 - n_2^2} \qquad (6.1)$$

where θ_c = the cone of acceptance angle
  n_1, n_2 = the refractive indices of the core and cladding, respectively

The numerical aperture decreases as the diameter of the core decreases. A typical value for a 50-μm core is 0.2, and it is 0.1 for a

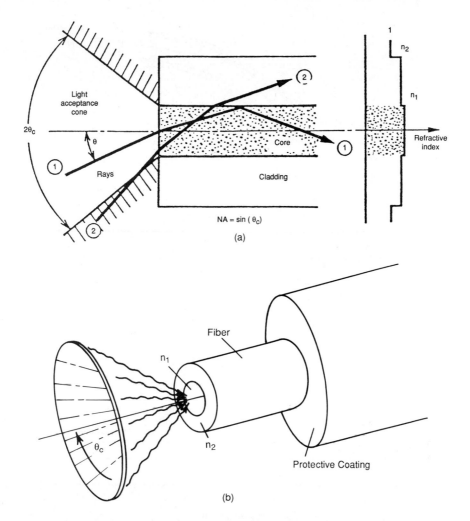

Figure 6.1 (a) Light-acceptance cone for a step-index fiber. (*Reproduced with permission from Yasuharu, S., Introduction to Optical Fiber Communications, © 1982, J. Wiley & Sons, Fig. 2.5.*) (b) Numerical aperture. NA = sin θ_c = $\sqrt{n_1{}^2 - n_2{}^2}$ = $n_1\sqrt{2\Delta}$. (*Reproduced with permission from IEEE, Keck, D. B., Fundamentals of Optical Waveguide Fibers, IEEE Communications Magazine, © May 1985, p. 19, Fig. 2.*)

10-μm core fiber. Figure 6.1*b* gives a three-dimensional view of the cone of acceptance.

6.2.2 Attenuation

Figure 6.3 shows the relationship between silica fiber attenuation and the wavelength of light. Rayleigh scattering is the main physical loss mechanism. The first fiber to be placed in service (about 1977) operated at 0.85-μm wavelength. The attenuation at 0.85 μm is about 3

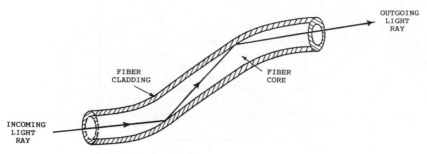

Figure 6.2 The optical fiber waveguide.

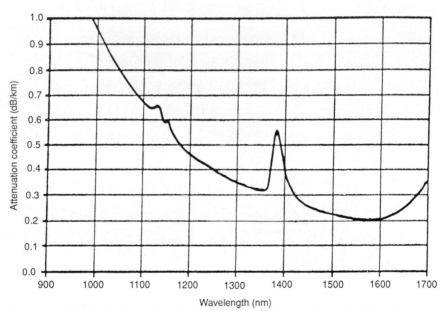

Wavelength (nm)

Figure 6.3 Graph of attenuation against wavelength for single-mode silica fiber.

dB/km. Since that time light sources have been developed for longer wavelengths, and now the standard wavelengths of operation are 1.3 and 1.55 μm. The lowest-loss silica fibers operate at 1.55 μm with a minimum attenuation of about 0.2 dB/km. This is close to the theoretical limit of about 0.16 dB/km at 1.55 μm for silica fibers.

Present research indicates that certain chloride and halide glass materials have the potential for attenuation values of 0.001 dB/km or less at longer wavelengths of several micrometers. Figure 6.4 graphically presents the theoretical attenuation against wavelength for several materials in consideration, and silica is included for comparison.

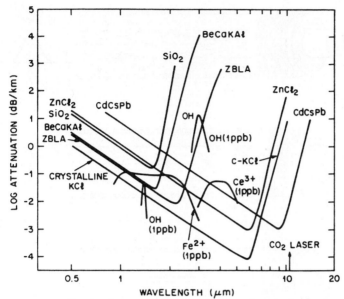

Figure 6.4 Graph of attenuation against wavelength for various materials. BeCaKAl are fluoride glasses made with those cations; ZBLA represents a zirconium barium lanthanum aluminum fluoride glass; CdCsPb is a chloride glass made from those cations. (*Reproduced with permission from IEEE, Nagel, S. R., Optical Fiber— The Expanding Medium, © IEEE Communications Magazine, Vol. 25, no. 4, April 1987, p. 37.*)

The best material shown is the crystalline KCl, which has an attenuation of 0.0001 dB/km at 6 μm, although the feasibility of using a crystalline material is very questionable. The attenuation values of these materials are extremely low, particularly when making a comparison with our most familiar type of glass (i.e., window glass, which has an attenuation of about 50,000 dB/km).

There are two types of optical fiber: (1) multimode and (2) single-mode. The distinction between the two types is simply determined by the size of the fiber core. If the core diameter is made small enough, approximately the same size as the wavelength of light, only one mode propagates. Figure 6.5 shows the typical relative physical sizes of multimode and single-mode fibers; Fig. 6.5*a* and *b* are multimode fiber, whereas Fig. 6.5*c* is a single-mode fiber.

The 8- to 10-μm core used for single mode has a smaller aperture than the 50-μm core used for multimode operation. At 1.3 μm, both multimode and single-mode systems are in existence. The attenuation is lower for the single-mode fiber and also the maximum bit rate is higher than for multimode. The 1.55-μm "window" of operation is used exclusively for single mode. The attenuation peaks in Fig. 6.3 are

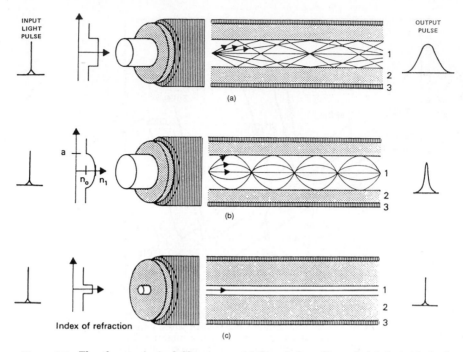

Figure 6.5 The three principal fiber types. (*a*) Step index; (*b*) graded index; (*c*) single mode. (*Reproduced with permission from ITU, Optical Fiber for Telecommunications, CCITT, Geneva 1984, Fig. 1.*)

caused by the presence of impurities such as OH⁻ ions. Above 1.55 μm, far infrared absorption causes most of the pure glass losses. Although infrared absorption is the limiting factor for the oxide glasses such as silica, other nonoxide glasses such as $ZnCl_2$ have infrared absorption peaks at much longer wavelengths: 6 to 10 μm. This is the reason for the optimism in obtaining extremely low-loss fiber at wavelengths above 2 μm in the future.

6.2.3 Dispersion

Dispersion causes optical pulses transmitted along a fiber to broaden as they progress down the fiber. If adjacent pulses are broadened to a point where they severely overlap each other, detection of the individual pulses at the receiver is not easily possible (Fig. 6.6). This misinterpretation of pulses, ISI, leads to a poor BER performance. So, dispersion limits the distance of transmission and the transmission bit rate. The dispersion is defined mathematically by a term which takes into account the broadening of a pulse over a distance traveled along the fiber, that is,

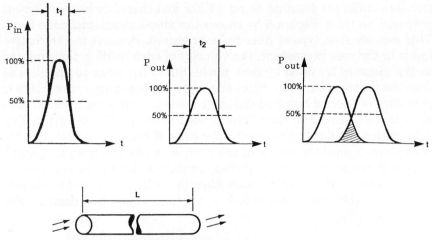

Figure 6.6   Pulse broadening due to dispersion.

$$\text{Dispersion} \approx \frac{\sqrt{t_2^2 - t_1^2}}{L} \quad \text{ns/km} \tag{6.2}$$

or, more precisely, at any wavelength, a pulse will be delayed by a time t_d, per unit length L, according to the equation

$$\text{Material dispersion} = \frac{1}{L}\frac{dt_d}{d\lambda} \quad \text{ps/nm} \cdot \text{km} \tag{6.3}$$

The causes of dispersion are as follows.

Spectral width of the transmitting source.   The optical source does not emit an exact single frequency but is spread over a narrow band of frequencies. The laser has a narrower linewidth than the LED and consequently is said to produce less *intramodal* or *chromatic* dispersion than the LED. If the light sources could emit only one frequency, there would be no dispersion problem.

Characteristics of the fiber.   The total fiber dispersion is the sum of three components:

1. Modal dispersion
2. Material dispersion
3. Waveguide dispersion

Modal dispersion is dependent only on the fiber dimensions or, specifically, the core diameter. Single-mode fibers do not have modal dispersion. Multimode fibers suffer modal dispersion because each mode travels a different distance along a fiber and therefore has a different propagation time. Figure 6.5a shows the step-index multimode fiber. This was the first type of fiber to be produced. Because the refractive index in the core is constant, the velocity of each mode is the same, so as the distance traveled by each mode differs from one to another, so does the propagation time. Since the light in these larger core fibers is composed of several hundred different modes, a pulse becomes broader as it travels along the fiber. The graded index profile shown in Fig. 6.5b causes the light rays toward the edge of the core to travel faster than those toward the center of the core. This effectively equalizes the transit times of the different modes, so they arrive at the receiver almost in phase. The graded index fiber pulse broadening (i.e., dispersion) is significantly improved over the step-index design. For multimode fiber:

$$\text{Total dispersion} = \sqrt{\text{intermodal}^2 + \text{chromatic}^2} \qquad (6.4)$$

The single-mode fiber of Fig. 6.5c, when operating in a true single mode, has no other interfering rays, so it does not suffer modal dispersion pulse broadening. The single-mode fiber does, however, suffer from *material dispersion*. This is because of the frequency dependence of the refractive index (and therefore the speed of light) of the fiber material. For silica, the material dispersion drops to zero at 1.3 μm, as indicated in the graph of Fig. 6.7. Unfortunately, at this wavelength, the attenuation is about 0.35 dB/km, which is not the minimum attenuation wavelength, so a dispersion-shifting technique has to be

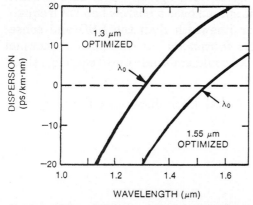

Figure 6.7   Material dispersion against wavelength for single-mode fiber designs. (*Reproduced with permission from IEEE, Nagel, S. R., Optical Fiber—The Expanding Medium, © IEEE Communications Magazine, Vol. 25, no. 4, April 1987, p. 41.*)

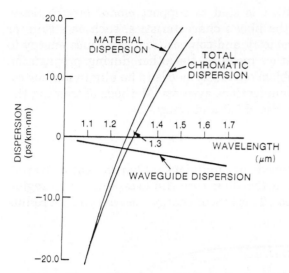

Figure 6.8   Waveguide and material dispersion in single-mode fibers. (*Reproduced with permission from IEEE, Nagel, S. R., Optical Fiber—The Expanding Medium,* © *IEEE Communications Magazine, Vol. 25, no. 4, April 1987, p. 35.*)

employed to fabricate the material for zero material dispersion at the lower attenuation wavelength of 1.55 μm. *Waveguide dispersion* is another form of dispersion which, compared to the material dispersion (about 0.1 to 0.2 ns/km), is generally negligible in multimode fibers. However, for single-mode fibers waveguide dispersion may be significant. Figure 6.8 shows how the waveguide dispersion combines with the material dispersion to shift the total chromatic dispersion curve. The zero dispersion wavelength shifts from about 1.26 to 1.30 μm. By modifying the fiber refractive index profile using precise fabrication techniques to affect the waveguide dispersion characteristic, the fibers can be designed to have the zero dispersion wavelength shifted to the lowest-loss wavelength of 1.55 μm. Manufacturing tolerances will unavoidably cause a fiber to have a small amount of dispersion even at the so-called zero value. For example, a 1.55-μm zero dispersion-shifted fiber will typically be rated at less than 3 ps/nm · km.

6.2.4   Polarization

Although single-mode fibers cure many of the problems associated with multimode fibers, they do introduce a problem of their own. Single-mode fibers having circular symmetry about the core axis can propagate two almost identical modes at 90° to each other (a TE mode and a TM mode). The fiber is said to support *modal birefringence*. Small imperfections in the fiber's characteristics, such as strain, or variation in the fiber geometry and composition, can cause energy to move from one polarization mode to the other during propagation along the fiber. This problem obviously needs to be eliminated for coherent optical fiber communications systems. Methods of treating the problem are discussed in Sec. 6.5.3 and Chap. 7.

6.2.5   Fiber bending

Whenever either a multimode or single-mode fiber is bent or bowed, some of the energy within the fiber core can escape into the region outside the fiber (Fig. 6.9a). This loss of energy increases as the radius

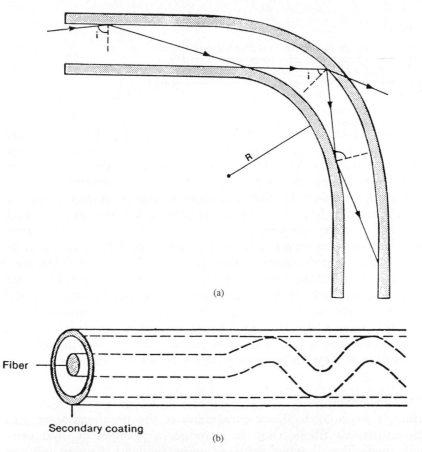

(a)

(b)

Fiber

Secondary coating

Figure 6.9   (a) A bow in an optical fiber; (b) microbending in an optical fiber.

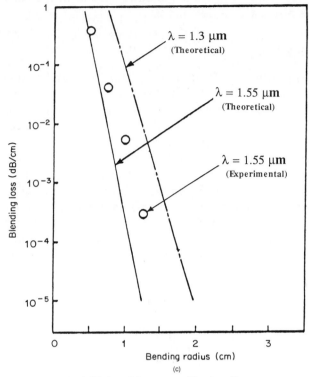

Figure 6.9 (c) Additional loss caused by bending.

of curvature of the bend decreases. If the radius of curvature approaches the fiber radius, very high losses occur, called *microbending attenuation* (Fig. 6.9b). Depending upon the circumstances, fiber bending can be either useful or detrimental. It is certainly necessary to fabricate the cable so that microbends are not created by impurities or defects in the fabrication process. Also, care must be taken during the installation of fibers to ensure no small radii bends are present at the splice boxes. Figure 6.9c shows the extent of bending loss incurred for 1.3-μm and 1.5-μm fibers, depending upon the radius of curvature. As one can see, this loss depends on the refractive index profile and the operating wavelength. The bending loss is very small for a 2-cm radius bend (i.e., less than 0.00001 dB/cm). As the radius of curvature becomes smaller than 1 cm, the loss increases rapidly.

In addition to the extra attenuation caused by bending, it can also be a source of unwanted surveillance of information, denying users their secrecy. Although this could be a problem for users wanting to maintain confidential communications, it opens up an interesting means of noninvasive testing, which is particularly useful in the fiber splicing procedure, as discussed in Sec. 6.4.2.

6.3 Design of the Link

The optical transmission link has many similarities to other types of transmission systems such as microwave systems. In its simplest form there is the usual transmitter and receiver with an intervening medium, which in this case is an optical fiber cable. As previously stated, one of the major objectives in optical fiber transmission is to ensure that the repeater separation is maximized to minimize costs. This is done by optimizing the properties of the transmitter, cable, and receiver. The transmitter output power should be as high as possible. The optical fiber attenuation should be as small as possible, and the receiver sensitivity should be maximized. The fiber cable in the past has been the major factor in determining the repeater separation. Today's optical fiber cables at the operating wavelength of 1.3 μm are readily available at 0.35 and 0.20 dB/km at 1.55 μm. In addition to the attenuation there is also the dispersion characteristic of the cable which can limit repeater separation. In other words, when the dispersion becomes too large, the distance is limited by intersymbol interference, causing incorrect detection of 1s and 0s. There is, however, a small loss penalty to be paid for shifting this dispersion value to 1.55 μm. It is quite small and only increases the attenuation to approximately 0.22 dB/km.

The transmitter, which uses an LED or laser diode (LD), should have the highest output power possible. The maximum transmitter power at present is limited by semiconductor physics technology. Even if the present output power values increase dramatically in the near future, there is a theoretical limit to the transmitted power level because of Remain and Brillouin scattering within the fiber. However, recent studies have established that the nonlinearities caused by scattering can be used to our advantage. Pulse compression occurs when transmitting at higher power levels (greater than 10 dBm), which, as described later, can drastically improve the repeater spacing and capacity of optical systems. The early LDs had a Fabry-Perot style of construction (see Sec. 6.5.1), which have an output frequency that is relatively broad in linewidth. These diodes have even been referred to as optical noise sources. The linewidth of the laser should be minimized in order to ensure that laser dispersion does not limit repeater separation. Recent developments have resulted in the distributed feedback LD, which provides a very narrow linewidth device, typically 0.2 nm or less. Considerable impetus has been given to transmitter technology, driven by the potential improvements obtained using coherent detection. The coherent detection mechanism, otherwise known as heterodyne or homodyne detection, requires optical sources to have extremely small linewidths. The linewidth should be 10^{-3} to 10^{-4} times the transmitted bit rate. The typical power transmitted by a laser in an optical fiber communication system is at present less than 10 mW.

Concerning the receiver, the two main types of semiconductor device in use are the avalanche photodiode (APD) and the PIN diode followed by an FET or high electron mobility transistor (HEMT) amplifier. The development of APD and PIN diodes has proceeded in parallel and there is little to choose between the two in terms of performance up to about a 2-Gb/s transmission rate. The PIN FET combination is the simplest to produce (see Sec. 6.5.2 for more detail).

An optical transmission link usually has an objective of transmitting x number of channels over a specific distance. The distance between repeaters for monomode cable is presently in the region of 40 to 300 km, depending upon the system design. This has enabled junction routes (interexchange traffic) to be constructed without the need for repeaters. However, for long haul, the repeater spacing distance is extremely important for economic considerations. For example, consider the requirement of transmitting approximately 8000 voice channels over a distance of several hundred kilometers. This link could be designed by using one pair of fibers for each direction at the 565-Mb/s rate, or it could be done using four pairs of fibers for each direction at the 140-Mb/s rate. The important difference in cost for these two systems would be due to the different repeater requirements. For a relatively cheap and simple, unamplified, 140-Mb/s system, repeaters would be required on each pair of fibers at typically 40-km intervals, whereas the 565-Mb/s system would require only one-quarter of this quantity of repeaters (provided the fiber dispersion characteristics allow 40-km repeater spacing). The higher-quality dispersion characteristic fiber would have a slightly higher cost, but this is more than offset by the reduced number of repeaters. The other economic factor to be considered is the price of the terminal equipment for the respective systems. At higher bit rates the equipment becomes considerably more expensive since more recent technology is used in such terminals. Over a relatively short span (link length) the increase in the number of repeaters for the 140-Mb/s system is offset by the higher price of the 565-Mb/s terminal equipment. If the desired link length is several hundred kilometers, the number of repeaters is considerably less for the 565-Mb/s system and it is therefore much cheaper than the 140-Mb/s system. A great effort is consequently being devoted to increasing the repeaterless span length and increasing the transmitted bit rate.

Theoretical limits to the *unamplified* repeaterless span using existing 1.55-μm, 0.22-dB/km fiber are in the region of 300 km, for a 5-mW optical transmitter output power with a 1-Gb/s bit rate, PSK homodyne modulation/coherent detection scheme, and a repeater sensitivity of about −57 dBm. *Note that there is approximately 6 dB of improvement in the receiver sensitivity for coherent detection systems (PSK homodyne) compared to present-day direct detection systems, which include optical preamplification.* Intensive research work is presently being done on chloride and halide fiber materials to achieve theoretical attenuation values as low as 0.001-0.0001 dB/km. If this becomes a reality, transatlantic repeaterless spans would be possible. At this level of attenuation, the splice losses would be a very significant portion of the overall link attenuation. At present, the splice loss objective is 0.1 dB or less. For some of the high-accuracy automatic fusion splicers, the splice loss can be reduced to 0.05 dB or better.

6.3.1 Power budget

The power budget is the basis of the design of an optical fiber link. Obviously,

$$\text{Total gain} - \text{total losses} \geq 0$$

Therefore $(P_t - P_r) - (\alpha_f + \alpha_c + \alpha_s + F_m)L \geq 0$. So,

$$L = \frac{P_t - P_r}{\alpha_f + \alpha_c + \alpha_s + F_m} \tag{6.5}$$

where P_t = transmitted power
P_r = receiver sensitivity (minimum received power)
α_f = fiber attenuation
α_c = connector attenuation
α_s = total splice losses
F_m = fiber margin
L = distance between repeaters

The *total length* of the link, the *bit rate,* and *error performance* are the three parameters which determine the design of a link. The total length can vary from a few meters in a local area network to thousands of kilometers in a cross-country trunk network. There are applications between these two extremes; for example, many countries are introducing optical fiber links for interconnecting exchanges within cities (5 to 20 km). The bit rate requirement is established by the number of telephone, video, or data channels proposed for the link. From the total length and bit rate, decisions can be made concerning the fiber characteristics and the type of transmitter and receiver to be

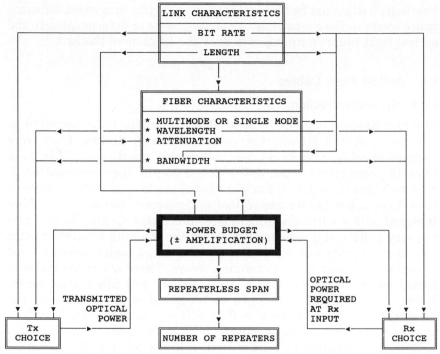

Figure 6.10 Decision process for the design of an optical fiber link.

used. As always, minimizing costs is essential, but the quality (error performance) of the link must satisfy the international standards.

Figure 6.10 shows the decision process for the choice of components for an optical fiber link. The starting point is to establish the capacity (i.e., how many channels are needed between points A and B). This automatically defines the bit rate, which has a bearing on whether the light source is an LED or LD and whether the receiver detector is an APD or PIN device. Next, is the link a short-, medium-, or long-haul communication system? The answer to this question decides what type of fiber is to be used. For example, the long-haul system must definitely use single-mode fiber, whereas a short-haul system (LAN) could use silica multimode or even the cheaper plastic fiber. Again, this decision has a bearing on the choice of device for the light source and detection devices. The overall length of the link determines whether or not repeaters are required. For medium-haul systems (junction routes) repeaters are not needed. For long-haul systems (trunk) repeaters are usually required. All of these interrelating factors are optimized in the "power budget" calculation, which must be done for every link design. Also, the fiber margin F_m of several decibels (e.g., 6 dB) must be factored into the analysis to account for extra splice losses in the event of future cable breaks or deterioration in the optical light source output power over the lifetime of the link.

6.4   Optical Fiber Cables

6.4.1   Cable construction

The first fibers commercially available in the late 1970s operated at 0.85 μm and had attenuation values of 3 to 5 dB/km. They were grouped in cables containing relatively small quantities of fiber. Less than 10 years later, in the mid-1980s, 1.55-μm dispersion-shifted fibers with less than 0.3 dB/km were being made with 2000 or more fibers in one cable. In all cases of fiber manufacture, the core of the fiber is coated with a silica cladding to offer protection against abrasion. A primary coating of plastic is placed over the cladding to provide extra structural rigidity and protection. In addition, a buffer jacket is used to protect the fiber against bending losses. There are three styles of buffer jacket: the tight buffer, loose buffer, and filled loose buffer jacket, as in Fig. 6.11. The tight buffer jacket is in contact with the primary coating, and since it is 0.25 to 1 mm in diameter, it stiffens the fiber so that it cannot easily be bent to the point of breaking. In practice it is possible to bend a tight buffered fiber to the breaking point, but this should not occur during normal handling procedures. Loose buffering allows the fiber to move within an oversized, hard plastic, extruded tube. Although it has an outside diameter larger than the tight buffering, it does offer better protection against excessive bending. Finally, at an added expense, the loose buffer tube is often filled with a silicone-based gel which provides moisture protection.

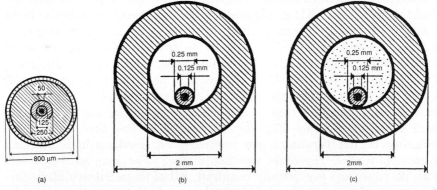

Figure 6.11   Optical fiber buffering. (*a*) Tight buffer jacket; (*b*) loose buffer jacket; (*c*) filled loose buffer jacket.

The gel-filled loose buffer fiber is the preferred choice for long distance links, whereas the tight buffer fiber is mainly used for LANs or short-to-medium distance links in the 1 to 10 km range.

Optical fibers are presently used in pairs (i.e., one for GO and the other for RETURN). Future wave division multiplexing (WDM) can reduce this to one fiber for both GO and RETURN (see Sec. 7.3.1). The economics of cable construction are such that it is usually not worthwhile buying a two-fiber cable. A four-fiber cable is considerably less than twice the cost of a two-fiber cable. Considering future capacity needs and spare fibers in case of failure, a six- or eight-fiber cable is usually the minimum considered for any interexchange or long distance link. There are many possible configurations for organizing optical fibers in a cable. All configurations use a strength member made of steel or a very strong material such as Kevlar (the bulletproof jacket material) in order to prevent longitudinal stress. This is particularly important for cables which are to be used in overhead installations. Figure 6.12 shows typical styles of packaging for 6, 12, 50, and 2000 fibers. The cables can be made with no metal whatsoever, if the application warrants such a construction. Other designs incorporate one or several pairs of copper wires to enable a current to feed remote regenerators. There are other equally acceptable methods of cable construction. One is illustrated by the ribbon matrix style in Fig. 6.13. Rows of fibers are stacked to form a ribbon matrix. The ribbon style has the potential benefit of reducing the splicing time when a large number of fibers are to be spliced. Note that the strength members are in the outer casing for this design instead of in the center.

Finally, submarine optical fiber cables require some special characteristics. For long-term transmission stability the fiber should be protected against water penetration and excessive elongation. The very high pressure exerted at depths up to 8000 m means that the cable construction is generally more rugged than land-based cables. Again, a variety of designs can be used, with a general preference for placing the fibers at the center with the strength member surrounding or external to the group of fibers. An example of a submarine cable is shown in Fig. 6.14.

6.4.2 Splicing, or jointing

One of the most essential requirements of any cable is the ability to join (splice) two pieces together to provide a low-loss connection that does not appreciably deteriorate with time. The optical fiber obviously cannot be treated in the same manner as ordinary metallic cable. Two techniques are currently in use for splicing optical fibers: (1) fusion splicing and (2) mechanical splicing. Both methods involve fiber-end

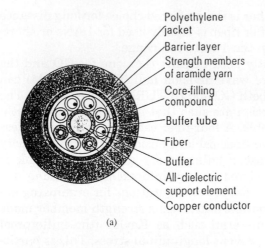

Polyethylene jacket

Barrier layer

Strength members of aramide yarn

Core-filling compound

Buffer tube

Fiber

Buffer

All-dielectric support element

Copper conductor

(a)

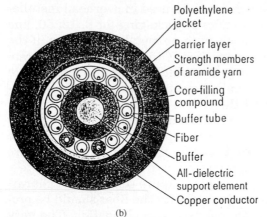

Polyethylene jacket

Barrier layer

Strength members of aramide yarn

Core-filling compound

Buffer tube

Fiber

Buffer

All-dielectric support element

Copper conductor

(b)

Figure 6.12 Optical fiber cable configurations. (a) 6 fibers; (b) 12 fibers.

preparation, alignment of the fibers, and retention of the fiber in the aligned position.

The fusion splice is done by placing two fibers end to end and applying an electrical arc at the point of contact for a short period of time. The glass momentarily melts at that point causing the two fibers to fuse together, resulting in a low-loss joint that should last for the lifetime of the fiber. This is considered to be the better method for fiber splicing, but it has the disadvantage of taking a relatively long time for each splice and requires very sophisticated and expensive equipment. The other method of splicing is to mechanically hold the two fibers end to end and to use a gel at the interface. The refractive index of the gel is the same as the fiber in order to provide continuity. The very-long-term (20 years) survival of such splices is unknown since they were introduced only in the early 1980s. Initially this method was used for temporary restoration of damaged links, but some admin-

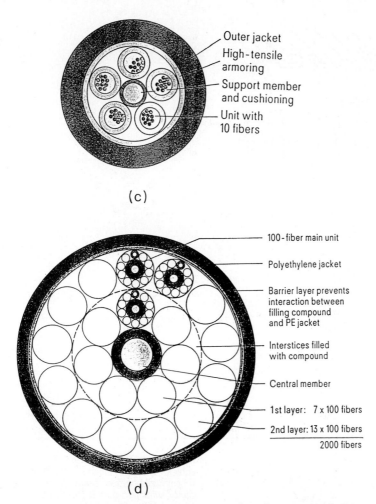

(c)

Outer jacket
High-tensile armoring
Support member and cushioning
Unit with 10 fibers

100-fiber main unit
Polyethylene jacket
Barrier layer prevents interaction between filling compound and PE jacket
Interstices filled with compound
Central member
1st layer: 7 x 100 fibers
2nd layer: 13 x 100 fibers
2000 fibers

(d)

Figure 6.12 Optical fiber cable configurations. (c) 50 fibers; (d) 2000 fibers.

istrations are now using it for permanent splices. This method will probably be dominant for fiber in the subscriber loop.

For either style of splicing, there are several aspects that are common to both. First, it is imperative that the two fibers to be joined have surfaces that are smooth, clean, and at 90° to the length of the fiber. Figure 6.15 emphasizes these requirements by indicating two unacceptable presplicing conditions. If a splice were made in either of these two situations, a high splice loss would be the result. Severe splice loss is always encountered unless extreme care and rigid splicing practices are observed. The ends of each fiber are always "cleaved" to provide a new, smooth, clean, 90° surface, whether the fiber is to be fusion or mechanically spliced.

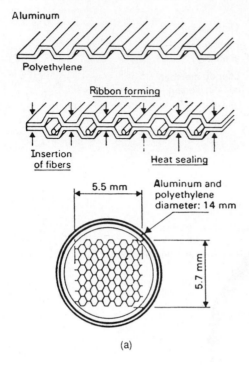

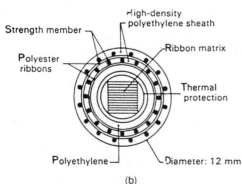

(b)

Figure 6.13 Ribbon optical fiber cables. (a) Loose ribbon (50 fibers); (b) tight ribbon fiber cable. (*Reproduced with permission from ITU, Optical Fiber for Telecommunications, CCITT, Geneva 1984, Fig. 8, p. 23 and Fig. 11, p. 24.*)

Fusion splicing. In the early days of optical fiber installation, use of the fusion splicing instruments required considerable skill and dexterity to ensure good low-loss splices. Fusion splicers now available are much easier to operate, and reliable, repeatable, very low-loss splices are the norm. This is mainly attributable to the automated aspect of these machines. Because they are built to operate within extremely fine tolerances, they are very expensive. The ends of the two fibers to be spliced are first cleaved, then cleaned in an ultrasonic cleaner. Next they are clamped in the alignment mechanism and brought into

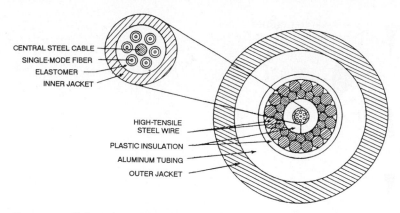

CENTRAL STEEL CABLE
SINGLE-MODE FIBER
ELASTOMER
INNER JACKET

HIGH-TENSILE
STEEL WIRE
PLASTIC INSULATION
ALUMINUM TUBING
OUTER JACKET

Figure 6.14   Submarine optical fiber cables.

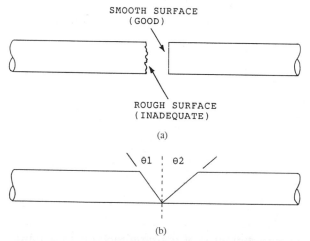

SMOOTH SURFACE
(GOOD)

ROUGH SURFACE
(INADEQUATE)

(a)

$\theta 1$ $\theta 2$

(b)

Figure 6.15   Fiber splicing preparation. (*a*) One fiber has an inadequately rough surface. (*b*) Both fibers have inadequate end surface perpendicularity.

position for the prefusion. The fiber ends are brought close to each other but not touching. The electric arc is then activated to melt one of the fiber ends to make a smooth rounded end. A microscope is used to give a clear view of the procedure, as illustrated in Fig. 6.16. The other fiber is treated the same way. The two fibers are then butted up against each other, and the arc is established for a few seconds to melt the two ends, causing fusion. The early, cheaper fusion splicing machines necessitated manual alignment and arc control. The more modern machines couple light into one fiber and use an optical feedback procedure to automatically align the two fibers to maximize the light transmission (i.e., minimum splice loss). The alignment accuracy required for single-mode fibers is considerably greater than for multi-

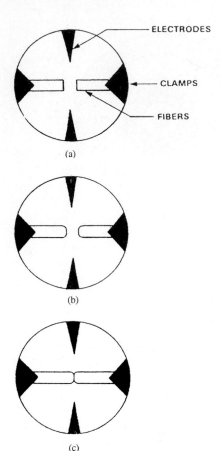

(a)

(b)

(c)

Figure 6.16   Fiber splicing steps. (*a*) Position of fibers before prefusion. (*b*) Condition of fibers after prefusion. (*c*) Position of fibers before fusion. *Note:* Fibers 1 diameter apart.

mode fibers because the core diameter of a single-mode fiber is much smaller (i.e., about 8 to 10 μm compared to about 50 μm for multimode). In one method, the two fibers to be spliced are brought to within a few micrometers of each other in an approximately aligned position. Light is then coupled into one fiber by bending the fiber and injecting an optical signal at about −35 dBm into the core. The level is again measured by bending the other fiber and using a photodetector to measure the level. Using a step-by-step procedure of moving the fibers relative to each other, the maximum light transmission position is the best aligned position. The main disadvantage of this method is the fact that the fiber has to be bent to a very small radius (less than 10 mm) to achieve adequate light input coupling and output decoupling. Another method in use which does not incorporate fiber bending is the *cladding leakage light* method (Fig. 6.17). In this case, light from the terminal transmitter is incident on the splice point, and

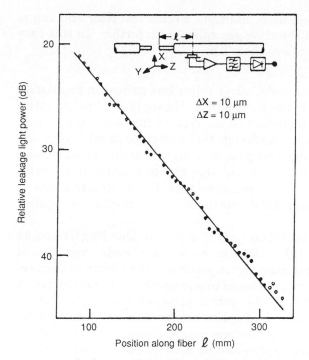

Figure 6.17 Leakage method of fiber alignment. (*Reproduced with permission from Ref. 134, © 1986 IEEE.*)

a small air gap is intentionally maintained between the two partially aligned fibers. Light is consequently coupled into the second fiber in the usual guided mode, plus some light is coupled into the cladding. This mode is confined only to the cladding over the region of fiber stripped back for splicing. When the light reaches the primary coated region, it also propagates in the coating because of its high refractive index relative to the cladding. Light tends to leak out of the coating, and this leakage can be used for alignment by orienting the two fibers until the minimum leakage is detected. On completion of the splice, the fused region is then protected by a reinforcement member and fixed in place by either epoxy and/or heat-shrinkable tubing.

Although the fusion splicer is used almost exclusively during installation, it is not always used for restoration of damaged fibers. If the break occurs at a busy time, the faster mechanical splice may be more appropriate, followed by resplicing with the fusion technique at a low-traffic time of the day. The time taken to fusion splice one pair of fibers obviously varies considerably depending upon the location and experience of the personnel, but 15 to 40 min is typical. This is more than twice as long as the mechanical splice time. When the ribbon constructed fiber cable is used, multiple mechanical fiber splicing is possible, which reduces the time per splice even further. In this case 24 splices can be done in less than 1 h.

Mechanical splicing. The mechanical splice has gained in popularity since it was introduced in the early 1980s. The early mechanical splice techniques included a score-and-break type of fiber cleaving to produce flat 90° end surfaces. Although this technique is still used, the use of the more repeatable and generally higher-quality grinding and polishing procedure is becoming widespread. For individual mechanical splices, the alignment is provided by a guiding insert structure as indicated in the successful GTE "elastomeric" or "fastomeric" splices shown in Fig. 6.18.

Figure 6.19 shows the ribbon type of splice. In this illustration 24 fiber ends of two 12-fiber ribbons have their ends ground and smoothly polished simultaneously. A portion of the ribbon is then removed and the bare fibers are placed in a grooved holding substrate. A matching gel is placed over the points at which the fibers butt up against each other. This type of splice can be done in 20 to 30 min, which is the usual time taken for one pair of fibers spliced by the fusion technique. The ribbon splice can be used for either multimode or single-mode fibers.

Whether the splicing is done by the fusion or mechanical method, the completed splices have to be placed in a *splice organizer*. As shown in Fig. 6.20, in a typical splice organizer, the excess length of each spliced fiber is coiled and carefully placed in position. This form of semipermanent containment provides protection of the splices against the elements, and can be entered at a later date for further work, if necessary.

Splicing problems. To appreciate the very tight tolerances which must be enforced to ensure good splices, various potential problems will now be addressed. First, high splice loss can be a result of splicing two fibers which have different fabrication characteristics. Figure 6.21a graphically illustrates the extra loss that can be expected if there is a mismatch in the *numerical apertures* of the two fibers. Notice how a difference of only about 5 percent can increase the loss by more than 0.5 dB. Similarly, if the *diameters* of the cores of the two fibers differ by only 5 percent, an additional loss of more than 0.5 dB can be anticipated (Fig. 6.21b). Second, extra loss can be incurred during the splicing procedure if there is an *offset* between the two fibers such that the core of each fiber is not precisely aligned with respect to the other. Figure 6.22a indicates how the misalignment of fibers increases the

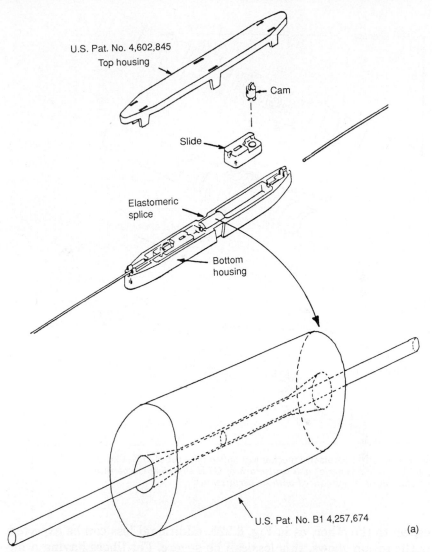

U.S. Pat. No. 4,602,845
Top housing

Cam

Slide

Elastomeric
splice

Bottom
housing

U.S. Pat. No. B1 4,257,674

(a)

Figure 6.18 (a) Elastomeric mechanical splice (*reproduced with permission from GTE Products Corporation*).

splice loss. A very small misalignment causes a large loss. For example, a misalignment of only 0.1 core diameters results in about a 0.5-dB loss. This is only about 0.1 μm for a single-mode fiber. The automatic alignment fusion splicers now available alleviate this problem. They can align the cores of the two fibers even if the core of one or both is eccentric in relation to the cladding. If one fiber is tilted with

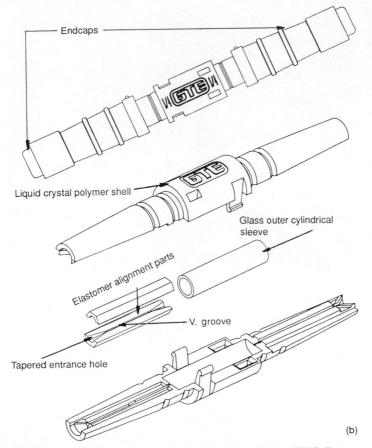

Endcaps

Liquid crystal polymer shell

Glass outer cylindrical sleeve

Elastomer alignment parts

V. groove

Tapered entrance hole

(b)

Figure 6.18 (b) Fastomeric mechanical splice (subject of U.S. Pat. no. 5,005,942) ("Fastomeric" is a trademark of GTE Products Corporation and the figure is reproduced with its permission).

respect to the other, as in Fig. 6.22b, additional loss can be expected. As the graph shows, this loss can be severe. For fibers having a numerical aperture of 0.2 (multimode), an angular tilt of only 2° causes a loss of about 0.5 dB. The loss is less severe for a single-mode fiber (i.e., the numerical aperture is about 0.1). Finally, if there is an air gap between the two fibers, a relatively small gap produces a large loss. For example, the graph in Fig. 6.22c indicates, for single-mode fibers, an additional loss of about 0.5 dB for a gap of 0.5 of a core diameter. This is only about 5 μm for a single-mode fiber. An air gap is unlikely to occur in a fusion splice but could occur in a mechanical splice. However, in normal circumstances, the index-matching gel should eliminate this problem.

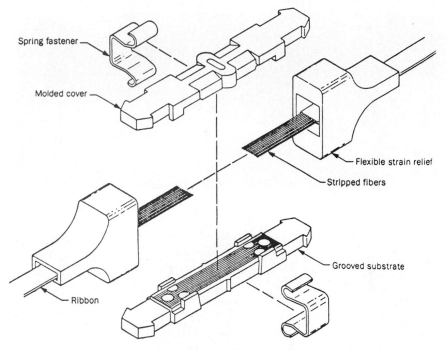

Figure 6.19 Ribbon mechanical splices. (*Reproduced with permission from Ref. 136,* © *1986 IEEE.*)

The splice loss should ideally be kept to 0.1 dB or less. Because of the problems stated above, this is not always possible. The environmental conditions at the site of the splicing are perhaps the greatest cause of high splice loss. Ideally, the splicing should be done in an air conditioned truck especially set up for the splicing procedures. This luxury is not always possible, and splicing often has to be done in manholes or at roadside locations. In such circumstances roadside dust is the worst enemy of the fiber splicing personnel. The presence of dust during a fusion or mechanical splicing procedure can easily produce splice losses in the 0.5- to 1-dB region. Such high losses are usually unacceptable. In very dusty regions it may be necessary to do one splice many times over until a satisfactorily low splice loss is established.

6.4.3 Installation problems

The intention of this section is not to relate the procedural details of placing the fiber cable but to discuss the advantages and disadvantages of the different types of installation. The installation methods

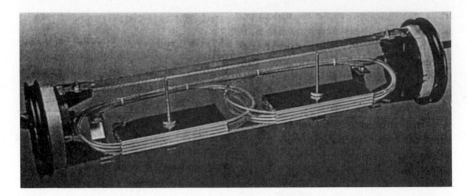

(a)

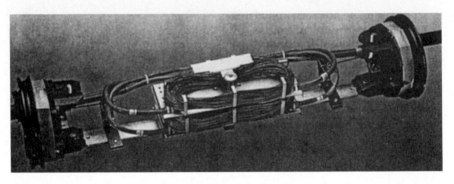

(b)

Figure 6.20 Typical splice organizers with (a) eight splicing modules or 80 fiber splices and (b) two splice mounts and, all together, 48 single splices. (*Reproduced with permission from Siemens Telcom Report 6 (1983), Special Issue, "Optical Communications," p. 56, Figs. 3 and 4.*)

can be simply split into two categories: (1) overhead and (2) underground. The choice is not always an easy one.

Invariably, the overhead installation is the cheaper one. However, the maintenance costs following installation can easily offset this initial advantage. Placing fiber on existing telephone cable poles makes it very vulnerable to extreme weather conditions such as hurricanes, tornadoes, or flooding. Fortunately these types of disasters rarely occur in most parts of the world, but in some countries they are an annual event. Even in areas that do not experience such severe conditions, heavy storms can knock poles down or accidents can occur such as a truck hitting a pole. In places of political instability, deliberate disruption of the communication system is very easy with overhead cables. Even internal labor unrest could lead to sabotage of cables, in

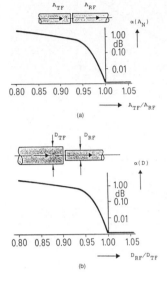

Figure 6.21 (a) Extra splice loss due to numerical aperture differential; A_{RF} = numerical aperture of receive fiber, A_{TF} = numerical aperture of transmit fiber. (b) Extra splice loss due to diameter differential; D_{RF} = core diameter of receive fiber, D_{TF} = core diameter of transmit fiber.

which case overhead cables are easy targets. A more secure overhead type of installation is to use existing electrical power transmission lines for suspending the cable. This location presents a high-voltage hazard to deter saboteurs, but it is also a severe hazard for the personnel who have to install and maintain the cables.

Underground cables are not without their problems. First, some regions are so rocky that very expensive blasting would be necessary before laying underground cables. In many cities the underground ducts are full to capacity, and in such cases it could be very expensive to lay more ducting. Then there is the question, Is ducting required for all underground fiber cables? If ducting is not used, the cable could be damaged by rodents if it is not placed deeper than about 1.5 m. Also, if ducting is not used, the cable is more susceptible to damage from construction workers or agricultural machinery.

In conclusion, if local conditions are favorable, and installation cost is secondary to security or preferable to later high maintenance costs, fibers should be placed underground in ducts set in concrete.

6.5   Fiber Optic Equipment Components

6.5.1   Light sources

As stated earlier, the two light sources available are the semiconductor LD and the LED. Both devices have small physical dimensions, which make them suitable for optical fiber transmission. As the term

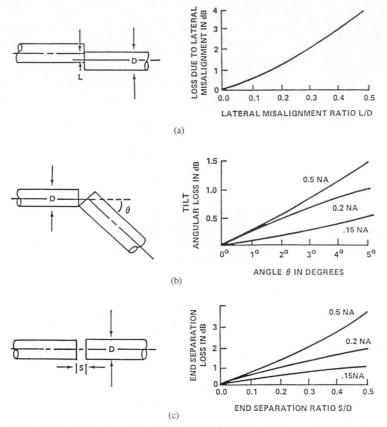

Figure 6.22 (*a*) Extra splice loss due to lateral misalignment; (*b*) extra splice loss due to angular tilt; (*c*) extra splice loss due to end separation. *D* = core diameter.

diode suggests, the LDs and LEDs are *pn* junctions. Instead of being made from doped single crystals, they now have exotic combinations of two or more single-crystal semiconductor materials. These hetero-junctions are consequently called heterostructures.

The fundamental difference between an LD and an LED is the fact that the light from an LED is produced by spontaneous emission whereas light from an LD is made by stimulated emission. This re-sults in the laser having an output which is coherent and therefore has a very narrow spectrum, whereas an LED has an incoherent out-put and a wide spectrum. The selection of LD or LED for an optical transmission system depends upon the following factors:

- Required output power
- Coupling efficiency

- Spectral width
- Type of modulation
- Linearity requirements
- Bandwidth
- Cost

LED. The semiconductor LED can be used in the surface-emitting or edge-emitting mode depending upon the type of fabrication, as indicated in Fig. 6.23. The surface-emitting style has good temperature stability and low cost. However, the coupling efficiency into the fiber is limited by its wide active area. The light power coupled into the fiber is typically less than 100 μW for multimode fiber and only several microwatts for single-mode fiber. Also, the light power output is incoherent (i.e., over a wide spectrum of about 40 nm). The operational bit rate is limited to several hundred megabits per second by the parasitic capacitance within the LED.

The edge-emitting LED has an improved performance compared to the surface-emitting type. The structure can achieve a higher coupling efficiency into a single-mode fiber, and the narrower active layer compared to the surface-emitting style has a smaller capacitance, which allows higher bit rate operation. The low cost and improved temperature characteristics of the edge-emitting LED compared to the LD has stimulated a lot of research to improve the devices so that they can be used in interexchange (junction) routes.

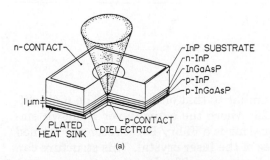

(a)

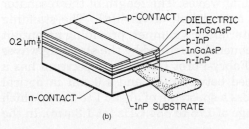

(b)

Figure 6.23 Structure and emission modes of a light-emitting diode. (*a*) Surface-emitting type; (*b*) edge-emitting type. (*Reproduced with permission from IEEE, © 1988, Nakagami, T., et al., IEEE Communications Magazine, vol. 26, Fig. 3.*)

For an LED to achieve a reasonable transmission distance (more than 10 km) at high bit rates (more than 565 Mb/s), single-mode fiber, operating at the zero dispersion wavelength, must be used. Multimode fiber operation significantly reduces the bit rate–distance product.

Laser diode.   The LD has evolved extremely quickly over the past decade. The development of the LD is central to the present-day long distance capability of optical telecommunications. The electrical characteristics of an LD are illustrated in Fig. 6.24. When the current density within the active region of the diode reaches a certain level, the optical gain exceeds the channel losses and the light emission changes from spontaneous to stimulated (i.e., lasing). The threshold current at which this occurs is quite low in the double heterostructure semiconductor lasers and is typically 50 to 150 mA. Figure 6.24 also shows how the threshold point shifts with temperature. This is a very undesirable characteristic because it means that the drive current must be increased as the temperature increases in order to maintain a constant output power. The internal power dissipation within the diode itself contributes to an increase in temperature, so a runaway situation can occur if some form of temperature control is not used. Also note how aging deteriorates the laser performance. Furthermore, the wavelength of the optical output is also temperature dependent. To counteract these problems, the LD is usually mounted on a Peltier-effect thermoelectric cooler with a feedback circuit to stabilize the temperature, and another circuit is included to maintain a constant drive current.

There are two major types of LD:

1. Fabry-Perot

2. Distributed feedback

The fabrication of the LD is similar to that of the LED. The main difference is clarified in Fig. 6.25, where the active layer is shown embedded in an optical resonator. This is a Fabry-Perot resonator formed from two opposite end-surfaces of the laser crystal. This structure can support multiple optical standing waves. The length of the resonator is an integer multiple of the optical half-wavelength. The standing waves add in a constructive interference manner, providing an output spectrum that has multiple frequencies defined by the standing waves as shown in Fig. 6.25a. This Fabry-Perot LD output spectrum has a number of spectral lines spaced between 0.1 and 1.0 nm in optical wavelength and spreading across a spectrum of about 1 to 5 nm, which corresponds to a frequency range of 176 to 884 GHz (at 1.3 μm). In the

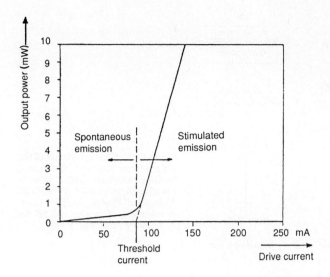

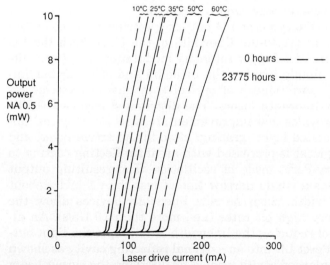

Figure 6.24 Characteristics of the laser diode. (*Reproduced with permission from Ref. 145, © 1988 IEEE.*)

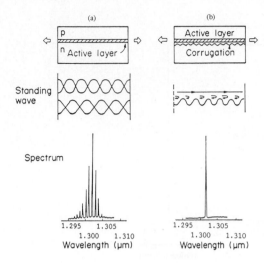

Figure 6.25 Characteristics of the Fabry-Perot (*a*) and distributed feedback (*b*) laser diodes. (*Reproduced with permission from IEEE, Nakagami, T. et al., © IEEE Communications Magazine, vol. 26, no. 1, January 1988, Fig. 2.*)

literature the term *linewidth,* ω, is often used; it is the width of the output spectrum at the 3-dB power points. For the Fabry-Perot LD ω is about 2 nm, which, in terms of frequency, is equivalent to more than 350 GHz at a center wavelength of 1.3 μm.

The width of the Fabry-Perot LD output spectrum is considerably narrower than the about 40-nm linewidth of the LED. Both the LD and LED devices are suitable for intensity modulation (i.e., transmitting 1s and 0s by simply turning the diode on and off). Further improvements in the performance of optical transmission systems require very narrow linewidth lasers. The distributed feedback (DFB) LD of Fig. 6.25*b* provides that improvement. It is achieved by the fabrication of a corrugated layer (grating) close to the active layer, and the light-emitting facet is processed with an antireflecting coating to suppress the Fabry-Perot mode of oscillation. The resulting output power spectrum has a single narrow line of less than 1 MHz (about 5×10^{-6} nm) in width. Such narrow linewidth devices allow the transmission of very high bit rates (i.e., more than 10 Gb/s). An alternative method of reducing the linewidth is to couple the light output from a Fabry-Perot LD into an external reflecting cavity as shown in Fig. 6.26. The enlarged cavity formed by directing the output beam to a diffraction grating about 20 cm from the laser and reflecting it back to the laser is a resonant structure. This technique can produce a linewidth of a few kilohertz.

Finally, there are on-going laboratory research efforts being directed toward reducing the linewidth and increasing the output power of semiconductor laser diodes. Without going into detail, an example of a recent success is the "corrugated-pitch-modulated, multiple quan-

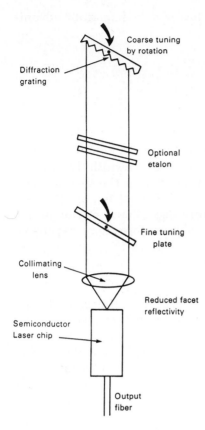

Coarse tuning
by rotation

Diffraction
grating

Optional
etalon

Fine tuning
plate

Collimating
lens

Reduced facet
reflectivity

Semiconductor
Laser chip

Output
fiber

Figure 6.26 The external cavity semiconductor laser. (*Reproduced with permission from IEEE, Stanley, I. W., © IEEE Communications Magazine, vol. 23, no. 8, August 1985, Fig. 5.*)

tum well, distributed feedback laser." A 170-kHz linewidth at 25-mW output power (late 1990) indicates the fast pace at which the field is advancing.

Summary for the LD and LED sources. The LED has the following advantages when compared to the LD:

- Higher reliability
- Simpler drive circuit
- Lower temperature sensitivity
- Immunity to reflected light
- Low cost

These characteristics make the LED suitable for short- to medium-distance applications. They are particularly attractive for LANs and subscriber loops where economy is a very important factor.

When compared to the LED, the LD has several important advantages:

- High output power
- High coupling efficiency
- Wide bandwidth
- Narrow spectrum

Figure 6.27 compares the output spectra for the LED, Fabry-Perot LD, and DFB LD: It is clear that the narrower the linewidth, the higher the bit rate that can be transmitted and the greater the transmission distance before dispersion causes serious ISI. One of the main disadvantages of LDs is the large temperature dependence of the output power. Also, they are more complex and therefore more expensive than LEDs.

6.5.2 Light detectors

The light emerging from the end of an optical fiber link must be detected and converted into electronic pulses for further processing so that the transmitted information can be received. There are two types of detector: (1) APD and (2) PIN diode. The progress of the performance of these two types of optical detector has been following a "cat and mouse" chase over the past decade with both devices now having multigigabits-per-second capability.

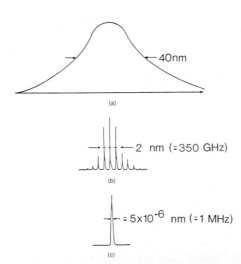

(a)

2 nm (≈350 GHz)

(b)

$\approx 5 \times 10^{-6}$ nm (≈1 MHz)

(c)

Figure 6.27 Spectrum comparison of semiconductor light sources. (*a*) Light-emitting diode; (*b*) Fabry-Perot laser diode; (*c*) distributed feedback laser diode.

40nm

APD. The early APDs were made from silicon and were used extensively in the 850-nm optical systems. Since silicon is effectively transparent at wavelengths greater than 1100 nm, another material was necessary for 1300- and 1550-nm systems. Germanium fulfilled the next generation of APD device requirements operating up to 140 Mb/s. It was soon realized that gigabit-per-second data rates would rapidly become a reality in the 1990s, and a new class of APDs was investigated using III–V semiconductor compounds such as GaAs, InP, GaInAs, InGaAsP, etc. Although such APDs operate at a few gigabits per second, the hunger for ever-improving performance has led to other combinations of material such as $Al_xGa_{1-x}Sb$ for use in the 10-Gb/s region.

The operation principle of the APD can be described with assistance of the simplified diagram of Fig. 6.28, which shows the APD is a semiconductor diode structure having a p^+-doped region, followed by an n-doped region, followed by an n^+-doped region. There are many variations on this basic structure incorporating other doped layers of III–V compound semiconductors, but they all operate essentially as follows. The diode is negatively biased with a voltage in excess of 100 V. When light from a fiber is incident on this diode, electron-hole pairs are generated. If the applied electric field is strong enough, accelerated free electrons generate new electron-hole pairs, and the process of multiplication continues, producing an avalanche effect. For each incident photon, many electron-hole pairs can be generated. As the multiplication factor is increased, the S/N decreases, so the multiplication factor chosen should not be too high. Typical values of multiplication factor for low-noise operation are 10 to 20 (10 to 13 dB), using a reverse bias of a few tens of volts to over 100 V. Furthermore, the multiplication factor is temperature dependent, which means some form of compensation is required to stabilize the device. In addition, the high reverse bias causes a small current to flow even in the absence of

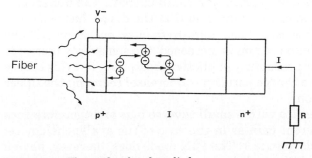

Figure 6.28 The avalanche photodiode.

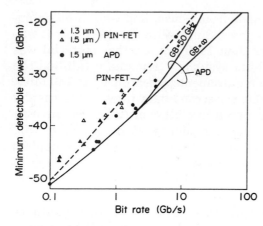

Figure 6.29 Comparison of APD and PIN receiver sensitivities. (*Reproduced with permission from IEEE, Nakagami, T. et al., © IEEE Communications Magazine, vol. 26, no. 1, January 1988, Fig. 5.*)

incident light. This so-called dark current is undesirable and must be minimized since it limits the minimum detectable received power.

The most important performance characteristic of an optical detector forming the receiver of an optical transmission system is its *receiver sensitivity*. This is defined as the minimum received power that will produce a BER of 10^{-9} at a particular bit rate. Figure 6.29 shows calculated and measured receiver sensitivities for direct detection using APDs and PIN-FETs. As expected, at the higher bit rates, the minimum detectable power increases (gets worse).

PIN diode.   Again, silicon and germanium were used in the early PIN device designs but are being superseded by III-V semiconductors. As illustrated in Fig. 6.30, the device is basically a *pn* junction with an intervening "intrinsic" region usually known as a charge depletion layer. When light from an optical fiber is incident on the *p* region of the reverse biased diode, electron-hole pairs are generated in the depletion layer. The electric field causes the electrons and holes to travel in opposite directions and so produce a small current. The dimensions of the depletion region can be chosen so that the device has good sensitivity and short rise time (i.e., high-frequency operation). If the thickness of the depletion region is increased, the sensitivity increases because of the higher probability of photon absorption, but this causes the travel time of the charge carriers to increase, reducing the upper frequency of operation.

The PIN diode operates with a small reverse bias and therefore does not contain any inherent gain as in the case of the avalanche device. Its maximum possible gain is 1. The PIN diode does, however, have a wide bandwidth capability well in excess of 10 GHz. This means that in order to take full advantage of this wideband characteristic, a low-

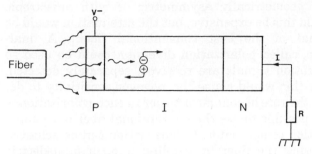

Figure 6.30   The PIN photodiode.

noise amplifier has to be placed after the PIN diode. The amplifier device originally used was the GaAs FET, but this has recently been superseded by the HEMT. This relatively new semiconductor device has lower noise and higher gain characteristics than the FET in the microwave region of operation (i.e., 1 to 100 GHz).

The minimum detectable power (receiver sensitivity) of the PIN-FET diode is shown in Fig. 6.29, as a comparison with the APD. The performance comparison of a PIN diode and an APD is a little obscure in that the APD has inherent gain whereas the PIN diode does not. Gain-bandwidth product (GB) is a necessary term to use in connection with APDs since the receiver sensitivity deteriorates rapidly above a few gigabits per second for low GB values. APDs have now been fabricated with a GB in excess of 100 GHz.

In conclusion, it is fair to say that the rate of improvement in the performance of optical photodetectors is keeping up with and perhaps ahead of the bit rate and bandwidth requirements for telecommunications systems.

6.5.3   Polarization controllers

A perfectly symmetric, circular, monomode fiber allows two orthogonally polarized fundamental modes to propagate simultaneously. In reality, the fiber is not perfectly circular or symmetric, since manufacturing irregularities cause the geometry and internal strain to be slightly imperfect. These imperfections make the power in each mode unequal. The polarization stability is such that the power in each of the two modes does vary over a period of time, but this is over minutes or hours and not just a few seconds.

For a satisfactory communications performance, it is necessary for only one mode to be present during optical transmission. Although polarization-maintaining fiber may be cheaply fabricated in the future, other methods are presently used. There are several techniques that can be applied to achieve this goal. The fibers could be designed specifically slightly geometrically asymmetric or with anisotropic strain. Not only would this be expensive, but the attenuation would be at least double that of the best conventional fibers. A dual-polarization receiver, called polarization diversity, could be used so that the two polarization signals are received, separately detected, and then added, but this would cause the receiver sensitivity to degrade by 3 dB. Since the rate of change of power in each polarization is relatively slow, it is possible to use the conventional fiber in conjunction with a polarization compensator. Compensation can be achieved by physically squeezing the fiber or winding it on a piezoelectric drum. However, the more attractive method is to use an opto-electric technique. Figure 6.31 shows a lithium niobate ($LiNbO_3$) polarization controller. First, the phase shifter adjusts the phase shift between the TE and TM modes so that they are 90° apart. The mode converter then increases the ratio of power in the TE compared to the TM mode, after which the second filter restores the phase relationship between the two modes. The polarization converter of Fig. 6.31 is fabricated on "Z-cut" $LiNbO_3$, and a voltage of about 90 V is required to produce greater than 99 percent mode conversion. On X-cut $LiNbO_3$, the operating voltage of the TE-to-TM mode converter can be reduced to about 10 V because of the increased overlap between the optical and electrical fields. Furthermore, interdigitated electrodes can improve the mode conversion efficiency compared to finger electrode structures of Fig. 6.31. The $LiNbO_3$ crystal style of mode converters is attractive

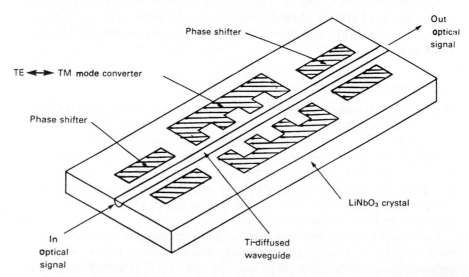

Figure 6.31 The polarization controller. (*Reproduced with permission from IEEE, Stanley, I. W., © IEEE Communications Magazine, vol. 23, no. 8, August 1985, Fig. 8.*)

because they can be easily incorporated with other devices onto a single crystal to form opto-electronic integrated circuits.

Other polarizers are being investigated in the form of depositing thin metal films such as chromium onto a fiber close to its core. This allows the propagation of what are rather exotically called *surface plasmon-polaritons*. Low-loss, high-conversion efficiencies (or extinction ratios) can be achieved by this technique.

6.5.4 Amplifiers

There are two main categories of optical amplifier: (1) semiconductor amplifiers and (2) doped fiber amplifiers. In either case, the objective is to increase the repeater spacing for a system that is limited by fiber attenuation. If the distance is limited by dispersion, that is a different problem, which will be addressed later. Both types of amplifiers can be used as in-line optical amplifiers (repeaters) to boost the optical signal without resorting to the regenerative style of repeater that requires the conversion of the optical signal to an electrical signal, pulse regeneration, and then conversion back to optical. Furthermore, the semiconductor amplifier can also be used as a preamplifier to boost the signal just prior to detection in the receiver. This technique increases the receiver sensitivity and therefore the repeater spacing. The application of amplifiers now gives a second definition to the term *repeater*. The optical amplifier is a repeater, but it is a much simpler, smaller, and cheaper repeater than the regenerative repeater. The optical amplifier does not eliminate the need for regenerative repeaters; it just increases the distance between them. This is because the optical amplifier introduces noise into the system, as all amplifiers do. However, the number of amplifiers that could be used before the noise causes an excessive BER problem translates to an unregenerated distance of thousands of kilometers. Of course, the system would be dispersion limited long before such a transmission distance were reached. Even the dispersion limit is now being challenged by the *soliton* transmission described in Sec. 7.5.2.

Semiconductor amplifiers. An LD used as a light source is simply an amplifier with enough positive optical feedback to cause oscillation. This feedback is provided by making reflective facets at each end of the semiconductor chip. So, by removing the feedback using facets with anti-reflecting coatings, the laser oscillator can be converted to an optical amplifier. Two types of semiconductor amplifier can be made, depending upon the reflectivity of the coatings on each facet. First, when the facet reflectivities are lower than for a laser oscillator, but still allow some light to be reflected back into the active region,

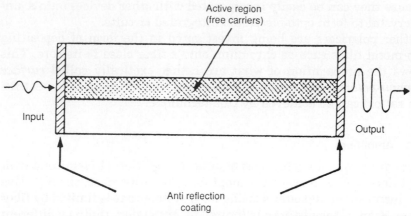

Figure 6.32   The semiconductor optical traveling-wave amplifier. (*Reproduced with permission from Ref. 265, © 1989 IEEE.*)

the amplifier is called a *resonant*, or *Fabry-Perot, amplifier*. Second, if the facet reflectivities are very low, light entering the device is amplified by a single pass, and the amplifier is called a *traveling wave amplifier* (Fig. 6.32). In practice, even with the best anti-reflective coatings there is a small amount of facet reflectivity, which means that most semiconductor amplifiers operate somewhere between a Fabry-Perot and traveling wave amplifier. Gain values in the 25- to 30-dB range are readily achievable, with an output power in excess of 5 dBm for the traveling wave amplifier. The single pass gain is different for each of the TE and TM polarization modes, implying that polarization-controlling devices may be necessary for this type of amplifier. The gain difference between polarization states is worse for the Fabry-Perot amplifiers than for the traveling wave amplifiers. The traveling wave amplifier is expected to have wider future application than the Fabry-Perot amplifier.

Doped fiber amplifiers.   This type of amplifier is fabricated using a length of conventional single-mode silica fiber, doped with a rare earth metal such as erbium (Er). A concentration of only 100 ppm is sufficient to provide high gain. If an LD is coupled into the fiber, it acts as a "pump" to increase the energy (i.e., signal level) of an incoming signal as it travels along the fiber. The pump light excites the Er atoms to higher energy levels than their normal state, so when the signal light encounters the Er atoms, stimulated light emission causes the signal power to gradually increase as the signal travels along the fiber. Figure 6.33 shows the typical construction of an Er-doped amplifier. The signal to be amplified is at 1.56 μm and passes through

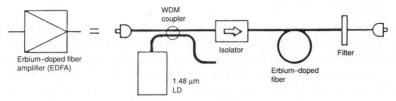

Figure 6.33   The Er-doped fiber optical amplifier. (*Reproduced with permission from Ref. 220, © 1990 IEEE.*)

the direct path of a 3-dB coupler to the 15 m of Er-doped fiber. Experiments have been made with various lengths of doped fiber up to more than 100 m. The 1.48-μm DFB laser diode is the pump source which is launched into the coupler at a power level of about 30 to 50 mW. An internal gain of about 25 dB is established between the low input level of −40 dBm and the amplifier output level of −15 dBm. In practical systems, isolators are usually used after each amplifier to stop any reflections back into the amplifier from causing low-level lasing action. This would cause noise to be superimposed on the signal, thereby degrading the BER. Also there is loss in the coupler and output filter. These losses can be up to about 5 dB, which gives a fiber-to-fiber gain of about 20 dB compared to the internal gain of about 25 dB. Depending upon the Er concentration, doped fiber length, pump power, etc., the early amplifier designs started to saturate with an input power of −15 dBm to 0 dBm.

If the Er-doped fiber amplifier pump source is at a lower wavelength, less than 1 μm, a lower noise figure and higher gain can be achieved. Not only does the gain increase with pump power, but also the gain saturation level increases. Recently (mid 1992), about 150 mW of pump power at 0.98 μm produced a doped fiber amplifier with a gain of about 40 dB which saturated close to 20 dBm.

At present, both Er-doped and semiconductor amplifiers operate over a relatively narrow band compared to the 1.25- to 1.35-μm and 1.45- to 1.65-μm optical bands available. For example, a 1.5-μm Er-doped amplifier design would typically be operable over about 1.53 to 1.56 μm (1991). An amplifier with a wider band is needed for high channel capacity WDM (see Sec. 7.3.1). Since Er-doped and semiconductor amplifiers are in their infancy, it can be anticipated that the bandwidth, gain, and saturation power level will all be gradually improved as the technology matures. Although many optical telecommunications system designers are already incorporating optical amplifiers into their new long distance system designs, it will probably be the mid 1990s before they are advanced enough for cheap, widespread implementation.

6.5.5 Modulators

Modulation of a light source can be done either by direct modulation of the dc current supplying the source or by using an external modulator following the source. The direct modulation method has the advantage of having fewer components than the external modulator. Direct modulation above 15 Gb/s is achievable but is restricted to noncoherent detection systems, whereas external modulators are readily suitable for coherent detection systems at very high bit rates. Also, for WDM, the direct modulation method requires a larger channel spacing than the external modulation method.

Direct modulation. The simplest form of direct modulation is to change the laser biasing current above and below the threshold value to turn the laser on and off to produce optical 1s and 0s. Unfortunately, this ON-OFF intensity modulation causes wavelength chirping (wavelength change with laser bias current) which results in a broadening of the optical spectrum relative to the information bandwidth. Unless the operating wavelength is very close to the zero-dispersion wavelength, severe system degradations occur. This type of modulation is clearly not suitable for WDM.

Advantage can be taken of the wavelength dependence of the laser bias current to frequency modulate the laser. If the injection current is varied while the laser is operating above threshold, the wavelength can be changed by up to about 2 nm. This laser tunability can be used for FSK modulation at well into the gigabits-per-second rate. Alternatively, the laser tunability could be used to create a multi-wavelength packet-switching system together with simultaneous intensity modulation. In these circumstances, modulating the bias current above and below the threshold value provides the intensity modulation and, by changing the value of the ON bias current, allows multiple channels to be created. Packets of ON-OFF data could consequently be distributed to several channels in a predetermined sequence. However, the frequency chirping associated with direct intensity modulation increases the linewidth for each channel. For multiple channel operation, this linewidth broadening defines the minimum channel spacing and therefore the total number of channels that can be used. A 2-nm tuning range allows at least 20 channels to operate, each at a several gigabits-per-second bit rate. Temperature effects, at present, tend to determine the maximum length of the packets.

Even multigigabits-per-second bit rate PSK can be accomplished by direct modulation, with a spectral width as narrow as the information bandwidth. Differential PSK has been used successfully without degradations due to thermal FM, by modulating the laser injection current with a bipolar signal. If an NRZ or RZ unipolar signal is used in the regular PSK mode, long strings of 1s or 0s cause large drifts in the optical frequency or phase due to temperature variations, resulting in severe distortion of the modulated signal. This problem is overcome by using a bipolar signal format. The NRZ signal is converted into a bipolar format because the bipolar pulses are better in terms of the thermal response of the laser. NRZ pulses cause a thermal FM within the laser, whereas the bipolar pulses are very short compared to the 1-μs thermal response time of the laser. Also, the polarity changes of the bipolar pulses allow a thermal improvement.

External modulation.   External optical modulators do not suffer the excessive chirp of direct intensity modulators, although a small amount of chirp still exists. For this reason considerable interest has been attracted to external modulators for application in high-bit-rate and long-haul optical telecommunication systems.

There are two types of external modulator presently in use which exploit the electro-optic property of certain materials. Lithium niobate has been established as a key material whose refractive index, and therefore phase, can be controlled by an applied electric field. As light propagates through the applied field region, it undergoes a phase change that is cumulative as it travels through the region. Figure 6.34 shows a simple lithium niobate modulator. The electrodes are

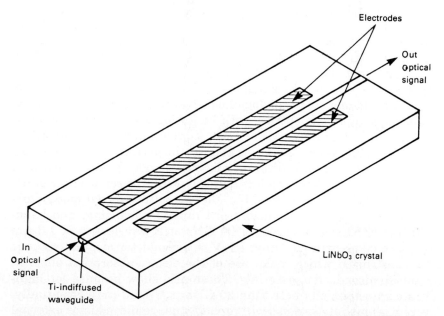

Figure 6.34   The lithium niobate phase modulator. (*Reproduced with permission from IEEE, Stanley, I. W., © IEEE Communications Magazine, vol. 23, no. 8, August 1985, Fig. 6.*)

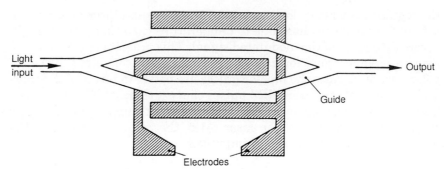

Figure 6.35 The Y-junction interferometric modulator. (*Reproduced with permission from Senior, J. M., Optical Fiber Communications, Fig. 11.24, © 1982, John Wiley & Sons.*)

about 2 cm long and the applied voltage is less than 10 V to produce a 180° phase change. By simply switching the voltage on and off a PSK modulation is achieved.

An intensity modulator can be formed by an interferometric technique. Figure 6.35 illustrates the modulator based on the Mach-Zender interferometer. This device has a Y-junction that splits the optical signal and a second Y-junction at a specific distance from the first which combines the two signals "in phase." Metal electrodes are arranged as indicated along the two arms of the interferometer. When an electric field is applied to the electrodes, because of the opposite field directions, a differential phase shift is obtained between the signals in each arm. The two signals are subsequently recombined with a 180° phase shift (i.e., in antiphase), giving the OFF state of the intensity modulator. A voltage of about 4 V is used to define the OFF state.

The lithium niobate technology has advanced to the stage where external modulators can operate with bandwidths in excess of 40 GHz. This should satisfy the needs of most telecommunication systems designers for many years to come. Although the performance is indisputably good, the material—lithium niobate—is not optimal because it does not allow complete circuit integration with the semiconductor devices such as lasers, detectors, etc. Electro-optic external modulators made from III–V semiconductors are rapidly advancing, and bandwidth performances in excess of 20 GHz at a wavelength of 1.3 μm have been demonstrated. Other III–V semiconductor modulators have been constructed which make use of the electro-absorption effect in certain semiconductor materials. These also show promise for high-bit-rate transmission (more than 20 GHz at λ = 1.3 μm). Eventually, future high-bit-rate systems will probably use semiconductor external modulators, monolithically integrated with laser diodes.

6.5.6 Couplers

Couplers are devices which tap off some of the signal power from the main transmission path, usually for power monitoring or feedback purposes. This definition implies that there are some special cases. For example, when 50 percent of the main path power is tapped off, the device could be called either a 3-dB coupler or a power splitter and combiner. This special-case coupler can be fabricated simply by forming a Y-junction (Fig. 6.36a), which produces a device which is inherently wideband and can be used across the present operation range of 0.8 to 1.6 μm. Such devices are often described as *achromatic*. The splitter and combiner of Fig. 6.36a is a very important component because it can be fabricated using the planar technology. In other words, standard photolithographic techniques are used to define the waveguide pattern which forms the Y-junction within a larger integrated circuit. This style of signal splitter and combiner can also be used for multiple splitting and combining by incorporating several Y-junctions as shown in Fig. 6.37.

The first optical couplers to be produced were discrete component devices. They had either a 3-dB coupling value or some other value and were constructed by taking two single-mode fibers and removing the glass down to the core over a small region and then fixing the two fibers together over this region. This has been called the proximity coupler in microwave circuitry. It produces a low-loss coupler, but its bulky size has led to the development of an integrated circuit component form of this coupler. This substrate fabricated device (Fig. 6.36b)

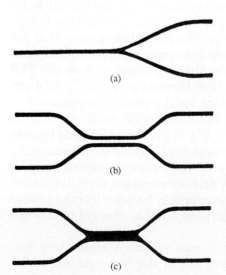

(a)

(b)

(c)

Figure 6.36 Various coupler configurations. (a) Y-junction coupler; (b) proximity coupler; (c) zero-gap coupler.

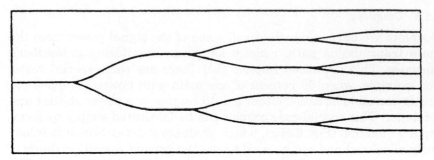

Figure 6.37   Multiple Y-junction splitter and combiner.

takes the same shape as the microwave equivalent, with the exception that the coupling region for the optical coupler is not a quarter wavelength as in the case of the microwave design. The proximity coupler is inherently wavelength dependent. Although this is a problem for some applications requiring wide bandwidth, it is a very useful, if not essential, characteristic for WDM applications (see Sec. 7.3). The coupling region can be symmetrical, in which case the wavelength transmission curve is periodic, or it can be asymmetrical, exhibiting a passband behavior. This wavelength (i.e., frequency) dependence illustrates the ability of the directional coupler also to act as a filter.

If the coupling region gap between the two waveguides of the proximity coupler is reduced, the coupling strength between the two waveguides increases. In the special limiting case of zero gap, the two individual waveguides become one two-mode waveguide (Fig. 6.36c). This device is called either the zero-gap coupler or the two-mode interference (TMI) coupler and has a periodic characteristic that is similar to that of the symmetrical coupler. The main difference is that the periodic passband peaks, which would correspond to the channel spacing when used for WDM, are less for the TMI coupler. Also the TMI coupler has the obvious advantage that there is no small, 3- to 4-μm gap to carefully control as is necessary for the proximity coupler.

When fabricated on an electro-optic material such as lithium niobate, with electrodes as shown in Fig. 6.38, the passbands become voltage tunable. This device has several applications such as modulators, switches, or WDMs. Tunability of WDM devices enables compensation for the fabrication tolerances and temperature sensitivity compensation. This is an important characteristic because a transmission wavelength shift can cause an increase in the channel-to-channel crosstalk.

Many high-speed LANs include a passive star coupler (see Sec. 7.5.3). Ideally, the input power to any one of the N inputs of a trans-

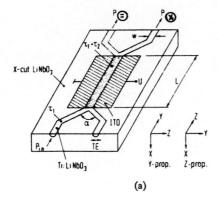

(a)

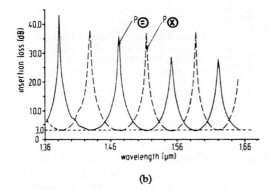

(b)

Figure 6.38 The TMI coupler (measured wavelength transmission $P_{\otimes}$, $P_{\ominus}$ of a dual-channel TMI device for TE polarization on X-cut, Y-propagating LiNbO$_3$). Fabrication parameters: $L = 12$ mm, w = 7 μm, $\tau_1 = 100$ nm, $\tau_2 = 40$ nm. Diffusion conditions: 1050°C, 10 h. (*Reproduced with permission from Ref. 258, © 1988 IEEE.*)

missive $N \times N$ star coupler is divided equally between its N outputs. Absorption and scattering detract slightly from this ideal. Figure 6.39*a* indicates how the 3-dB symmetric proximity coupler operates as a transmissive 2×2 star coupler and how this is the basic building block for higher-order star couplers. Figure 6.39*b* is a 4×4 transmissive star constructed from four 3-dB couplers.

6.5.7 Isolators

Isolators are essential components for optical communications systems. They are used at the output of LD sources to stop reflections from increasing the laser linewidth; 50 to 60 dB of isolation are required to ensure that reflections are adequately reduced. Such isolators have been fabricated using a bismuth-substituted iron garnet material which produces Faraday rotation. The classic operating principle of this isolator is to produce a unidirectional phase shift due to Faraday rotation so that any reflected signal is highly attenuated

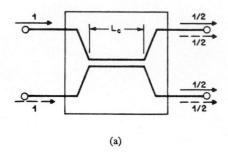

(a)

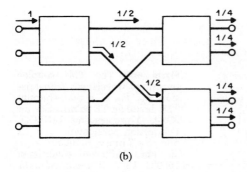

Figure 6.39 (*a*) A passive transmissive 2 × 2 star using a 3-dB coupler; (*b*) a transmissive 4 × 4 star from four 2 × 2 stars. (*Reproduced with permission from Ref. 234, © 1988 IEEE.*)

(b)

by anti-phase cancellation. Although these isolators produce the required isolation characteristics, they are relatively bulky devices which are not presently amenable to optical integrated circuits. Whether optical electronics will follow the path of microwave electronics by adopting the buffer amplifier for oscillator isolation remains to be seen.

6.5.8 Filters

Filters are essential components of all telecommunications systems, and optical systems are no exception. They are used in many places, in particular, in WDM and coherent receivers (see Sec. 7.2). There are two categories of filter: (1) fixed frequency and (2) frequency tunable. Fixed frequency filters are discussed in more detail in Sec. 7.3.

Frequency tunable filters. The important parameters of a tunable filter are channel bandwidth, channel spacing, and tuning range as described in the diagram of Fig. 6.40. Obviously, the channel spacing should be as small as possible to allow the maximum number of channels to be tunable within the tuning range. The channel spacing limitation is based on the necessity to minimize the crosstalk between

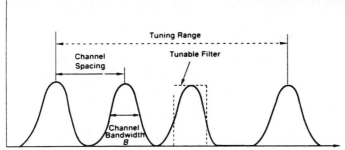

Figure 6.40 Tunable optical filter characteristics. (*Reproduced with permission from Ref. 223, p. 54, © 1989 IEEE.*)

channels, which entails keeping the crosstalk penalty to less than 0.5 dB. The channel spacing is determined by the transmitter source linewidth, the modulation scheme, and the filter lineshape and can vary from about 3 to 10 times the channel bandwidth, B.

Tuning speed, the time taken to tune between any two frequencies (channels), is another parameter of concern for tunable filters. In a video broadcast system, the random selection of video channels in a subscriber television set can be done in a tuning time of several milliseconds. Although the tuning of multiaccess data networks and cross-connect systems is done rather infrequently, the tuning speed required is in the microsecond time frame. For packet switching of data having packet lengths of a few microseconds, tuning speeds of less than 100 ns are required.

There are several types of frequency tunable filters. *Fabry-Perot* interferometric filters have been in existence for a long time. They work on the principle of partial interference of the incident beam with itself to produce periodic passbands and stopbands. Tuning requires mechanical movement of mirrors, so wavelength selection is relatively slow (i.e., more than milliseconds). Also, the Fabry-Perot filter tuning range is limited to about 10 percent of the center frequency. Typical single-stage filter insertion loss is about 2 dB.

Mode coupling tunable filters using acousto-optic, electro-optic, or magneto-optic effects have produced some useful tuning characteristics. As Fig. 6.41 indicates, the light enters the filter in the TE mode and is converted to the TM mode using either acoustic, electric, or magnetic fields. The mode coupling conditions are satisfied only by optical signals that have a very narrow wavelength. The two propagating modes are separated by a polarizer. The filter output has a bandpass characteristic without the inconvenience of multiple passband outputs generated by the Fabry-Perot filter. The filter output bandwidth is relatively wide (i.e., about 1 nm for both the acousto-optic and electro-optic filters). The acousto-optic filter can be tuned

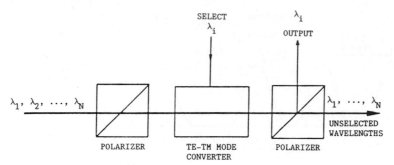

Figure 6.41 Mode coupling tunable optical filter. (*Reproduced with permission from Ref. 223, p. 57, © 1989 IEEE.*)

over the full 1.3- to 1.6-μm range, whereas the electro-optic filter has less than a 20-nm tuning range. The acousto-optic filter therefore has a much higher channel capacity than the electro-optic filter (i.e., several hundred versus about 10, respectively). Both types can be tuned relatively fast. Furthermore, the acousto-optic filter has a unique property. If multiple acoustic waves are applied to the interaction region, multiple passbands can be simultaneously and independently created. This allows multiple channel selection, which may have some interesting future applications. The electro-optic filter, however, is faster than the acousto-optic filter, with tuning speeds in the nano- and microsecond ranges, respectively.

Semiconductor laser structures form the third main category of tunable filters. Biasing a resonant semiconductor laser structure below its lasing threshold produces a resonant optical amplifier, with reasonable selectivity. While this would be a problem for wideband amplifier designers, it is very convenient for tunable filters. The Fabry-Perot laser structures have multiple passbands whose center frequencies are periodically spaced, depending upon the physical dimensions of the cavity. The passband spacing, known as the free spectral range (FSR), defines the maximum range over which incoming signals can be tuned. A typical 300-μm-long Fabry-Perot laser would have a FSR of about 150 GHz, or about 1 nm of tuning range. Since this is a very limited range, DFB laser structures are mainly used for tunable filters. In single electrode DFB amplifiers, the passband narrows and the gain increases as the bias current is increased toward the threshold value. In a multielectrode DFB amplifier, the resonant frequency and gain can be controlled independently.

The frequency tuning in both Fabry-Perot and DFB semiconductor filters is produced by the change of injection current as the bias is varied. The injection current determines the carrier density, which subsequently defines the refractive index of the region. The refractive index decreases with increasing carrier density. The resonant wavelength, λ, is proportional to the refractive index, so that $m\lambda = 2nL$, where m is an integer, n is the refractive index, and L is the grating pitch for the DFB structure or cavity length for the Fabry-Perot structure.

The DFB filter tuning range of about 4 nm is still relatively small compared to other filter techniques. The width of the filter passband is less than 0.05 nm. This allows a capacity of a few tens of channels. The tuning speed is very good (i.e., in the nanosecond region). Also, the filter has a more than 10-dB gain instead of a 2- to 5-dB loss.

Polarization control or polarization-maintaining fiber is needed since the gain for the TE orientation is about 10 dB more than for the TM orientation. Also, the input signal power level must be constant; otherwise the resonant frequency (filter passband) and gain are affected. There are additional problems, but one must remember that semiconductor tunable filters are in their infancy, and it can be anticipated that they will excel in the future as more research is devoted to improving their characteristics.

6.5.9 Photonic switches

Exchanges throughout the world are undergoing an upgrade from the slow electro-mechanical switches to the faster and more reliable electronic, stored program controlled (SPC) switches. When these switches are used in conjunction with optical fiber transmission systems, there must always be a transformation of the signal from electronic to optical and vice versa in order to perform the switching in the exchanges. As wideband connection to the subscriber becomes increasingly attractive and eventually becomes expected, the practice of switching at the optical level instead of at the electrical level will be essential. Optical switches will also be necessary for optical packet switching to produce time domain multiple access (TDMA) systems. Subpicosecond switching times are possible by switches that are activated by an optical signal. In a different but closely related field of technology, computers promise to expand in power and speed capability by using optical switching. Optical computing substitutes photons for electrons, which travel through free space 1000 times faster than electrons travel through electronic media. Light beams replace wires, and information is transmitted in three dimensions. Since light beams do not interact with one another, multiple signals can propagate along the same path, increasing the information transfer by at least 10^6.

At present, the most useful application of photonic switching is for LANs utilizing the star coupler arrangement of Sec. 6.5.6. Optical cross-connects could already find a useful application for rerouting multiplexed traffic or for use in optical protection switching in a $1 \times N$ protected optical transmitter system. The electro-optic directional coupler provides an adequate switching element for such purposes.

6.5.10 Solid-state circuit integration

The concept of integrating several optical devices and electronic circuits was first proposed in the early 1970s. It was not until the early 1980s that the technology on several fronts had matured enough to produce circuits with eye-catching performances. Monolithic integration of a variety of optical devices is called photonic integrated circuits (PICs), and they have only recently (late 1980s) started their technological advancement. The development of methods to produce very precise dimensions of semiconductor layers has been a key to this rapid advancement. The evolution of semiconductor growing techniques such as metal organic vapor phase epitaxy (MOVPE) and chemical beam epitaxy (CBE) has provided indispensable tools for PIC fabrication. In addition to optical device integration, for a complete system such as a coherent optical receiver, it is also necessary to integrate high-performance electronic devices such as laser drive circuits or HEMT amplifiers on the same substrate (i.e., monolithically). The resulting circuits are called opto-electronic integrated circuits, or OEICs.

Monolithic integration has already had phenomenal success for low-frequency electronics. The benefits are well known. The main obvious manufacturing improvements derived are lower cost, higher reliability, higher productivity, compactness, and simple assembly. Another very desirable quality is the small deviation in design characteristics from one circuit to another. Also, circuit performance is enhanced by minimizing parasitic capacitances and inductances.

Lightwaves and microwaves are both electromagnetic waves, the only difference being the frequency. So, it is not surprising that some of the techniques used in designing microwave integrated circuits can be applied to optical integrated circuits. Optical components such as couplers, modulators, phase shifters, switches, etc., which are the necessary building blocks for an optical telecommunications system, have been designed and constructed using the electro-optic material $LiNbO_3$, as described earlier. When each component is fabricated individually and then connected into a system, there are many losses incurred by the waveguide-to-fiber transitions and connectors at the input and output of each component. If many of the components are fabricated on a single $LiNbO_3$ substrate, losses are reduced substantially.

The $LiNbO_3$ material technology is probably an interim solution for optical integrated circuits. Eventually, as technology progresses, the more complex semiconductor materials such as GaAs/GaAlAs and InP/InGaAsP will probably be used not only as the device materials but also as the substrate for monolithic circuit integration. Much research work is currently being devoted to III–V semiconductor, and multiple-quantum-well (MQW) monolithic optical integrated circuits. A MQW is the solid-state physicist's terminology for a device constructed by sandwiching a low-band-gap material such as GaAs between two layers of high-band-gap material such as GaAlAs to produce what is called a thin heterostructure.

In summary, opto-electronics is progressing along the same road as the lower-frequency electronics industries by moving from discrete components to integrated circuits. At present, the optical technology is passing through the hybrid integrated circuit phase en route to the MIC maturity. Although the technology is still new, very rapid development is taking place.

An example of the present technology using (1) the lithium niobate substrate material and (2) the semiconductor substrate, will now be presented.

Lithium niobate integrated circuits. The $LiNbO_3$-technology-based opto-electronic integrated circuits had several years head start on the semiconductor-based PICs. This is because materials technology, which was blossoming in the 1980s, was still not mature enough by about 1985 to produce the excellent PIC performance required for long distance communications systems. During the early-to-mid 1980s, the more well-understood $LiNbO_3$ material technology received considerable research attention, which produced quick results, enabling it to provide the first commercially available optical integrated circuits in the late 1980s.

Optical waveguides are formed at the surface of a $LiNbO_3$ substrate by evaporating or sputtering titanium strips onto the surface of the substrate. The evaporated strips are typically 600 Å thick and 5 μm wide. The waveguide dimensions are then defined by standard photolithographic techniques. The titanium is then diffused into the $LiNbO_3$ substrate by heating it to about 1050°C for 8 h. The waveguide is formed by the diffused titanium changing the refractive index by a small amount in the diffused region compared to the bulk material. The resulting waveguides are relatively low loss (≤ 0.3 dB/cm at 1.3 μm).

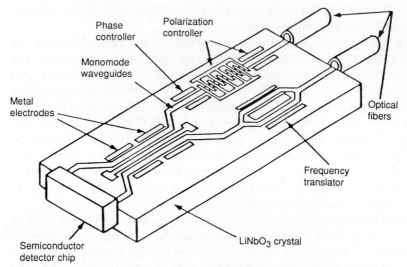

Figure 6.42 LiNbO$_3$ integrated circuit example. (*Reproduced with permission from IEEE, Stallard, W. A., et al., Integrated Optic Devices for Coherent Transmission, © IEEE LCS, July 1986, Fig. 3.*)

Figure 6.42 illustrates the combination of the optical waveguides on LiNbO$_3$ with several components to form the main circuitry of a coherent optical receiver. Two optical fibers feed in the received signal and local oscillator signal, respectively. The received input signal passes through the polarization controller, and the local oscillator signal passes through a frequency translator. The two signals are then mixed and split at the coupler. The resulting signals are detected by the semiconductor chip photodetectors, ready for bit stream regeneration. Notice that this LiNbO$_3$ integrated circuit does not contain the local oscillator or the photodetectors as part of the planar circuit fabrication process. They are discrete semiconductor devices which are mounted as extra chips or fed into the LiNbO$_3$ circuit using an optical fiber. By the low-frequency definition, this is a hybrid integrated circuit and not an MIC.

Semiconductor integrated circuits. It was only by about 1990 that semiconductor PICs and OEICs started reaching a satisfactory level of performance for serious commercial viability. It may be 1995 or later before a substantial quantity of these circuits is in widespread use in optical telecommunications systems.

GaAs is a well-developed material technology, but its use as a PIC substrate material is limited to the short optical wavelengths of operation (i.e., in the 0.8-μm region). The less-well-established InP material is the suitable substrate material for PIC operating at 1.3 and 1.5 μm.

Figure 6.43 is a diagram of a late-1990, laboratory-fabricated, balanced heterodyne receiver PIC. It contains a continuously tunable local oscillator, an adjustable 3-dB coupler, and two MQW waveguide detectors monolithically fabricated on an InP substrate. The MOVPE fabrication process was used, and as Fig. 6.43 illustrates, the multilayer structure is basically planar but takes on a three-dimensional character because of the various different layers used for each component. Low-loss transitions between the components can be achieved by selectively etching the layers at the interface regions between components.

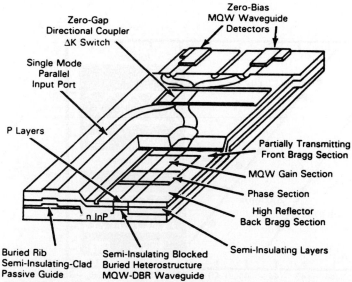

Figure 6.43 Semiconductor PIC example. (*Reproduced with permission from IEEE, Koch, T. L. et al., Photonics Integrated Circuits: Research Curiosity or Packaging Common Sense?, © IEEE LCS, November 1990, Fig. 4.*)

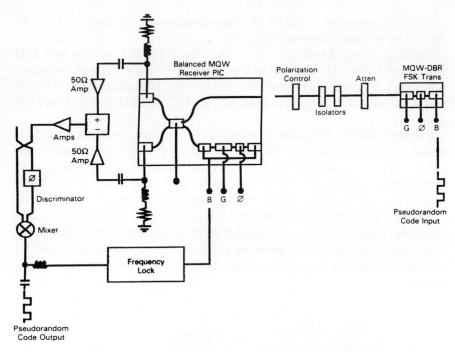

(a)

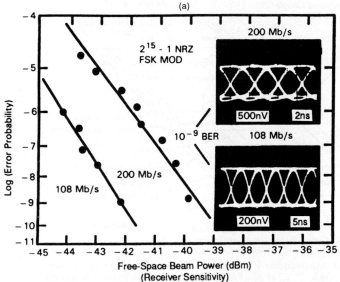

(b)

Figure 6.44   (*a*) Semiconductor PIC evaluation (*reproduced with permission from IEEE, Koch, T. L. et al., Photonics Integrated Circuits: Research Curiosity or Packaging Common Sense?, © IEEE LCS, Nov. 1990, Fig. 6*). (*b*) Semiconductor PIC performance (*reproduced with permission from IEEE, Koch, T. L. et al., Photonics Integrated Circuits: Research Curiosity or Packaging Common Sense?, © IEEE LCS, Nov. 1990, Fig. 8*).

To give an indication of the performance of this PIC, Fig. 6.44a shows the chip incorporated into a narrowband FSK receiver, with an automatic local oscillator frequency lock. The transmitter in this case is an FSK-modulated three-section MQW laser. Using a $2^{15} - 1$ pseudorandom NRZ bit stream first at 108 Mb/s and then at 200 Mb/s produced the receiver sensitivities shown in Fig. 6.44b. The sensitivity of -39.7 dBm at 200 Mb/s shows the growing maturity of PICs.

Optical Fiber Transmission Systems

Optical fiber is gradually finding application in all aspects of telecommunication systems. The three broad categories, which relate to short, medium, and long distance, are referred to as:

1. Local loop (for subscribers) and LANs (commercial, etc.)
2. Intercentral office (or interexchange) traffic
3. Long haul (intercity traffic)

The distance and capacity (bit rate) are the primary factors which influence these systems designs and the associated economic viability of constructing and operating them. This chapter will address the various technical factors which need to be taken into account to realize each of the above three system categories.

First, the optical fiber transmission systems technology is evolving along two different paths:

1. Intensity modulated systems
2. Coherent systems

Figure 7.1 shows the historical evolution of optical fiber transmission systems over the past 20 years. The early first-generation systems were primarily limited in repeater spacing and bit rate by high fiber loss and excessive chromatic dispersion in the fiber caused by the use of LEDs. The development of 1.3-μm wavelength fiber systems produced the improved repeater spacing and bit rates of the second-generation systems. These systems were still operating with multi-

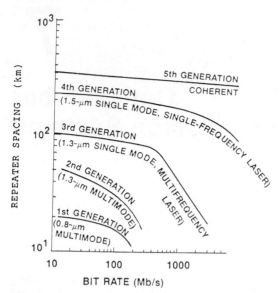

Figure 7.1 Five generations of optical fiber communication systems. (*Reproduced with permission from Ref. 285, © 1985 IEEE.*)

mode fibers, which limited the performance by interference between the propagating modes of the fiber (i.e., modal dispersion). The move to single-mode fibers operating at 1.3 μm gave the third generation a very impressive repeater spacing and bit rate performance. The fourth generation benefited from shifting the operating wavelength to 1.5 μm, which offers the minimum achievable attenuation point for silica fibers. So far, all of the systems described use the intensity modulation of the optical transmitter. The fifth generation uses phase modulation of the optical transmitter and a coherent detection scheme which improves the receiver sensitivity, thereby increasing the repeater spacing. The very narrow-spectrum distributed feedback diodes used in these systems allow very high bit rates of operation. In addition, the coherent optical communications systems allow FDM, which gives an astounding increase in the channel capacity because of the more efficient use of the available optical bandwidth. Not shown here is the sixth generation, which includes optical amplifiers, as discussed later.

The case in favor of using the coherent optical system in future designs appears to be clear-cut. In practice, there are many variables to take into consideration. First, what is the nature of the interconnection? As highlighted in Fig. 7.2, there are three categories of interconnection:

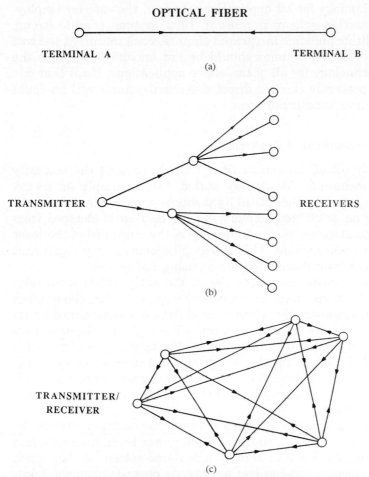

Figure 7.2 Optical fiber configurations. (*a*) Point-to-point: link (one transmitter and one receiver). (*b*) Point-to-multipoint: broadcast (one transmitter and many receivers). (*c*) Multipoint-to-multipoint: network (many transmitters and many receivers).

1. Point to point (link)
2. Point to multipoint (broadcast)
3. Multipoint to multipoint (network)

So far, the focus of attention has been on the point-to-point link, and it has been assumed that the ever-increasing hunger for more bandwidth inevitably leads to the necessity of coherent detection technology. Although coherent systems are technically more elegant than the structurally simpler direct detection systems, cost does not favor the coherent technology for all applications. In fact, the case for employing direct detection systems in present (1992) systems is quite strong. When monolithic photonic integrated circuits reach maturity, the cost of coherent detection systems should be low enough to make it the dominant technology for all of the above applications. Until that day arrives, the presently cheaper direct detection systems will no doubt continue to have widespread use.

7.1 Intensity Modulated Systems

Until recently all of the optical fiber links have used the intensity modulation technique. As already stated, this is simply an ON-OFF transmission, whereby the optical light source produces 1s or 0s by the source being on or off, respectively. As the current is changed from zero to its operation value, the frequency of the output from the laser changes by a small amount. This *chirping* becomes a very significant problem for coherent detection systems using PM or FM.

The intensity modulation schemes of the early 1980s were relatively narrow in bandwidth. The bandwidth and therefore the number of channels transmitted by intensity modulation is determined by (1) how fast the laser can be turned on and off and (2) the dispersion associated with the chirp within the laser. The appearance of the DFB laser revived the life of the directly (intensity) modulated approach. At 1.3 and 1.5 μm, systems have been demonstrated to operate well above 10 Gb/s using intensity modulation and over repeaterless spans in excess of 100 km.

The repeater spacing of an optical fiber transmission system depends mainly upon the transmitter output power level, fiber loss, and receiver sensitivity. For an intensity modulated system (ASK), which uses direct detection, the receiver sensitivity depends upon the minimum number of generated electron-hole pairs that can be translated into a detectable current. An ideal receiver would be able to observe the generation of a single electron-hole pair. For this ideal receiver one can calculate the minimum required pulse energy needed to achieve a specific error probability. This is called the quantum limit. Calculations show that the minimum number of photons that can be detected in an ideal direct detection receiver for a BER of 10^{-9} is 21. Unfortunately, the ideal receiver does not exist and early direct detection receivers had amplification following the photodetection, which added both thermal and transistor shot noise. These additional noise sources increased the received power level (average number of photons per second) to 13 to 20 dB above the quantum limit (i.e., 400 to 2000 photons per bit). However, the introduction of optical preamplification, with large gain values, prior to photodetection has been shown to improve the sensitivity to 38 photons per bit. In this configuration, the signal and amplifier spontaneous emission noise, which is approximately proportional to the gain, is dominant over other noise sources including the receiver thermal noise. FSK requires a higher number of photons per bit than ASK.

7.2 Coherent Optical Transmission Systems

In optical communications literature, the term *coherent* refers to any detection process that uses a local oscillator. Note that this differs from the radio literature definition of coherent detection, which specifically refers to the phase of the signal involved in the detection of the IF signal. Coherent optical transceiver systems have many similarities to their microwave transmission counterparts. For example, many present-day coherent optical system designs have a PM in the transmitter and a heterodyning technique in the receiver. The circuit block diagram of Fig. 7.3 highlights the main features of a coherent optical communications system.

In the coherent detection mechanism, a weak optical signal is mixed with a relatively strong local optical oscillator. In these circumstances, the calculation of receiver sensitivity is somewhat different from the radio system. Here, the electrical signal power is proportional to the product of the optical signal and local oscillator power, and the receiver shot noise is proportional to the sum of the optical signal and the local oscillator power. This means that with sufficiently high local oscillator power, the shot noise in the coherent detection system receiver will be large compared to the thermal and active de-

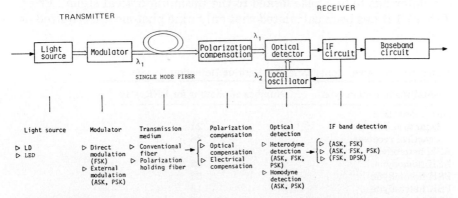

Figure 7.3 Schematic diagram of the coherent optical communication system with various configuration combinations. Homodyne $\lambda_1 = \lambda_2$, heterodyne $\lambda_1 = \lambda_2$. (*Adapted with permission from Ref. 254, © 1988 IEEE.*)

vice noise power of the receiver electronics. In comparison, for the direct detection system, the detector output power is proportional to the square of the incident optical signal power, and the receiver noise is mainly determined by the electronics following the detector. The coherent system should be able to closely approach the quantum limit performance.

Analyses have been made for several optical systems to establish which type of modulation produces the best receiver sensitivity performance. The results of these calculations are summarized in Table 7.1 in terms of the number of detector photons required to produce a BER of 10^{-9}. Figure 7.4 graphically shows how the receiver sensitivities degrade with increasing the bit rate. For coherent detection, these receiver sensitivities are the theoretical best possible (quantum-noise-limited) values. As a comparison, some practically measured receiver sensitivities are also plotted for direct detection. For these systems, there was no optical preamplifier incorporated prior to the photodetector.

As indicated in Fig. 7.4, PSK homodyne detection in the receiver is theoretically the best technique, and this is borne out by experiment. When used with PSK and specifically differential PSK in the transmitter, the best system performance is obtained. Good performance is achieved by using either the heterodyne or homodyne coherent detection technique in the receiver. The difference between the heterodyne and homodyne receiver is simply that the *heterodyne* receiver has a local oscillator whose frequency is *different* from that of the transmit frequency whereas the *homodyne* technique has a local oscillator whose frequency is *equal* to that of the transmit frequency. The homodyne technique theoretically produces a receiver sensitivity 3 dB better than heterodyne, but it is more complex to build since the local oscillator has to be phase locked to the incoming optical signal. From Table 7.1 it has been calculated that only nine photons are required at

TABLE 7.1   Receiver Sensitivities for Various Detection Schemes

Modulation/detection type	Number of photons for BER = 10^{-9}
Direct detection	
Quantum limit	21
Practical receiver	38
ASK heterodyne	72
ASK homodyne	36
FSK heterodyne	36
PSK heterodyne	18
PSK homodyne	9

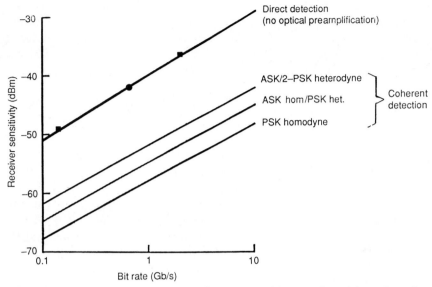

Figure 7.4 Graph of receiver sensitivity against bit rate for various detection schemes.

the receiver for perfect homodyne reception of 1 bit of transmitted information at a BER less than 10^{-9}, which is approaching the quantum limit.

When compared to intensity modulation with optical preamplification, the coherent system therefore has an improvement, but only about 6 dB at best. The question is, Is this alone sufficient to warrant the use of this more expensive technology? Today, the answer may be Yes only for long-haul very-high-capacity systems and probably No for all other systems. However, as technology improves and photonic integrated circuits mature, coherent optical systems will probably become the norm. The advantages of coherent systems reside not only in the improved sensitivity and therefore larger repeaterless distance but also in the numerous channels that can be multiplexed in an FDM style over the 25,000-GHz or so bandwidth available in the 1.45- to 1.65-μm band (Fig. 7.5). Within each channel, thousands of telephone channels can be established, which means that the capacity of this type of system is potentially enormous. The components required for coherent optical systems need considerably improved characteristics compared to intensity modulation systems.

The various possible combinations of system configuration for coherent optical fiber transmission are detailed in the block diagram of Fig. 7.3. Considering the *light source*, the laser diode FM-to-PM noise

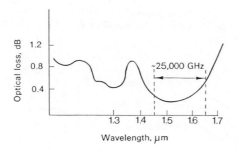

Figure 7.5 Bandwidth available for optical multiplexing.

must be minimized by making the linewidth as narrow as possible. Ideally it should have zero linewidth, which of course is impossible. The minimum acceptable linewidth requirements depend on the bit rate and the *modulation scheme,* as summarized in Fig. 7.6. This figure highlights the fact that the more sophisticated, higher-quality performance modulation schemes require as relatively small a light source linewidth to transmission bit rate ratio as possible. It is also desirable for the IF bandwidth to be equal to the bit rate, as is the case

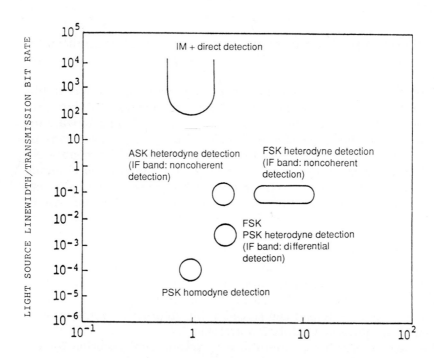

Figure 7.6 Linewidth requirements for various modulation schemes. (*Reproduced with permission from Ref. 254, © 1988 IEEE.*)

for homodyne PSK. Note that intensity modulation and direct detection is very good in this respect. The other modulation schemes such as heterodyne ASK, FSK, PSK, and homodyne ASK and PSK have also been included in this figure. Table 7.2 complements Fig. 7.6. The first systems to be investigated incorporated ASK, but these were soon superseded by the more sophisticated FSK and PSK systems.

The heterodyne ASK and FSK can tolerate an LD spectrum linewidth to bit rate ratio of 10^{-1}. In a 565-Mb/s system, for example, the required source linewidth of about 50 MHz is easily attainable with a basic DFB LD. The PSK homodyne requirement of 10^{-4} is a much more difficult specification to fulfill, especially at low bit rates. This requirement favors high-bit-rate systems because the allowable laser linewidth increases linearly with signal bit rate. For the 565-Mb/s system the source linewidth must now be about 50 kHz. This is a very narrow linewidth, but it can be achieved by present-day external cavity DFB LDs. At 5 Gb/s the linewidth can be relaxed to the more easily achievable value of about 0.5 MHz.

The *transmission medium* decision is whether to use conventional fiber or polarization holding fiber. If the polarization fiber can be purchased at a price and attenuation value not significantly higher than the conventional fiber, it is preferable. If conventional fiber is chosen, some form of *polarization compensation* is necessary, either at the optical or electrical level.

The *demodulation* obviously is the inverse of the modulation, with homodyne PSK being the best choice for high performance. An *IF band detection* is then necessary to convert the microwave IF signal down to the baseband bit steam. This can be done either by envelope detection or by the better-quality heterodyne detection.

Observing the modulation scheme in more detail, the most attractive technique, as indicated in Table 7.1, is the homodyne PSK detection system. It is theoretically superior to all other detection schemes and has considerable potential for future systems. An optical phase-locked loop (OPLL) is universally accepted as being essential for PSK homodyne detection systems. There are several problems associated with the practical implementation of PSK homodyne detection. First, the carrier component of the modulated transmit signal is suppressed, so what does the PLL lock on to? This suppression is evident by observing that in Fig. 7.7a the digital 1s and 0s give a net average value of zero. Second, for optimum homodyne detection, the data and the local oscillator signals must be locked in-phase (Fig. 7.7b). Unfortunately, a conventional PLL locks the local oscillator at 90° to the data (reference) signal, as in Fig. 7.7c. Third, the data signal and local oscillator signals are mixed at the photoconductor to produce the required beat frequency. The detector output current contains an addi-

TABLE 7.2 Comparison of ASK, FSK, and PSK Modulation Techniques

Modulation format	Optical modulation	Optical demodulation	IF band demodulation	Relative sensitivity	LD spectrum width/ transmission speed	IF bandwidth/ transmission speed
ASK	External modulator	Heterodyne	Envelope	0	$\approx 10^{-1}$	≈ 2
		Homodyne		3 dB	$\approx 10^{-4}$	≈ 1
FSK	Direct modulation	Heterodyne	Dual-filter envelope detection	3 dB	$\approx 10^{-1}$	$\approx 4-10$
			Differential detection	$\approx 4-6$ dB	$\approx 10^{-2} - 10^{-3}$	≈ 2
PSK	External modulator	Heterodyne	Differential detection	≈ 6 dB	10^{-3}	≈ 2
		Homodyne		9 dB	10^{-4}	≈ 1

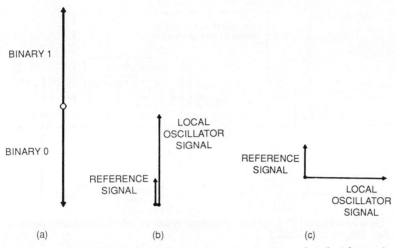

BINARY 1

BINARY 0

REFERENCE
SIGNAL

LOCAL
OSCILLATOR
SIGNAL

REFERENCE
SIGNAL

REFERENCE
SIGNAL

LOCAL
OSCILLATOR
SIGNAL

(a)   (b)   (c)

Figure 7.7   LO signal phase relative to data signals. (*Reproduced with permission from Ref. 302, © 1986 IEEE.*)

tional, unwanted dc signal which must be eliminated; otherwise it interferes with the PLL operation. These three problems can be solved by using either a nonlinear loop or a balanced PLL.

7.2.1 Nonlinear PLL

There are two main types of nonlinear PLL:

1. Costas loop
2. Decision-driven loop

In the Costas loop, the phase-lock current is multiplied by the signal current from the data photodetector, whereas in the decision-driven loops, the phase-lock current is multiplied by the output signal of the data receiver. The decision-driven loop, homodyne optical receiver is theoretically superior in performance to the Costas loop receiver.

Figure 7.8 illustrates a homodyne optical receiver based on the decision-driven loop. The received signal power is split by a 90° optical hybrid. One part of this signal is sent to the data detector and the other part is sent to the PLL. This implies an extra power penalty due to the loss in the 3-dB coupler. Since the average value of the phase-lock current from the loop photodetector is zero (as $\phi = 90°$), it must be processed nonlinearly prior to use in phase locking. This is achieved by multiplying the output from the 1-bit delay circuit (phase-lock current) by the output from the data detector (data output signal). Notice the RC circuits which are included for ac coupling to eliminate the dc

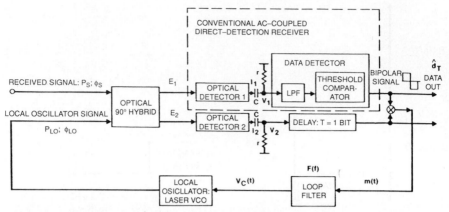

Figure 7.8 Decision-directed PLL optical homodyne receiver. (*Reproduced with permission from Ref. 302, © 1986 IEEE.*)

component. For this decision loop optical homodyne receiver circuit the required laser linewidth to bit rate ratio is calculated to be about 3×10^{-4}.

7.2.2 Balanced PLL

The block diagram for the balanced PLL receiver is shown in Fig. 7.9a. The optical power is split by the 3-dB coupler. Note this is a 180° coupler as opposed to the usual 90° coupler. In the other arm of the coupler is fed the optical output from the voltage controllable laser local oscillator. The signals are detected by the photodetector and passed to a difference amplifier whose output is fed back via the loop filter to the VCO, forming the PLL. The dc component is eliminated by the difference amplifier.

The linewidth to bit rate ratio for this circuit is calculated to be about 6×10^{-6} (for a 1-dB power penalty at a BER of 10^{-10}). This figure can be improved to 10^{-5} or even better than 10^{-4} by incorporating extra post-detection processing to reduce the data-to-phase-lock crosstalk (i.e., interference between the data detection and phase-lock branches of the receiver, Fig. 7.9b).

Comparing the performance of the decision loop receiver with the balanced loop receiver reveals that the balanced loop receivers suffer data-to-phase-lock crosstalk noise which must be canceled. This is in addition to the usual laser phase and shot noise. From the linewidth point of view, the decision loop receivers are better than balanced loop receivers without data-to-phase-lock crosstalk cancellation. If data-to-phase-lock crosstalk is canceled, the linewidth required by the balanced loop systems approaches that of the decision loop systems.

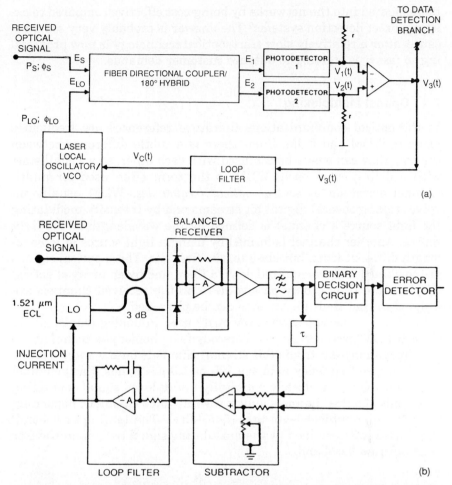

Figure 7.9 (a) Balanced PLL optical homodyne receiver (*reproduced with permission from Ref. 302, © 1986 IEEE*). (b) Improved balanced PLL optical homodyne receiver (*reproduced with permission from Ref. 296, © 1990 IEEE*).

Photonic and electronic integrated circuits should eventually neutralize any disadvantage one system may have over the other.

In summary, the laser source for homodyne PSK requires a minimum spectrum width to bit rate ratio which depends upon the receiver configuration. At present, the binary PSK homodyne detection, balanced OPLL receiver is fashionable, requiring the minimum linewidth to bit rate ratio of the order 10^{-4}. Homodyne PSK systems have experimentally verified the theory by exhibiting receiver sensitivities quite close to the quantum limit of nine photons per bit.

The only question that now remains is when will these systems be implemented into the networks by being cost effective compared to existing direct detection systems? The answer is probably very soon, because, after a relatively slow start, optical technology is now progressing so fast it is starting to outpace customer demands.

7.3 Optical Multiplexing

In the optical communications literature references are frequently made to WDM and FDM. Since there is a subtle difference between the two, they can easily be confused with each other, so we shall start with a definition of each. WDM is the term often used for multichannel operation at *several optical frequencies*. WDM usually involves taking several digital bit streams and by intensity modulating the light source a channel is defined by the wavelength of the light source. Another channel is formed by using a light source at a wavelength different from, but close to, the first, etc. These optical signals are combined and transmitted down a fiber, and by an array of optical couplers and filters (only passive devices) the received channels are isolated so that their bit streams can be processed.

FDM is a term which was previously used in analog transmission. Optical FDM systems use the heterodyne or homodyne technique to multiplex many electronic (bit stream) channels onto a *single optical carrier* whose frequency is in the 1.5- or 1.3-μm bands. Each channel to be multiplexed is already a digitally multiplexed signal containing thousands of voice channels and/or several video channels, depending upon the hierarchical level (e.g., 565 Mb/s). The total optical bandwidth used is determined by the modulating signal (i.e., microwave or even infrared baseband).

7.3.1 Wave division multiplexing

Figure 7.10 shows a typical WDM system, which in this case has just 18 channels. The DFB lasers are used to provide relatively narrow light sources. Each laser operates at a wavelength such that the channels are spaced apart by 2 nm (about 250 GHz at 1500 nm). The 1500-nm band of silica fiber offers over 100 nm of optical bandwidth. The 1300-nm band has a similar available bandwidth, providing over 200 channels in total. In practice, the number of usable channels would be less than 200 because the zero dispersion point occurs at only one wavelength, which means that as the channels move out from this wavelength, the dispersion will get worse. For example, if a system is designed to operate at the normal fiber dispersion minimum of 1300 nm with some of the multiplexed channels operating in the 1500-nm band, the dispersion at the 1500-nm channels would be approximately

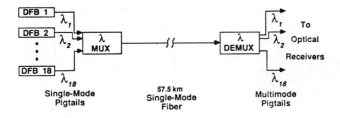

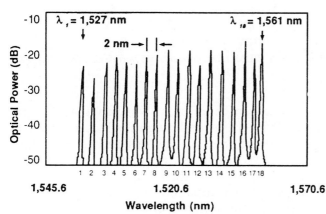

Figure 7.10   Optical WDM transmission. (*Reproduced with permission from Ref. 260, © 1989 IEEE.*)

15 ps/nm · km. This would severely degrade the maximum bit rate per distance performance available to these channels. If a channel spacing of 2 nm is used, good performance can be achieved if 50 channels are wave division multiplexed onto one fiber in the 1500-nm band and another 50 channels are wave division multiplexed onto one fiber in the 1300-nm band. The 2-nm channel spacing figure is determined by the temperature stability of the light sources and the filtering capability of the multiplexers. This spacing could be reduced if the cost of the system is increased to counteract temperature instability. In Fig. 7.10, the individual channel passbands are about 0.3 nm, which necessitates a temperature stabilization of each DFB laser to within 1°C. Each laser is modulated at 2 Gb/s, which is equivalent to more than 20,000 voice channels. So, if 18 optical channels are wave division multiplexed, the system transmission capability is more than 360,000 voice channels on one pair of optical fibers. An example of the optical multiplexers that could be used in such a system is shown in Fig. 7.11. For demultiplexing, the light enters the device through the fiber denoted as *line*. It is collimated with a lens, after which it is incident

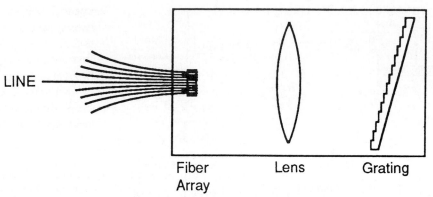

LINE

Fiber   Lens   Grating
Array

Figure 7.11   Optical multiplexer for WDM. (*Reproduced with permission from Ref. 260,* © *1989 IEEE.*)

upon a grating. The grating causes different wavelengths to be reflected at different angles. On returning through the lens, each wavelength is refocused to a different point on the fiber array. The output of the multiplexer is therefore a specific wavelength channel on each fiber. The device is reciprocal, meaning that it operates equally well as a wavelength multiplexer. When inputs to the fibers are at different wavelengths, the output on the single fiber contains the combined channels. The loss associated with each channel passing through this multiplexer/demultiplexer is about 4 dB. Channel spacings of 1 to 2 nm are possible with this style of wavelength multiplexer. To improve the channel spacing below this value and therefore incorporate more channels within the optical bandwidth available, one must use tunable optical filters and direct detection or coherent detection and/or FDM.

Another, simpler application of WDM is to use WDM to enable a full-duplex communication between two points using only one optical fiber instead of a pair of fibers. If, for example, one wavelength at 1.50 μm is used for GO and another at 1.60 μm for RETURN, only one fiber is necessary. This has interesting implications for future cost reduction of local loops utilizing fiber to the home (FTTH).

7.3.2   Frequency division multiplexing

FDM is considered to be superior to WDM because more channels can be multiplexed within the available optical bandwidth. FDM is often described as dense optical multiplexing. The channel spacing for FDM can be typically 5 GHz (or 0.04 nm at 1.5 μm), which is considerably less than the 250 GHz (or 2 nm) for WDM. Also, the losses in the components are less for FDM. Figure 7.12 illustrates a point-to-point FDM

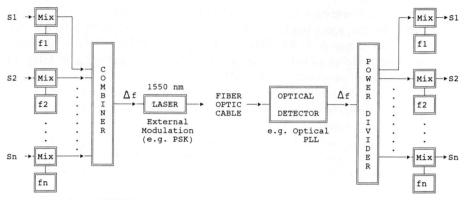

Figure 7.12 Optical FDM.

system. Here, the transmitted optical signal is built up in a manner similar to the AMR system. The total number of bit streams that can be multiplexed depends on the bit rate of each stream. Obviously more bandwidth is going to be consumed by heterodyning a 10-Gb/s bit stream than a 100-Mb/s stream. At the receiving end the optical signal is detected, power is divided, and the individual bit streams are recovered by IF heterodyning.

The full capability of optical FDM will be realized only if a baseband of up to 12,500 GHz or so can be used to modulate the LD. This corresponds to an optical $\Delta\lambda$ of about 100 nm, which could constitute the upper sideband of a PSK modulated optical carrier covering 1500 to 1600 nm (or 200,000 to 187,500 GHz). To allow a large repeaterless distance, it would be necessary for the zero dispersion to be almost flat across the full 100-nm optical bandwidth.

The 12,500-GHz baseband could be constructed by 2500, 5-GHz spaced, 2.488-Gb/s (STS-48 or OC-48) bit streams, totaling $30,720 \times 2500 = 76.8$ million voice channels (or about 180,000 video channels) on one pair of fibers. However, 12,500 GHz far exceeds the present and foreseeable future capability of microwave and infrared technology, not to mention present voice and video channel capacity requirements.

The following is an example of a realistic system that could be designed using present-day technological capability.

Example. In Fig. 7.12, consider $S_1 = S_2 = S_3 \cdots = S_n = 140$ Mb/s (i.e., *1920* voice channels per input). Also, consider 40 GHz to be presently the highest frequency for obtaining high-quality microwave components (e.g., for modulating the LD).

The number of FDM channels depends on the modulation technique, which defines the bandwidth efficiency; that is, if the bandwidth efficiency = 1 b/s/Hz, $f_1, f_2, \ldots, f_n$ would be typically 2.0, 2.2, 2.4, 2.6,..., 40 GHz (Δf = 38 GHz); → 190 CH → 1920 × 190 = *364,800* voice channels on one 1550-nm optical carrier using only 38 GHz, or about 0.3 nm, of *optical* bandwidth.

The 64-QAM modulation technique already provides a bandwidth efficiency of 4 b/s/Hz and 256-QAM will soon be available having 5 b/s/Hz. If the bandwidth efficiency = 5 b/s/Hz, $f_1, f_2, \ldots, f_n$ would be typically 2.05, 2.10, 2.15, 2.20,..., 40 GHz (Δf = 38 GHz); → 760 CH → 1920 × 760 = *1,459,200* voice channels on one 1550-nm optical carrier using only 38 GHz, or about 0.3 nm, of *optical* bandwidth.

Even present-day technology could use up to 50 of the 0.3-nm (1,459,600 voice channels) FDM signals in a WDM system having 2-nm channel spacing. This combination of FDM and WDM would provide about 73 million voice channels on one pair of fibers. Although this large quantity of voice channels far exceeds existing or even future capacity requirements, the corresponding 152,000 video channels is an interesting prospect. This number of video channels is calculated using 34 Mb/s per video channel. This is present TV quality video, which could be distributed to customers via the telecom network using an FDM-WDM system, provided optical fiber is installed in all sections of the network. That is presently a very large proviso.

7.4   Repeaters

Because of the distances involved and the point-to-point nature of the junction (interexchange routes), repeaters are not required. However, for the distances involved in long-haul links, repeaters are necessary. Many LANs also require repeaters because of multiple splitting of the light source power. Repeaters take two forms: (1) regeneration of the transmitted bit stream and (2) optical amplification. It is only recently that optical amplifiers have become commercially available, and they are now invaluable in both of the above applications. Optical amplifiers have a noise accumulation, which means that pulse regeneration is needed after a certain distance.

7.4.1   Regenerative repeaters

The block diagram of Fig. 7.13 shows a typical sequence of events in a conventional optical fiber regenerator. The optical signal is converted to an electrical signal by the optical detector (APD or PIN diode). The resulting electrical pulses are amplified, equalized, and then completely reconstructed by a standard pulse regenerator and timing cir-

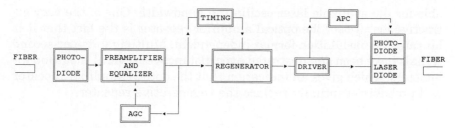

Figure 7.13   Block diagram of an optical regenerator.

cuit. A measure of the BER is established at this stage of the repeater. The clean pulses are then used to drive a light source (LD or LED) that is ready for onward transmission down the optical fiber. This method was used exclusively in almost all long distance systems installed in the 1980s. The regenerators were placed in manholes and were relatively small for a small fiber cable having only eight or ten fibers. For large cables having several hundred fibers, the regenerators would occupy several cubic meters, necessitating large manholes. Furthermore, the power required for the regenerators could be quite an engineering problem, particularly for submarine links. Since power is not usually available at the manholes, copper wires must be included in the construction of the cable to remotely power the regenerators from the nearest line terminal equipment. Note that in the usual (non-WDM) systems, each optical channel requires two regenerators, one for GO and the other for RETURN. The main point in favor of regenerative repeaters is the fact that the effects of fiber dispersion (intersymbol interference) are cleaned up at each regenerator. This is not the case for optical amplifier repeaters where dispersion effects accumulate with distance, regardless of the presence of the repeaters.

7.4.2 Optical repeaters

The commercial availability of optical amplifiers in the early 1990s has revolutionized the long distance communications designs and also made optical LANs economically viable. Optical amplifiers, which do not require the signal to be converted to an electrical signal and then back to optical, offer a smaller and cheaper approach to the regenerative repeaters. If the signal of a long-haul route is periodically boosted with optical amplifiers, the length of the link is determined by the cumulative optical noise introduced by each optical amplifier and cumulative fiber chromatic dispersion. The dispersion can, to a large extent, be eliminated by using dispersion-shifted fiber and transform-limited pulses. A transform-limited pulse is the shortest laser pulse achiev-

able for the available laser oscillating bandwidth. One of the very attractive features of the optical amplifier repeater is the fact that it is
bit rate and modulation format independent. Multigigabits-per-second
signals have been demonstrated over distances of several thousand kilometers, which gives an indication that the optical amplifier repeater
will probably eventually replace the regenerative repeater.

7.5   Systems Designs

Let us consider the three main categories of fiber transmission
systems: (1) long haul, (2) junction routes (interexchange traffic), and
(3) local (subscriber) loop. The technology is moving so fast that engineering designs for each of these categories are changing rapidly. As
usual, cost considerations dictate the design and components used for
a particular link. The junction routes are perhaps the most standardized designs which use well-proven repeaterless technology. A mas-
sive research effort is being made to increase the repeaterless distance
and therefore reduce the cost of long-haul links. Many engineers consider this to be the last major hurdle to overcome to enable global optical fiber connectivity.

Optical fiber in the local loop is the subject of intense debate in the
developed world. The cost of delivering fiber to every customer is enormous, but the services which could be provided and the subsequent
revenue generated is even greater. Some of the link designs presently
in use and proposed for the future will be discussed in the following
subsections.

7.5.1   Junction routes
(interexchange traffic)

The early optical fiber technology was ideally suited to interconnecting COs, which are 10 to 20 km apart, with medium to high capacity.
These links initially took the form of step-index 1.3-μm multimode fibers fed by LED light sources and detection at the receive end by
APDs or PIN diodes. As the technology improved in the 1980s, the
LEDs were replaced by LDs, and 1.3-μm multimode fiber was introduced. In fact, multimode optical fibers could still be used for this application today.

Today in the United States, the 45- and 90-Mb/s (34 and 140 Mb/s
elsewhere) 1.3-μm, *single-mode* optical fiber systems are widely
used to interconnect COs. It makes more sense to use single-mode
fiber instead of multimode to facilitate future capacity upgrades. A
typical 140-Mb/s optical fiber line terminal equipment (LTE) is
shown in Fig. 7.14. The 140-Mb/s system is chosen since this is the

TRANSMITTER UNIT

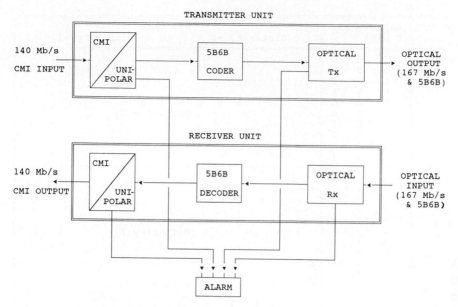

Figure 7.14   Block diagram of a 140-Mb/s line terminal.

closest bit rate to the 155 Mb/s STS-3 (STM-1) which will be prevalent in the synchronous digital hierarchy systems to be implemented in the near future. In the transmitter part of the LTE, the incoming signal from the digital multiplexer has the CMI code, which is first converted to the unipolar NRZ code. It is then sent to a 5B6B encoder which increases the bit rate by allowing the inclusion of omnibus order-wire, automatic switching protection information, and alarm information to be inserted. The bit rate is increased by the ratio of 5 to 6 (i.e., from 139.264 to 167.117 Mb/s). The composite signal, which is a stream of electronic pulses, is then converted to optical pulses in the optical transmitter. In the receiver section of the LTE, the incoming optical pulses are detected and converted into electronic pulses. Service information is extracted during the 5B6B decoding process. Finally the NRZ coded pulses are converted to the CMI line code and are ready for demultiplexing down to individual voice, data, or video channels.

Although the NRZ-to-CMI and CMI-to-NRZ converters were discussed in Chap. 2, the 5B6B line coder and optical transmitters and receivers were not. The 5B6B code conversion is shown in Fig. 7.15. The original bit stream is split up into 5-bit words which are converted into 6-bit words according to the following rules. If the original 5-bit word has two 1 bits and three 0 bits, the sixth bit of the 5B6B signal is a 1. If three bits are 1 and two bits are 0, the sixth bit is 0.

Original signal (5 bit)					5B6B code Disparity +2						Disparity -2					
1	1	0	0	0												
1	0	1	0	0												
1	0	0	1	0												
1	0	0	0	1												
0	1	1	0	0	x	x	x	x	x	1						
0	1	0	1	0			(Disparity 0)									
0	1	0	0	1												
0	0	1	1	0												
0	0	1	0	1												
0	0	0	1	1												
0	0	1	1	1												
0	1	0	1	1												
0	1	1	0	1												
0	1	1	1	0												
1	0	0	1	1	x	x	x	x	x	0						
1	0	1	0	1			(Disparity 0)									
1	0	1	1	0												
1	1	0	0	1												
1	1	0	1	0												
1	1	1	0	0												
1	1	1	1	1	1	1	1	0	1	0	0	0	0	1	0	1
1	1	1	1	0	1	1	0	1	1	0	0	0	1	0	0	1
1	1	1	0	1	1	0	1	1	1	0	0	1	0	0	0	1
1	1	0	1	1	1	1	1	0	0	1	0	0	0	1	1	0
1	0	1	1	1	1	1	0	1	0	1	0	0	1	0	1	0
0	1	1	1	1	1	0	1	1	0	1	0	1	0	0	1	0
1	0	0	0	0	0	1	1	1	0	1	1	0	0	0	1	0
0	1	0	0	0	1	1	0	0	1	1	0	0	1	1	0	0
0	0	1	0	0	1	0	1	0	1	1	0	1	0	1	0	0
0	0	0	1	0	0	1	1	0	1	1	1	0	0	1	0	0
0	0	0	0	1	1	0	0	1	1	1	0	1	1	0	0	0
0	0	0	0	0	0	1	0	1	1	1	1	0	1	0	0	0

Figure 7.15   The 5B6B code conversion table.

For five, four, one, or zero 1 bits, the 5B6B signal has four 1 bits and two 0 bits the first time (disparity +2) and two 1 bits and four 0 bits the second time (disparity −2).

A typical frame structure for the 140-Mb/s 5B6B optical signal is shown in Fig. 7.16. The frame is subdivided into 12 subframes, each containing 18 words. The last bit of the first 6-bit word in each 18-word subframe is used for overhead information such as frame alignment, supervisory and automatic protection switching, omnibus order-wire, service order-wire, etc. The first 5 bits of those words are uncoded.

The 5B6B, 167.1168-Mb/s NRZ signal, typically at ECL level, is then used to intensity modulate the LD light source (i.e., turn it ON or OFF) ready for transmission down the fiber. The electrical pulse width of ON must be wider than the OFF because the rising amplitude of light emission in an LD lags behind that of the electrical signal. The output

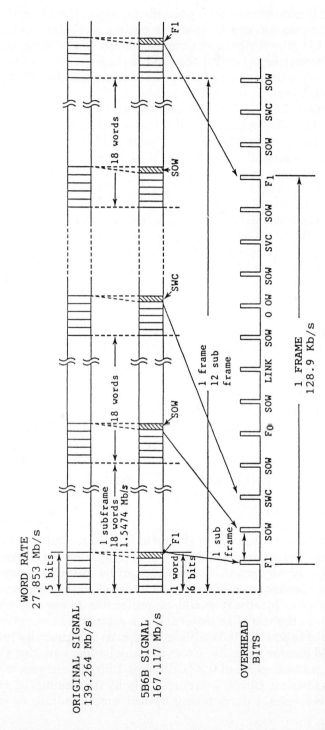

Figure 7.16 Frame structure for the 140-Mb/s 5B6B optical signal. F1,F0: frame alignment bit. SWC: supervisory and control bit for automatic protection switching. SVC: This bit is used for switching the SOW transmission lines (service channel 1) and the omnibus OW transmission lines (service channel 2). SOW: service order-wire signal. Omnibus OW: omnibus order-wire signal. LINK: This bit is used for the data link information transmission.

power from the LD is monitored by a photodiode behind the LD so that temperature compensation can be made to ensure a constant output power from the LD. The temperature compensation can be done simply by using a thermister to control the LD bias current. In the event of a cable break, a monitoring circuit is used to automatically shut off the LD so that restoration can be done safely.

The typical optical LTE of Fig. 7.14 also incorporates an optical receiver. The 167.1168-Mb/s optical signal is converted into an electrical signal, in this case, by a germanium avalanche photodiode (Ge-APD) which has a current amplification factor M of 1 to 20 at 1.3 μm. M varies with the APD bias voltage and is therefore used in an AGC circuit to maintain a constant electrical output from the APD for at least a 10-dB variation in optical input power.

The amplified electrical signal is then converted to the 139.264-Mb/s CMI signal by a 5B6B decoder according to the conversion rules; that is, six parallel signals are converted to five corresponding parallel signals. The overhead bits are also recovered to form the order-wire signal and other service channel signals. A circuit is also included to detect any violations of the code conversion and to raise an alarm if any violations occur.

In some optical line terminal transmitting equipment the signal is scrambled prior to 5B6B encoding. This has two purposes. First, it ensures that the LD is turned on and off in a manner that almost equalizes the distribution of 1s and 0s, thereby minimizing thermal variations in the LD. Second, it eliminates periodic sequences of signals so that the 5B6B decoder at the receive end can achieve word synchronization regardless of bit sequence. Also, clock recovery is facilitated in circumstances where there are long strings of zeros. Of course, descrambling must be done at the receiving end after the 5B6B decoding.

7.5.2 Long-haul links

There are several interpretations of the term *long-haul,* but any distance in excess of junction route distances is usually long haul. Long-haul links are specifically trunk routes that can be intercity connections, a country backbone network, an international link, or a transcontinental link. Ideally it would be preferable to span any of these links without the use of repeaters. If the attenuation of optical fiber cables ever gets down to 0.01 dB/km or less, as theoretically predicted, this would become a reality. From the link budget equation the present minimum attenuation of 0.22 dB/km at 1.55 μm dictates that the repeaterless distance can be increased only by increasing the optical source power level, incorporating optical amplification, or increasing the receiver sensitivity. The receiver sensitivity limit is fixed by the laws of physics, and laboratory experimental systems are already approaching that limit. The optical power sources are continually being improved as time goes by. LD technology is steadily improving the optical output power, but again, the laws of physics step in and supposedly limit the maximum power that can be used to a little more than 10 mW because of scattering phenomena. Recent developments dispute that initial proposal, and now it is believed that pulse compression can occur at transmitting power levels higher than about 10 mW. The benefits of that observation are far reaching and will be discussed in the next section, which is on *soliton* transmission.

Optical amplifiers promise to be so advantageous that in the future they will undoubtedly be indispensable. Conventional regenerative repeaters effectively reset the fiber dispersion effects at each repeater. When using optical amplifiers, the dispersion and optical noise accumulate along the entire length of the link and become transmission distance limiting factors. Figure 7.17 shows the distance limitation calculated for various 1.55-μm single-mode systems. The plot 1 is for a step index fiber having a dispersion of 17 ps/nm · km. Also, the transmitter has a directly modulated LD with a chirp of 2 Å. Plot 2 indicates a significant distance improvement by using the same step index fiber but externally modulating the laser source so that the signal is transform limited. This plot is perhaps surprisingly better than the system 3, which has a dispersion-shifted fiber with a dispersion value of 0.1 ps/nm · km but uses direct laser modulation (chirp = 2 Å). The

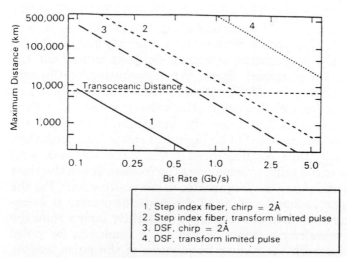

1. Step index fiber, chirp = 2Å
2. Step index fiber, transform limited pulse
3. DSF, chirp = 2Å
4. DSF, transform limited pulse

Figure 7.17 Distance limitation due to dispersion. (*Reproduced with permission from Ref. 185, © 1990 IEEE.*)

best performer is plotted in 4, which has the superior combination of dispersion-shifted fiber (0.1 ps/nm · km) and external laser modulation (signal transform limited). This system should be capable of transoceanic distances if amplifiers are used which do not set a distance limit due to optical noise. Amplified spontaneous noise emission is a wideband phenomenon which can be minimized by using optical filters between each amplifier in the chain and also in front of the receiver. Calculations indicate that relatively cheap Er-doped amplifiers having a 19-dB gain, 4-dB noise figure, 2-nm optical bandwidth, and 0-dBm launch power and spaced 100 km apart could span a repeaterless link of over 7000 km before optical noise starts to degrade the BER performance. The calculations indicate that even at 10,000 km the noise contribution from the amplifiers degrades the BER only to about 10^{-9}. Long-haul optical fiber systems appear to have a very bright future, although practical verification of these large-distance postulations has not yet been made.

Soliton transmission.   Perhaps the most interesting innovation in optical communications to appear so far is the optical soliton transmission. Although this method of transmission has been theoretically known for some years, it was only in about 1989, when optical components such as Er-doped fiber amplifiers became readily available, that efforts to achieve soliton transmission succeeded. What are solitons? They are dispersionless pulses of light. They only exist in a nonlinear medium, where the refractive index is changed by the varying intensity of the light pulse itself. There are at least two classes of solitons: (1) temporal and (2) spatial. The temporal soliton is the candidate for ultralong distance optical communication systems. A soliton is a transmission mode of an optical fiber which is the only stable solution to the fundamental propagation equation. Solitons cancel out the chromatic dispersion of optical fibers. When operating at the zero dispersion-shifted value of 1.55 μm, a pulse will be spectrally broadened because of silica's nonlinearity. The mere presence of a light pulse which has a varying light intensity over its envelope causes the refractive index of the material to be higher at the pulse's peak than at its lower intensity tails. As illustrated in Fig. 7.18a, the peak is retarded compared to the tails, which lowers the frequencies in the front half of the pulse and raises the frequencies in the trailing half. For the soliton to propagate, a small amount of chromatic dispersion is necessary. If the system operates at a wavelength slightly longer than the zero-dispersion wavelength, there would be a tendency for pulse spreading to occur with the higher frequencies of the pulse leading and the lower frequencies being retarded (Fig. 7.18b). As one can see, the effects of the fiber nonlinearity are balanced by the fiber's chro-

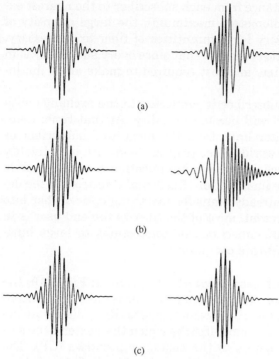

(a)

(b)

(c)

Figure 7.18 Soliton formation. (*a*) Nonlinearity; (*b*) dispersion; (*c*) soliton. (*Reproduced with permission from Ref. 216, © 1990 IEEE.*)

matic dispersion, producing the soliton of Fig. 7.18c. A soliton is therefore truly nondispersive in both the frequency and time domains. Furthermore, it has already been demonstrated that solitons remain stable after passing through many amplifiers provided the distance between amplifiers is short enough so that the pulse's shape is not disturbed. A practical demonstration of this theory was recently successfully completed over a 9000-km-long optical fiber with a 4-GHz signal in a recirculating loop incorporating Er-doped amplifiers spaced 25 to 30 km apart.

7.5.3 Local area networks and subscriber loops

Optical fiber deployment in the telecommunications network is progressing rapidly. Although the subscriber loop is the shortest distance in the network, it will be the last part of the optical system to be connected. This is obviously because of the enormous quantity of subscribers. Although the distance from each subscriber to the nearest exchange is small (a few kilometers maximum), the large quantity of subscribers means that very large quantities of fiber are necessary. But, even more problematic is the fact that since every subscriber is in a different location, the time and cost required to make all of the installations is formidable.

It is only when every subscriber is connected to the exchange with fiber that a truly B-ISDN will become a reality. At that time, there can be interactive computer links from the home to a subscriber or business anywhere in the world. Video transmission with high-fidelity sound will be commonplace. The telecommunication services possible will be limited only by the imagination. In several countries in the developed world, there are already plans for installing optical fiber into the subscriber loop. At present, most of the fiber to the end user is in the form of experimental connections or connections to large businesses that utilize high-bit-rate computers.

Subscriber networks (local access network).   First, as in Fig. 7.19, the CO (exchange) is connected to its remote nodes (or remote line concentrators, or remote line units). These remote nodes (RNs) distribute cables out to the service access points (SAPs) which then extend the service cables the last few meters to the customer premises (CP). The objective is eventually to use optical fiber cable for all three of these segments, which is often called FTTH. Economic considerations indicate that the last segment (i.e., from the SAPs to the customer premises), will be the last to be installed. The phrase *fiber to the curb* has been coined to describe the interim solution which will probably exist before the local loop becomes completely optical fiber. There are several interconnection options, described as topologies, when considering the local loop, and each segment may have a different topology. Figure 7.20 shows the three categories of topology (i.e., the star, bus, and ring). The star configuration is generally considered to be the best topology for the SAP-to-CP segment since it provides the highest possible bandwidth to the customer while maintaining a high degree of privacy and security. In the feeder and distribution segments (1) the

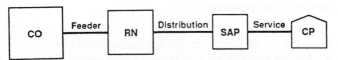

Figure 7.19   Subscriber loop (local access). (*Reproduced with permission from Ref. 250, © 1989 IEEE.*)

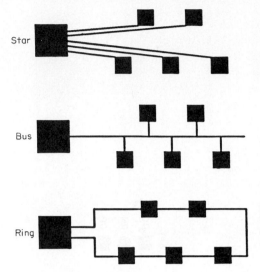

Star

Bus

Ring

Figure 7.20 Interconnection configurations. (*Reproduced with permission from Ref. 250,* © *1989 IEEE.*)

ring or star topologies can be used for voice and data services, (2) bus or star topologies can be used for distributive video, (3) the star topology is preferred for the point-to-point, high-bandwidth, switched video services. There are many possible combinations of star, bus, and ring topologies for the local loop (local access network) and several examples are illustrated in Fig. 7.21. The star-star configuration offers the highest flexibility for supplying voice, data, and video (broadcast or interactive) services. At increased expense and complexity, the reliability can be enhanced by supplying route diversity as in Fig. 7.21*b*. The bus-star and bus-bus topologies of Fig. 7.21*c* can be used for video program distribution as well as voice and data communications. Additional measures would be required to ensure security and privacy. Figure 7.21*d* is a ring-star configuration, where the ring in the feeder segment offers some protection against equipment failure or cable cuts by effectively providing route diversity. Figure 7.21*e* combines the bus-star for video distribution and star-star for voice and data communications.

Having decided upon the topology or topologies to use for the interconnections between the CO and the subscribers, the next technical problem to solve is the *style of transmission*. Should the system incorporate WDM, FDM, or heterodyne/homodyne (coherent) detection? Figure 7.22 shows an early optical fiber systems application which used WDM to provide each customer with a dedicated transmit and receive wavelength. This technique allows a single optical fiber feeder to be used between the CO and the remote node or SAP. No wave-

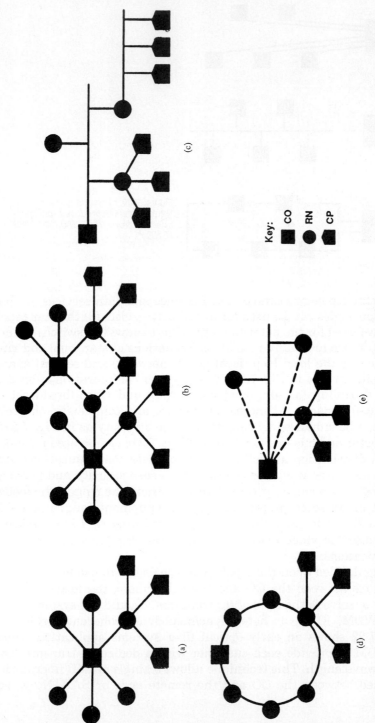

Figure 7.21 Subscriber loop (local access) topologies. (a) Star/star; (b) star/star with route diversity; (c) bus/star and bus/bus; (d) ring/star; (e) overlay of bus/star and star/star. (Reproduced with permission from Ref. 250, © 1989 IEEE.)

Key:
■ CO
● RN
⬠ CP

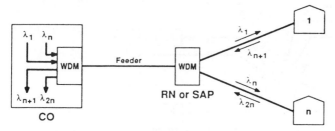

Figure 7.22 WDM in the subscriber loop. (*Reproduced with permission from Ref. 250, © 1989 IEEE.*)

length filtering is necessary at the customer premises since each customer receiver has only one allocated wavelength specifically assigned by the CO. The number of subscribers that can be connected by this system is relatively small. For a given available optical bandwidth (e.g., 1200 to 1300 nm) the number of usable channels is limited by the filter characteristics of the WDM devices. If the filter passband is typically 0.4 nm, this dictates the minimum value of channel separation (i.e., about 2 nm). Narrower channel spacing would cause intolerable adjacent channel crosstalk. Therefore the maximum number of usable channels in this case would be about 50. FDM uses a heterodyne multiplexing technique at the transmitting end to form a composite signal to be transmitted. The heterodyne/homodyne receiver is an appropriate choice in these circumstances. Since this can enhance the channel spacing to about 5 GHz (0.04 nm), the number of usable channels is therefore increased to 2500 for an optical passband of 100 nm (e.g., 1.2 to 1.3 μm); except as mentioned earlier, the microwave and infrared technology is not yet sufficiently developed to exploit such a large number of channels.

Examples of coherent systems are shown in Fig. 7.23. In Fig 7.23*a*, this multichannel network could be used for CO interconnections or LAN to metropolitan area network (MAN) computer interconnections. Each transmitter has a specific optical frequency and is connected to the passive star coupler, which combines the transmitted signal and distributes the composite signal to all receivers. Each receiver can be tuned to any of the transmitted frequencies, thereby allowing communication between any of the transmitters and receivers within the network. Figure 7.23*b* is a distribution network in which many optical signals are combined by a star coupler and then passively split to a large number of subscribers. Each subscriber can tune to one of the many transmitted signals using a coherent detection receiver. This network has applications for TV distribution or B-ISDN.

Local area network. LANs, which are primarily associated with computer interconnectivity, are evolving at a fast pace. These are net-

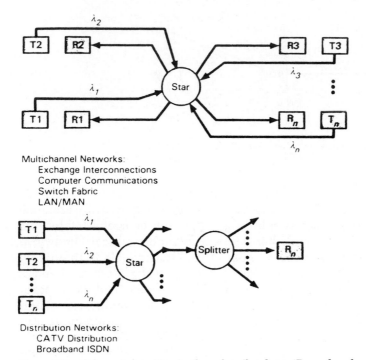

Multichannel Networks:
   Exchange Interconnections
   Computer Communications
   Switch Fabric
   LAN/MAN

Distribution Networks:
   CATV Distribution
   Broadband ISDN

Figure 7.23 Coherent detection in the subscriber loop. (*Reproduced with permission from Ref. 314, © 1990 IEEE.*)

works within a building or group of buildings such as business premises, factories, educational institutions, etc. The bit rate required for each application can differ significantly. For example, personal computer interconnection requires about 1-Mb/s. Higher-performance workstation and file-server interconnection requires about 10 Mb/s. In the 100-Mb/s range are networks such as the fiber distributed data interface (FDDI). For supercomputer and complex graphics, bit rates in excess of 1 Gb/s will be necessary. The LAN has some extra complexities not experienced in the long distance point-to-point systems. Optical fiber cable is not simply a direct substitute for copper cable to upgrade the capacity of a wire-based LAN. While the fiber-based systems have abundant transmission bandwidth, they also have a severe signal power limitation. The primary objective of most LANs is for any one of the network users to be able to communicate with any, or all, of the other users simultaneously. This is a significant increase in system complexity compared to the point-to-point connections predominant in most telecommunication systems. The LAN has an interconnectivity which is both point to point and simultaneous broadcast in nature; this is accomplished without the use of switches.

There are presently two favored topologies for optical fiber LANs: the star network and the unidirectional bus network (Fig. 7.24). In order to fully exploit the available bandwidth of an optical fiber system, a LAN must not be (1) power limited or (2) suffer an electronic bottleneck. The power transmitted by one user is split many times by, for example, an N × N star coupler, in order to have the broadcast capability. Depending upon the topology, number of users, distance between users, etc., it may be necessary to incorporate amplifiers to boost the power at each receiver. The electronic bottleneck problem is particularly severe in the ring topology, where each node must process all of the traffic for the network. The electronic processing capability of the entire network is therefore limited by the electronic processing capability of one node. The obvious way of surmounting this problem is to use multiple wavelengths instead of just one. WDM or the more sophisticated FDM techniques are being successfully applied to large user LANs.

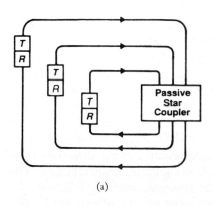

(a)

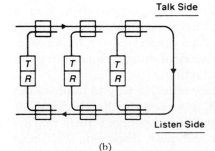

(b)

Figure 7.24 LAN topologies. (*a*) Star network; (*b*) unidirectional bus. (*Reproduced with permission from Ref. 265, © 1989 IEEE.*)

7.5.4 BER improvement

There are several countermeasures that can be applied to optical communication systems to improve the link BER, such as:

- Optical equalization of fiber chromatic dispersion
- Equalization of polarization dispersion
- Forward error correction

Fiber chromatic dispersion is a major system limitation which causes intersymbol interference and therefore errors, depending upon the unregenerated distance and bit rate. As already discussed, this problem can be alleviated by operating systems at, or very close to, the zero dispersion wavelength of 1.3 μm, or the lower-loss, zero dispersion shifted wavelength of 1.55 μm. Consider a conventional single-mode fiber design optimized for zero dispersion operation at 1.3 μm. This fiber would have a dispersion of about 15 to 20 ps/km · nm at 1.55 μm. This dispersion can, to a large extent, be equalized by placing a Gires-Tournois interferometric equalizer at the receiving end prior to the optical to electrical conversion. Such an interferometer is a reflective device which returns light from the input port that has undergone a frequency-dependent phase shift. This type of equalization results in a significant improvement in the receiver sensitivity for a given BER, which translates into an increased repeaterless distance or a higher bit rate for a specified maximum BER value. In addition, this equalization allows "dual window" operation, which is a very useful feature that has many applications.

Polarization dispersion is a bit rate–distance-limiting factor in high-capacity long-haul optical fiber systems. It causes signal degradation in optical amplifier systems at bit rates of 8 Gb/s and above. The main signal degradation is due to the received signal having different relative delays and amplitudes for the two orthogonal polarizations. The two signals add in optical power at the receiver, causing a linear distortion in the electrical signal. The problem can be alleviated by the inclusion of either an adaptive electrical analog tapped delay line, a decision feedback equalizer, or a nonlinear canceler. Without going into detail, it can be said that polarization dispersion can be satisfactorily corrected by operating on the electrical output signal after the optical to electrical conversion (i.e., detection).

FEC, as described in Chap. 5, is a very effective way to improve the BER of an optical fiber link. The FEC also acts on the electrical bit stream. FEC improves the BER for power-independent BER impairments such as mode partition noise or chromatic dispersion. FEC also improves BER floors that are induced by reflection, phase jitter, laser chirp, etc. By using a highly efficient code such as [224,216] FEC with only 3.7 percent overhead, 565-Mb/s or 1.2 Gb/s systems can be sig-

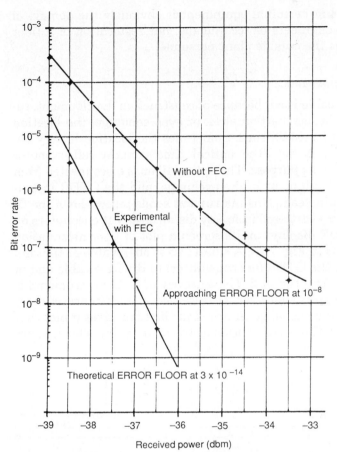

Figure 7.25   BER improvement for a 565-Mb/s system using FEC.
(*Reproduced with permission from IEEE, Grover, W. D., © May 1988 IEEE, J. Lightwave Technol., vol. 6, no. 5, p. 648.*)

nificantly improved. Figure 7.25 shows the measured performance of such a 565-Mb/s system operating at 1.3 μm. The BER versus receive power was established by inserting increasing amounts of fiber dispersion into the link while maintaining a large power margin. The improvement between the curves with and without FEC is very significant.

7.6 Optical Fiber Equipment Measurements

The measurements made on telecommunications equipment are usually clearly divided into external plant equipment and exchange equipment. In the case of optical fiber transmission, for some measurements, cooperation is usually required between the outside plant engineers or technicians and transmission engineers or technicians.

The optical fiber line terminal equipment is obviously the domain of the transmission engineers or technicians, but restoration of a fiber fault also requires the outside plant personnel.

7.6.1 Cable measurements

If an optical fiber cable is cut because of construction development, severe weather conditions, earthquakes, or even sabotage, the location of the break must be established. Sometimes the location is very obvious, but often it is not. The optical time domain reflectometer (OTDR) is used for this purpose. The optical fiber is disconnected from the line terminal equipment in the exchange, and the OTDR instrument is connected instead. This instrument sends pulses down the fiber, and light is backscattered from any discontinuity it encounters on its path. The OTDR displays on its screen a plot of fiber attenuation against distance as in Fig. 7.26. A small loss is observable at the connector interfacing the OTDR (or transmitter) to the fiber cable and at each splice point along the link. The accuracy of these instruments is so high that a 0.01-dB splice can be observed at a distance of more than 40 km from the exchange. A break in the fiber gives a plot that suddenly drops to infinite attenuation at the break point. The distance

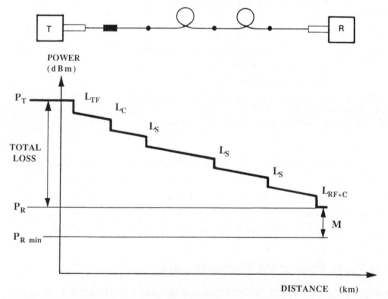

Figure 7.26 Attenuation versus distance using an OTDR. L_{TF} = transmit fiber loss, L_C = connector loss, L_S = splice loss, L_{RF} = receive fiber + connector loss. Margin M = received power − receiver sensitivity = $P_R - P_{R\,min}$.

from the exchange to the break can then be very accurately measured to within less than 1 m.

When repairing the break, it is useful to have a voice link-up between the person operating the OTDR and the person at the site of the repair. In remote locations a pair of very high-frequency (VHF) radios may be necessary. When the fiber has been spliced, or even just prior to splicing or mechanically fixing it in position, the OTDR operator can communicate whether or not the fiber alignment is looking good. On completion of the fusion or mechanical splice, the OTDR operator can indicate to the splicer whether the splice loss is adequately low or whether the splice needs to be redone. Fiber restoration and then measurement of the received power at the distant exchange could also be done to establish whether the splice is good or bad by comparing the power received after the splice with the value obtained just after cable installation. Incidentally, during initial cable installation, the OTDR is invaluable for establishing good splices.

7.6.2 Line terminal measurements

The optical fiber LTE is very simple to operate and maintain compared to other transmission systems such as microwave radio. After the usual power supply voltage checks, the following main measurements to be made are:

1. Optical output power

2. Optical received power

3. Optical output pulse waveform

4. BER

5. Output pulse waveform

6. Alarm indication and protection switching

The *optical output power* is simply a periodic check to ensure that the LD is healthy. The transmitter output is disconnected from the optical fiber line and a hand-held power meter can be connected to the transmitter output. Obviously, if the output power is rated too high for the power meter, an attenuator must be inserted to protect the power meter. A typical output power value is 0 to 10 dBm. A similarly simple measurement of *optical received power* is made by disconnecting the fiber into the receiver to check that the anticipated received power level is correct and there has been no degradation over a period of time (weeks or months). Incidentally, on very short links, the received power may be too high for the photodetector (typically more than −12 dBm) and an attenuator may need to be permanently installed to protect the photodetector.

With an electrical input to the LTE from a pattern generator, the *optical output pulse waveform* measurement is performed as in Figure 7.27a. With the attenuator adjusted to ensure that the optical power level is below 0 dBm, the waveform monitor converts the signal from optical to electrical, which is then displayed on the oscilloscope. The

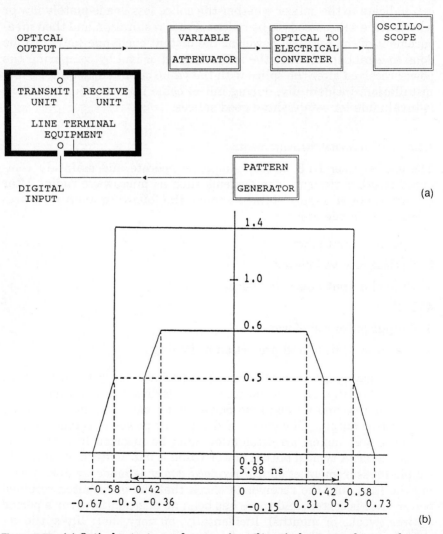

Figure 7.27 (*a*) Optical output waveform monitor; (*b*) optical output pulse waveform.

waveform should conform to the mask shown in Fig. 7.27*b* or similar specification.

The *BER* measurement of the LTE is performed as indicated in Fig. 7.28*a*. The optical output is looped back to the optical input and an attenuator is included in the loop to simulate the fiber link and also to

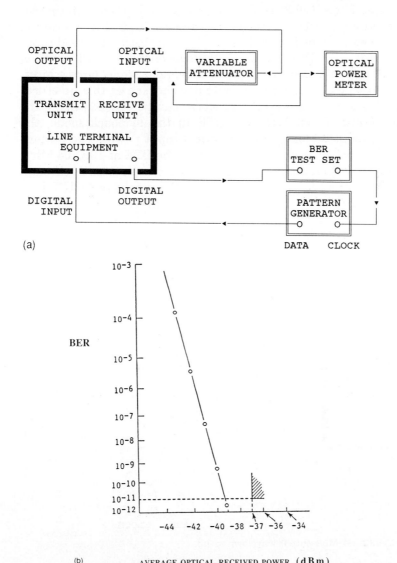

(a)

(b)   AVERAGE OPTICAL RECEIVED POWER (dBm)

Figure 7.28   (*a*) BER measurement; (*b*) typical graph of BER against optical received power for APD direct detection.

prevent damage to the optical detector. The pattern generator provides a PRBS ($2^{23} - 1$), CMI, 139.264-Mb/s signal into the LTE, and an error detector measures any errors. In order for the overall system BER performance to conform to CCITT Rec. G.821, it is necessary to ensure that there is the lowest possible BER for the back-to-back LTE (i.e., Tx to Rx loopback) without the presence of the link optical fiber and repeaters. For a 139.264-Mb/s, intensity modulated, 1.3-μm, LD transmitter and an APD detector in the receiver, a good line terminal BER should typically be better than 10^{-11} for a received optical input signal of about −37 dBm. Figure 7.28*b* shows the desired graph of BER against optical received power level. Some systems have a built-in error monitor which can be conveniently checked at this juncture.

The 139.264-Mb/s *output pulse waveform* can be checked using the test setup of Fig. 7.29*a*. With the LTE in the loopback mode, the attenuator is adjusted for −37-dBm optical input power. The pattern generator feeds an all 1s or all 0s signal into the LTE at 139.264 Mb/s, and the output waveform is observed after a complete path through the LTE. The 1s and 0s should conform to the masks of CCITT Recommendation G.703 as indicated in Fig. 7.29*a* and *b*.

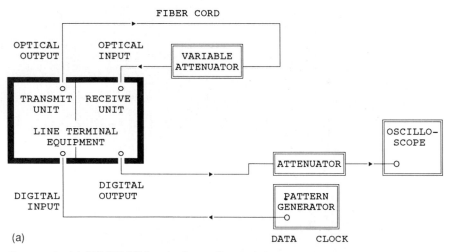

Figure 7.29   (*a*) 139.264-Mb/s output waveform setup.

Pulse Shape	Nominally rectangular and conforming to the masks shown in Figures 19/G.703 and 20/G.703
Pair(s) in each direction	One coaxial pair
Test load impedance	75 ohms resistance
Peak-to-peak voltage	1 ± 0.1 V
Rise time between 10% and 90% amplitudes of the measured steady state amplitude	≤ 2 ns
Transition timing tolerance (referred to the mean value of the 50% amplitude points of negative transitions)	Negative transitions: ± 0.1 ns Positive transitions at interval boundaries: ± 0.5 ns Positive transitions at mid-interval: ± 0.35 ns
Return loss	≥ 15 dB over frequency range 7 MHz to 210 MHz
Maximum peak-to-peak jitter at an output port	Refer to section 2 of Recommendation G.823

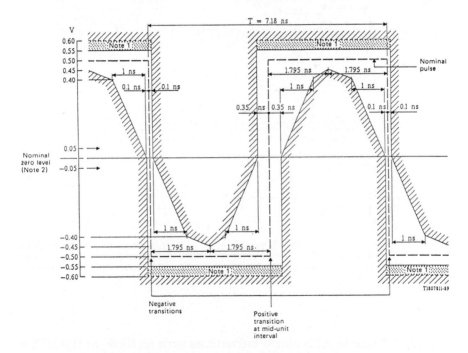

Note 1 — The maximum "steady state" amplitude should not exceed the 0.55 V limit. Overshoots and other transients are permitted to fall into the dotted area, bounded by the amplitude levels 0.55 V and 0.6 V, provided that they do not exceed the steady state level by more than 0.05 V. The possibility of relaxing the amount by which the overshoot may exceed the steady state level is under study.

Note 2 — For all measurements using these masks, the signal should by AC coupled, using a capacitor of not less than 0.01 μF, to the input of the oscilloscope used for measurements.

The nominal zero level for both masks should be aligned with the oscilloscope trace with no input signal. With the signal then applied, the vertical position of the trace can be adjusted with the objective of meeting the limits of the masks. Any such adjustment should be the same for both masks and should not exceed ± 0.05 V. This may be checked by removing the input signal again and verifying that the trace lies within ± 0.05 V of the nominal zero level of the masks.

Note 3 — Each pulse in a coded pulse sequence should meet the limits of the relevant mask, irrespective of the state of the preceding and succeeding pulses. For actual verification, if a 139 264 kHz timing signal associated with the source of the interface signal is available, its use as a timing reference for an oscilloscope is preferred. Otherwise, compliance with the relevant mask may be tested by means of all-0s and all-1s signals, respectively. (In practice, the signal may contain frame alignment bits per Rec. G.751.)

Note 4 — For the purpose of these masks, the rise time and decay time should be measured between −0.4 V and 0.4 V, and should not exceed 2 ns.

Figure 7.29 *(b)* Mask of 139.264-Mb/s pulse (binary 0) *(reproduced by permission from ITU, CCITT Blue Book, vol. III, Fascicle III.4, General Aspects of Digital Transmission Systems; Terminal Equipments, IXth Plenary Assembly, Melbourne, November 1988, Geneva 1989, p. 70, Fig. 19/G.703).*

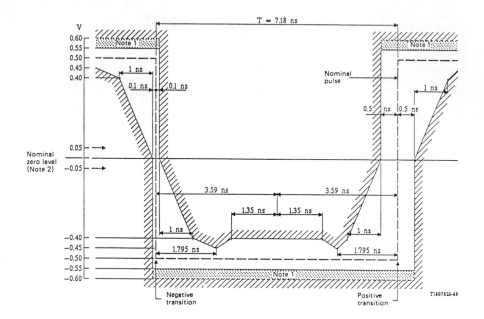

Figure 7.29   (c) Mask of 139.264-Mb/s pulse (binary 1). *(reproduced by permission from ITU, CCITT Blue Book, Vol. III, Fascicle III.4, General Aspects of Digital Transmission Systems; Terminal Equipments, IXth Plenary Assembly, Melbourne, Nov. 1988, Geneva 1989, p. 71, Fig. 20/G.703).*

Finally, the usual LTE alarm indications such as those in the following typical list are checked for correct operation:

1. Power supply failure

2. Fuse failure

3. Loss of incoming 139.264-Mb/s signal

4. Loss of outgoing 139.264-Mb/s signal

5. Loss of incoming optical signal
6. Loss of outgoing optical signal
7. Loss of frame alignment
8. Degradation of BER above 10^{-3}
9. Degradation of BER above 10^{-6}
10. CPU failure
11. Increase in LD drive current
12. Detection of AIS signal
13. Remote alarm detection

Mobile Radio Communications

8.1 Introduction

From the subscriber's point of view, mobile systems are perhaps the most exciting telecommunications development since the invention of the telephone. The developments of optical fiber technology sound very impressive and the statistics involved are mind boggling to the average person in the street, but that person does not usually fully appreciate how he or she benefits personally. Such developments are mainly occurring in the wings or backstage and are not really tangible service improvements to the subscriber. In fact, good service is expected.

The pocket telephone, on the other hand, is a revelation that is centerstage and whose benefits can be instantly appreciated by a person who decides to invest in one of these devices. However, despite the many advantages of being in contact with business associates or friends at all hours of the day, no matter where one may happen to be within a city, there are some sociological disadvantages to the pocket telephone. For example, the antisocial aspects of receiving calls in a quiet restaurant have already prompted some establishments to require customers to leave their pocket phones at the reception desk.

Cellular mobile telephone systems are difficult to classify. They could be considered to be part of the local loop since they extend out to the subscriber handset. Because of the distances traveled between a fixed subscriber and the mobile subscriber (or mobile-to-mobile subscriber), they could be called long-haul circuits. The technology incorporates some of the most advanced radio transmission techniques. In addition, the call processing requires high-level digital switching techniques to locate the mobile subscriber and setup and to maintain calls while the mobile subscriber is in transit.

The portable telephone was only recently made possible by the min-

iaturization resulting from VLSI electronic circuitry. Even today
there are still some technological problems to solve, such as increasing
the time between recharging the batteries of portable telephones.
More serious is the fact that vehicular mobile telephones and portable
radio telephones have some severe technical incompatibilities. These
take two forms. First, the systems developed in different parts of the
world (e.g., Europe, North America, and Japan) do not yet even have
the same operating frequency, and the system designs vary quite considerably. However, the ITU is currently coordinating Recommendations with all major parties to enable equipment compatibility for the
next generation of cellular mobile radio systems. Second, one can say
there are three generations of cellular radio systems: analog FM,
narrowband TDMA, and wideband CDMA. The analog has been
around for a number of years, narrowband TDMA is currently being
put into service, and wideband CDMA is possibly on the horizon.
There are significant equipment incompatibilities when moving from
one system to the other or trying to incorporate two or three of these
into the same network. Many of these problems stem from the difference in power outputs. Vehicle radios transmit at relatively high
power levels in the region of 1 to 10 W, whereas portable units transmit relatively low power levels of 1 to 10 mW. While this is fine for the
customer, it makes coexistence of the two systems a network planner's
nightmare.

In summary, cellular telephony is the culmination of several technologies which have progressed in parallel over the past two decades.
In fact, the progress has been so rapid that the ITU has had a problem
in organizing CCITT/CCIR meetings fast enough to determine standards that are consistent with the new technology.

8.2 Cellular Structures and Planning

The mobile telephone system uses a hexagonal "honeycomb" structure
of cells, with a base station at the center of each cell which gives radio
coverage to that cell and connects into the public telephone network
(Fig. 8.1). The hexagonal cell pattern arises from the best method of
covering a given area, remembering that radio coverage is ideally ra-
dial in nature. Figure 8.2 illustrates three possible methods of covering a particular area.

A quick observation of Table 8.1 reveals that the area of coverage of
a hexagon is twice that of a triangle, with a square midway between
the two. The area of overlap is calculated for a completely surrounded
cell (i.e., by six cells for the hexagon, four for the square, and three for
the triangle). The hexagon has a small overlap compared to the triangle. To cover an area of three hexagonal cells, or $7.8r^2$, would require

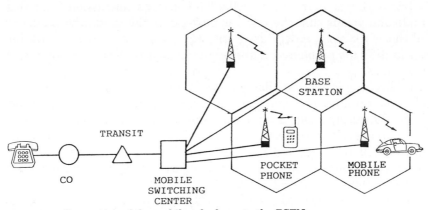

Figure 8.1 Connection of the mobile telephone to the PSTN.

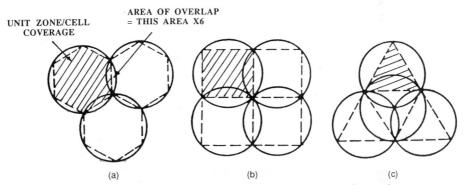

Figure 8.2 Cell structure possibilities. (*a*) Regular hexagon zoning; (*b*) regular square zoning; (*c*) regular triangle zoning.

TABLE 8.1 Cell Characteristics for the Three Main Cell Types

Cell type	Center-center distance	Unit zone coverage	Area of overlap	Width of overlap	Min. number of frequencies
Triangle	r	$\approx 1.3r^2$	$\approx 3.7r^2$	r	6
Square	$r\sqrt{2}$	$2r^2$	$\approx 2.3r^2$	$0.59r$	4
Hexagon	$r\sqrt{3}$	$\approx 2.6r^2$	$\approx 1.1r^2$	$0.27r$	3

six triangular or four square cells. The obvious conclusion from this simple analysis is that the regular hexagon is the most advantageous and therefore most widely used structure, with the triangle suitable only in difficult propagation areas which require deep overlapping of radio zones.

Initially, the limited available power transmitted by the mobile subscriber determined the cell size. As the subscriber moves (*roams*) between cells during a journey, the communication with the base station of the departing cell ceases and communication with the base station of the entering cell commences. This process is known as *handoff* in North America and *handover* in Europe. Each adjacent base station transmits a frequency that is different from its neighbor. The handoff is accomplished when the received signal from the base station is low enough to exceed a predetermined threshold. At the border between two cells the subscriber is under the influence of two or even three base stations, and the link could pass back and forth between base stations as the moving subscriber receiver experiences a fluctuating field strength depending upon the immediate environment, such as being surrounded by tall buildings. Some further intricacies of handoff are described in Sec. 8.8.1. In the real world, there are no true circular cells. The signal strength contour for each cell does not produce the pleasing precise pattern of Figs. 8.1 or 8.2 but can be very distorted and is usually more like the strange jigsaw puzzle shapes of Fig. 8.3. One can say that only the *nominal* cells are hexagonal. The nominal cell diameter also varies, depending on the traffic density. Typically, the center of a city is the most populated, with the suburbs gradually decreasing in population. This leads to the cells in the center having a small diameter with a gradual increase in diameter when moving outward. An example of a realistic, nominal city cell plan is shown in Fig. 8.4, which indicates the location of the base

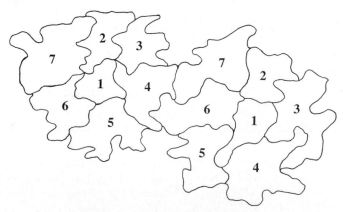

Figure 8.3 Cell structure in the "real world."

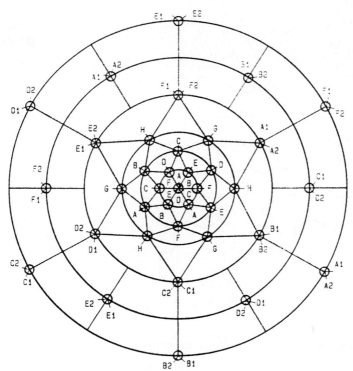

Figure 8.4 Example of the concentric nominal cell plan. (*Reproduced with permission from Ericsson Review, No. B, 1987, Lejdal, J., and Lindquist, H., p. 11, Fig. 1.*)

stations. The letters represent the choice of frequency groups. This is a concentric circle model, where the base stations are located at the corner points of the hexagons (six per circle). Variations on this nominal plan would be necessary to account for irregular field strength contours caused by buildings and irregular terrain. So, while the hexagon is the hallmark of mobile radio technology, it is only the design starting point, and the final real-life cellular structure may bear little resemblance to hexagons.

To enable the available bandwidth to be used efficiently and thereby increase the number of users, a frequency reuse mechanism is built into the cellular structure. In the diagram of Fig. 8.5, the cells are clustered into groups of seven, each group having the same pattern of seven base station frequencies. The distance between different base stations using the same frequency is D, which in this case is 4.6 times the cell radius R. A considerable effort is made to check that signal strength contours for each cell within the whole area of the system are not distorted to the point which would allow two calls to in-

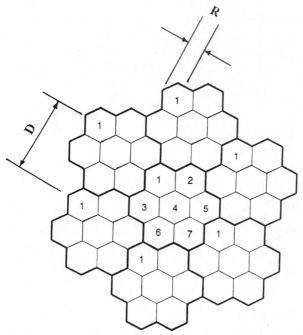

Figure 8.5 Frequency reuse using the hexagonal cell structure. D/R = 4.6.

terfere with each other. Note that each cell can use only one-seventh of the channels available within the system.

There is a trade-off here between interference and channel capacity. If the number of cells in the pattern was increased beyond seven, clearly the distance between the base stations of identical frequency would be greater. The cochannel interference would correspondingly decrease. Unfortunately, the system capacity would drop because the number of channels per cell is simply

$$\frac{\text{Total number of channels per repeat pattern}}{\text{Number of cells (base stations) within the repeat pattern}}$$

For the hexagonal cell, the number of cells within a regular repeat pattern can be only 3, 4, 7, 9, 12, and multiples of these values. Seven is the most usual choice for cellular radio systems, as a compromise between degradation due to cochannel interference and high channel capacity.

A cellular structure with smaller and smaller cells is evolving for two main reasons: first, the increasingly limited available power which can be transmitted by the smaller and smaller mobile subscriber telephone sets, and second, increased capacity. Whereas the mobile base station can have a large power output and therefore cover a radius of 50 km or more, the move to smaller, pocket-sized mobile telephones restricts the return path distance to only a few hundred meters. For example, a transmitted output power of 5 mW can reliably work over a distance of about 300 m. The small cell size of less than 1-km radius is called a *microcell* structure and has the major benefit of providing orders of magnitude improvement of system capacity over the large (20-km) cell size. The increase in channel capacity is approximately inversely proportional to the square of the microcell radius. For example, reducing the cell radius from 8 km to 150 m increases the network capacity by a factor of more than 2500. In a rapidly expanding market, keeping up with the ever-increasing customer demand for access to the mobile network is a major problem. It is currently believed that the microcell structure will satisfy that demand.

Capacity can be enhanced even further by cell sectorization. As cell sizes decrease, the distance between base stations of identical frequency also decreases. This can be offset by careful control of the power radiated from either the base station and/or the mobile station. That process is described later, but another way to reduce the cochannel interference coming from the six surrounding cells of a seven-cell structure is to use several directional antennas at each base station. If three antennas each cover 120°, the cell is split into three sectors, each having its own channel set (Fig. 8.6a). In a seven-cell repeat pattern, the three sectors allow 21 channel sets instead of 7. Further sectorization using six antennas (Fig. 8.6b) results in a four-cell repeat pattern (instead of seven), providing a total of 24 channel sets. Notice that greater sectorization has diminishing returns because the cell pattern becomes repeated more often (over a shorter distance) as the number of sectors increases. Sectorization may appear to be a disadvantage because there are fewer channels per sector and therefore a

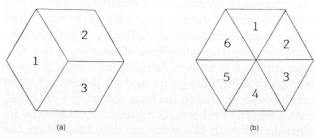

(a)   (b)

Figure 8.6   Cell sectorization. (*a*) Three sectors; (*b*) six sectors.

worsening of the trunking efficiency, which means reduced traffic capacity for a given level of blocking. In fact, sectorization results in an *enhanced* capacity because the cell sizes can be greatly reduced.

As most telecom organizations have followed the path from analog FM to digital microcell mobile radio, there have been some formidable technical and economics problems. When the system becomes saturated for a given cell size, it becomes necessary to add new radio base stations for more channels, and consequently the cell size reduces and frequency reuse increases. This must be done without impairment to the voice quality. Just as important, the infrastructure cost per subscriber must not exceed critical limits which would preclude affordability. In addition, there is the major technical problem of the coexistence of mixed analog and digital technologies during the transition from the old analog to the new digital systems. Perhaps just as difficult is the problem of coexistence of relatively high-power vehicular mobile radio systems and the low-power portable systems. The frequencies of operation must be different for the analog and digital systems; otherwise severe cochannel interference would be intolerable.

From the planning point of view, there are some problems which differ quite considerably from the usual network planning. For example, the traditional telephone network generally expands as the population of urban areas expands. Demographic data usually provide a starting point for the network planner and projected population increases allow an estimation of future demand. The customers for a mobile telephone system live mainly in towns and consist of up to millions of people. A large percentage of homes may already possess a telephone. The question is, How many of them want to "go mobile" and have a pocket telephone? If the mobile service could be provided without any additional cost to the customer, the answer to the question is, probably, *all of them*. The mobile network planner must be acutely aware of how much extra, on average, the subscriber is prepared to pay for the privilege of mobility. The potential demand is already known to be enormous, if the price is right. Maximizing system economy while maintaining voice transmission quality is the difficult balancing act. If the requirement of data transmission is brought into the picture, there is an added dimension of difficulty. Furthermore, the newly mobile population, by the definition of mobile, is not predictably at specific locations for known lengths of time. Suppose 50,000 people decide to go to a football game and 25 percent of them want to use their portable telephones to make a call at the interval. This obviously presents a serious problem. Similarly, cities are the focus of population migration during the working day, and when the people return home at the end of the day, they cause the mobile channel requirements to shift to the city suburbs.

The evolution of the vehicular mobile telephone is substantially different from the portable telephone. In the initial start-up phase, in a new area, the vehicular mobile system has only a few customers and therefore only a few cell sites which are widely separated to provide a large area of coverage while minimizing the cost per customer. As the number of users increases, the density of base stations increases accordingly. The portable telephone, in contrast, provides service to a large number of customers within a relatively densely populated area. Such areas, or islands, can then be interlinked to provide a wider area of coverage.

Mobile location.   For a mobile subscriber to receive a call, the subscriber's precise location must obviously be known. There are several possible ways to track the movement of a mobile station. A convenient method is to split up the whole cellular network into a number of location areas, each having its own ID number. Each base station within a particular area periodically transmits its area ID number as part of its system control information. As the mobile station moves from the control of one base station to another, eventually it will move across to a new ID number region, and the network is updated with the new area in which it can be found. This is all done using signaling information transfer.

8.3 Frequency Allocations

As usual, in the absence of global standards, mobile radio system designs in different parts of the world have evolved using frequencies in the bands made available by their national frequency coordinators. Inevitably, the frequencies chosen are different from one country to another. While this may not be a problem in a large country like the United States, it renders the mobile telephone useless to many European users who travel across the borders of several countries to conduct their normal daily business. In the 1970s the Federal Communications Commission (FCC) in the United States allocated the frequencies shown in Table 8.2 for land-based mobile radio telephones. The accompanying diagram, Fig. 8.7, illustrates how the allocated frequency spectrum is used.

The 800-MHz band has been allocated for the use of cellular mobile radio, and the original systems used analog modulation. In the not too distant future, this band will be probably completely digital, although, at present, it contains both analog and digital. Full-duplex op-

TABLE 8.2   Comparison of Spectrum Usage

	Base transmit (MHz)	Mobile transmit (MHz)	CH spacing (kHz)	Total bandwidth (MHz)	Number of channels
U.S.	Analog + digital				
	870–890	825–845	30	20 + 20	666
Europe	Analog modulation				
	935–950	890–905	25	15 + 15	600
	935–960	890–915	25	25 + 25	1000
	Digital modulation				
	950–960	905–915	200	10 + 10	400
	935–960	890–915	200	25 + 25	1000
	(Eventually)	(Eventually)			
Japan	Analog modulation				
	870–885	925–940	25	15 + 15	600–1000
	Digital modulation				
	860–885	915–940	12.5	25 + 25	2000

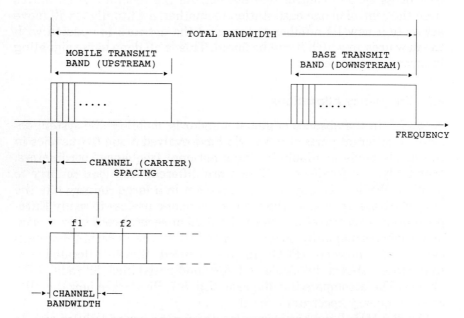

Figure 8.7   Cellular mobile radio frequency structure. f1, f2,..., fn are carrier frequencies.

eration is made possible by using 20-MHz upstream and 20-MHz downstream carriers, separated by 45 MHz; 870 to 890 MHz is the base station transmit frequency band, and 825 to 845 MHz is the mobile station transmit frequency band. The carrier spacing is 30 kHz.

Personal communications network (PCN) is a term now associated with microcellular technology (see Sec. 8.9) which specifically uses portable or pocket-sized telephones. The trend for these systems is to use higher frequencies than previous mobile systems, notably the 1.7- to 2.3-GHz band. Although most of this band is already allocated, the FCC is under considerable pressure to allocate exclusively, or at least up to 200 to 300 MHz of this band for PCNs.

In Europe, the frequencies in Table 8.2 have been adopted for land-based mobile radio telephones. Initially, the AM systems operated in the 935- to 960-MHz (base transmit) and 890- to 915-MHz (mobile transmit) bands. On introduction of the digital modulation systems, they were allocated the upper 10 MHz of upstream and downstream bands, but eventually the analog systems will be displaced and the full 2 × 25 MHz will be for digital.

In Japan, yet another set of frequencies has been chosen, as indicated in Table 8.2. As in the United States, the newer digital systems operate in frequency bands that are common to the analog systems. Upstream and downstream frequencies are separated by 55 MHz.

8.4 Propagation Problems

The following discussion is primarily related to moving vehicles but is equally applicable to portable telephone users, where the vehicle is just one of the propagation situations encountered and perhaps the most difficult to quantify. Radio propagation is one of the most fundamental problems in mobile communication engineering. Unlike fixed point-to-point systems, there are no simple formulas that can be used to determine anticipated path loss. By the nature of the continuously varying environment of the mobile subscriber, there is a very complicated relationship between the mobile telephone received signal strength and time. The situation is not so bad if the mobile unit remains stationary for the duration of the call. If a particularly poor reception point is encountered, a short move down the road may significantly improve the reception. For a moving subscriber, as in the typical call placed from a moving car, the signal strength variation is a formidable problem which can be approached only on a statistical level. For land-based mobile communications the received signal variation is primarily the result of multipath fading caused by obstacles such as buildings (described as "clutter") or terrain irregularities.

The obstacles (clutter) can be classified in three areas, as follows:

1. *Rural areas,* in which there are wide open spaces with perhaps a few scattered trees but no buildings in the propagation path
2. *Suburban areas,* including villages where houses and scattered trees obstruct the propagation path
3. *Urban areas,* which are heavily built up with large buildings or multistory houses or apartments and greater numbers of trees

The terrain conditions can be categorized as follows:

1. *Rolling hills,* which have irregular undulations but are not mountainous
2. *Isolated mountains,* where a single mountain or ridge is within the propagation path and nothing else interferes with the received signal
3. *Slopes,* where the up- or downslopes are at least 5 km long

8.4.1 Field strength predictions

Perhaps a little surprisingly, experiments have shown that it is in the *suburban* areas that the received field strength suffers the widest fluctuations.

A typical plot of the received field strength measured while traveling is shown in Fig. 8.8. When moving at any specific fixed radius from the base station between a rural area and an urban area, a 30- to 40-dB variation of field strength can be expected. The average field strength is lowest in urban areas, followed by suburban, then rural. The rate of variation of field strength is proportional to the product of the frequency of the carrier wave and the speed of the mobile subscriber. Incidentally, for high-velocity mobile units the doppler effect must be taken into account in the equipment design.

For urban areas located on relatively flat terrain, Fig. 8.9 shows a graph of the statistically predicted median attenuation relative to free space plotted against frequency at various distances from the base station having an effective antenna height of 200 m and the mobile

900-MHz   BAND

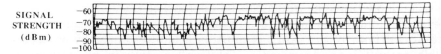

SIGNAL
STRENGTH
(dBm)

Figure 8.8   Typical field strength of a moving mobile station.

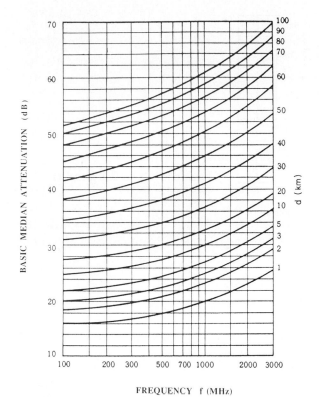

BASIC MEDIAN ATTENUATION (dB)

FREQUENCY f (MHz)

Figure 8.9 Graph of urban areas median attenuation against frequency for quasi-smooth terrain. Base station antenna height = 200 m; mobile station antenna height = 3 m. [*Reproduced with permission from Ref. 342, p. 110 (from Ref. 6), © 1982 McGraw-Hill.*]

antenna height of 3 m. These curves were derived by Okumura et al. based on empirical data. For example, at 900 MHz over a quasi-smooth terrain distance of only 1 km, the attenuation in addition to that of free space is approximately 20 dB. Figure 8.10 gives the predicted correction factors which must be included in the calculation depending upon whether the mobile unit is traveling in an urban area in the direction of propagation or at 90° to the direction of propagation. The difference can be substantial, particularly at small distances. For example, at a distance of 5 km, there is more than 11-dB difference in the attenuation values.

In suburban areas, where the effects of obstacles are less severe than in urban areas, the field strength is generally higher. Figure 8.11 indicates the predicted correction factor which must be sub-

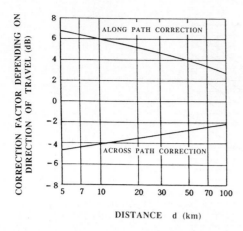

Figure 8.10 Correction factor for direction of travel.

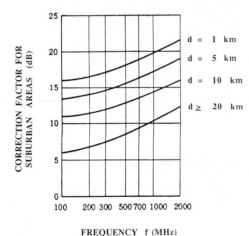

Figure 8.11 Correction factor for suburban areas.

tracted from the Fig. 8.9 attenuation values for suburban areas. Notice that the correction value is larger, signifying more improvement, at the higher frequencies.

Figure 8.12 shows the predicted correction factor for rural areas. The corrections to be subtracted from the Fig. 8.9 attenuation values are even greater, which indicates an even larger received field strength than in suburban and urban areas.

Clearly, the antenna height of the base station is a very important parameter in determining the field strength. Figure 8.13 shows how the estimated field strength is increased or decreased depending on whether the base station antenna height is above or below 200 m.

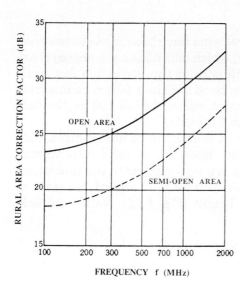

Figure 8.12 Correction factor for rural areas. [*Reproduced with permission from Ref. 342, p. 111 (from Ref. 6), © 1982 McGraw-Hill.*]

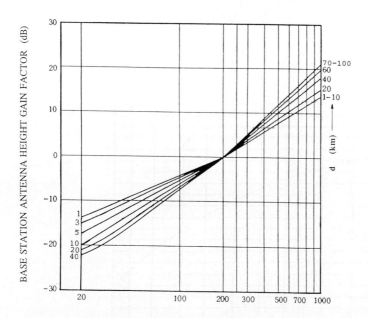

Figure 8.13 Multiplication factor for urban area antenna heights other than 200 m.

8.4.2   Effects of irregular terrain

For a rolling hilly terrain, measurements have been made to establish the field strength correction factor, which depends on the degree of undulation of the hills. Figure 8.14 illustrates this correction factor calculated for the 900-MHz frequency band within a 10-km radius from the base station. For a peak-to-trough rolling hill of 100 m, the field strength suffers an extra loss of about 7 dB.

Operating within the vicinity of an isolated ridge or mountain, field strength measurements have been made which take into account knife-edge diffraction effects. They are very similar to the theoretical correction factors depending on the distance from an isolated ridge or mountain of nominally 200 m in height (Fig. 8.15). Notice the field

DEFINITION OF Δh

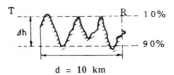

d = 10 km

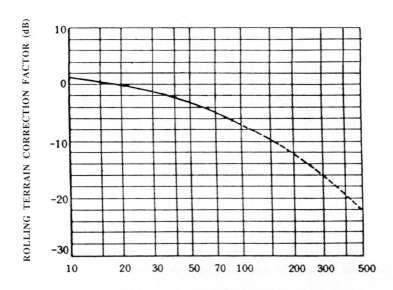

THEORETICAL TERRAIN UNDULATION HEIGHT Δh (m)
(900-MHz BAND)

Figure 8.14 Field strength correction factor for rolling hill terrain.

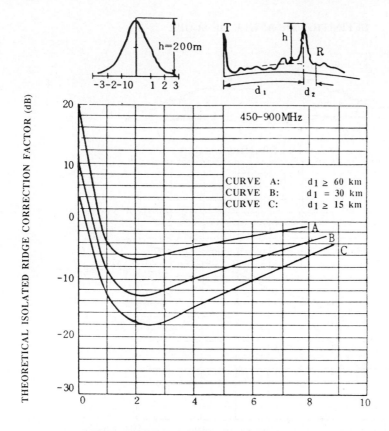

DISTANCE FROM ISOLATED RIDGE TOP d_2 (km)

Figure 8.15 Correction factor for an isolated ridge.

strength *enhancement* on the top of the ridge and the maximum attenuation values in the immediate shadow of the ridge.

Finally, Fig. 8.16 shows the measured and interpolated correction
factors for terrain which slopes for a distance of at least 5 km. A positive slope obviously enhances the signal strength, whereas the negative slope decreases it.

There are other factors that also come into the picture—for example,
tree foliage can cause as much as a 10-dB attenuation variation between the winter and summer months. Tunnels, bridges, subways,
and close proximity to very high skyscrapers can produce highly unpredictable signal strength variations.

The propagation evaluations are at best an approximation with a
relatively high degree of uncertainty. Nevertheless, they do provide

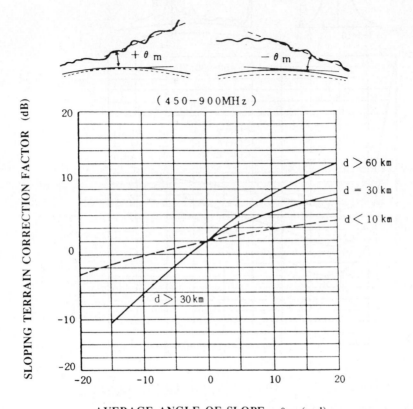

Figure 8.16 Correction factor for sloping terrain.

guidelines for designers who are deciding where to place base stations in order to provide the best signal coverage for the terrain to be covered.

8.5   Antennas

Previously described microwave antennas have all been parabolic in nature. Those were highly directional, since it was necessary to focus the energy in one direction. Conversely, mobile station radios require antennas to be omnidirectional *in the horizontal plane* (looking from above) but to have very little upward radiation. This is because, at any time, the mobile unit could be at any point around the full 360° range of the base station antenna. Note that these are not isotropic antenna qualities, which would require equal radiation in *all* direc-

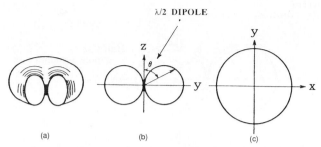

Figure 8.17 Dipole antenna radiation pattern. (*a*) Radiation pattern; (*b*) vertical plane; (*c*) horizontal plane.

tions. There are several styles of antenna that can be used for this purpose. Since these antennas are generally variations on the dipole antenna, a few comments on the dipole antenna are necessary. The dipole is the simplest of all antennas as far as physical construction is concerned (Fig. 8.17), but even this antenna has some rather elaborate mathematics associated with it. The radiation pattern for a short piece of metallic rod is shown in Fig. 8.17, which indicates that it is indeed nondirectional in the horizontal plane, although there is some directivity in the vertical plane. When the dipole is fed from the center of the metal rod antenna such that the total length is one-half wavelength, as one would expect, this is referred to as the half-wave dipole antenna, as in Fig. 8.18. The gain of a half-wave is about 2 dB (i.e., relative to the isotropic antenna).

The impedance of antennas is important because the idea is to match the impedance of the transmitting device to that of free space so that 100 percent energy transfer takes place. If there is a bad mismatch, reflections from the antenna back into the transmitter cause severe interference. In the analogous case of the TV antenna, the interference caused by mismatch is observed as "ghosting" on the picture.

8.5.1 Base station antennas

There are several types of antenna frequently used for the mobile base station, and three are illustrated in Fig. 8.19.

1. *Bent or folded dipole antenna (Fig. 8.19a).* This is constructed as a bent or folded conductor whose horizontal dimension is one-half wavelength.

2. *Ground plane antenna (Fig. 8.19b).* The coaxially fed antenna is physically convenient for many applications. The finite ground plane tends to incline the radiation pattern maximum slightly upward in-

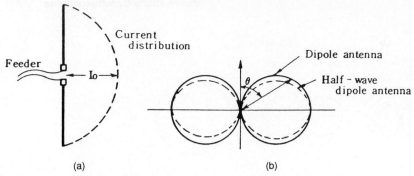

(a)   (b)

Figure 8.18 The half-wave dipole antenna. (*a*) Current distribution; (*b*) vertical pattern.

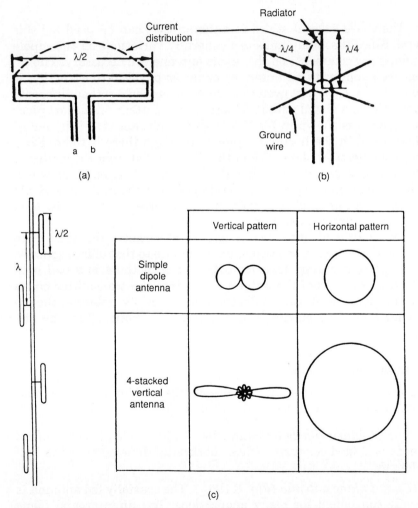

(c)

Figure 8.19 Base station antennas. (*a*) Bent or folded dipole antenna; (*b*) ground plane antenna; (*c*) stacked antenna.

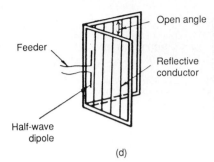

Figure 8.19 Base station antennas. (*d*) Corner reflector antenna.

stead of horizontally. This is not usually desirable and can be circumvented by several techniques for improving the ground plane.

3. *Stacked antenna (Fig. 8.19c).* A stack of several half-wave dipole antennas reduces the radiation in the vertical direction and effectively increases the omnidirectional horizontal gain. For example, a stack of four folded dipoles increases the gain to about 6 dB.

Variations on the stacked antenna can be used if the gain in the horizontal direction needs to be asymmetrical to illuminate preferentially some areas of a cell which may have some geographical or building screening problems.

4. *Corner reflector antenna (Fig. 8.19d).* For the case of cell sectorization, the base station antenna must radiate only over a specific angle (e.g., 60°). For this antenna, a half-wave dipole is placed in the corner of a V-shaped wire plane reflector at 0.25- to 0.65-λ spacing from the vertex. The wires are typically spaced less than 0.1 λ apart. With an angle of 90°, a typical gain of at least 8 dB relative to a half-wave dipole can be achieved. As the angle is made smaller, the directivity (gain) increases.

In order to provide coverage to some difficult areas, it may be necessary to use an antenna that has a high directivity or preferred orientation of radiation. A corner reflector may not have enough gain, in which case the high-gain YAGI could be used.

8.5.2   Mobile unit antennas

The antennas used for car mobile radios must be omnidirectional and as small as possible and must not be adversely affected by the car body. The location of the antenna is usually on the roof, trunk (boot), or rear fender (bumper). The whip antenna of Fig. 8.20a is one possibility. This is a λ/4 vertical conductor with the body (preferably the

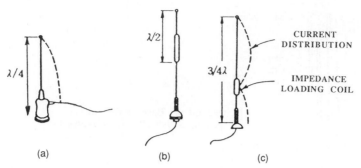

Figure 8.20 Mobile station antennas. (*a*) Whip antenna; (*b*) λ/2 coaxial sleeve antenna; (*c*) high-gain whip antenna.

roof) acting as a ground plane. The compact size makes this a very popular style. The center of the roof is the best location so that the car body least affects the signal. Figure 8.20*b* is a λ/2 coaxial antenna. Figure 8.20*c* is a high-gain whip antenna with a loading impedance coil used to adjust the current distribution.

Portable radio antennas are usually either short (λ/4 or less) vertical conductor whip antennas or normal-mode helical designs. There are a variety of options. Some are coaxial fed and retractable for compactness, whereas others are detachable. Still others are encased in a plastic or rubber material for ruggedness. Ideally, for maximum convenience, the antenna should not even protrude from the body of the portable radio. The antenna used for portables is a difficult problem. First, there is no ground plane as is the case for automobiles, so its efficiency is reduced. Second, the user may not be pointing the antenna in the optimum direction, or, even worse, it may be held almost horizontally. In either case the antenna may not be in the best orientation for the correct polarization reception. Third, the user's head may cause disturbances by mismatching the antenna impedance. Incidentally, the long-term health effects of holding a portable telephone close to the head for extended periods of time are at present unknown.

8.6   Types of Mobile Systems

Historically, the technology has lagged behind the design of mobile telephone systems, and it was only as recently as 1983 that the first good-quality systems were put into operation. Since those early, low-capacity pioneering systems, the subscriber demand has mushroomed, despite the significantly increased cost of calling from a mobile or portable telephone. The different types of mobile systems, in terms of frequency spectrum usage, are summarized in Fig. 8.21. They differ primarily in modulation technique and carrier spacing.

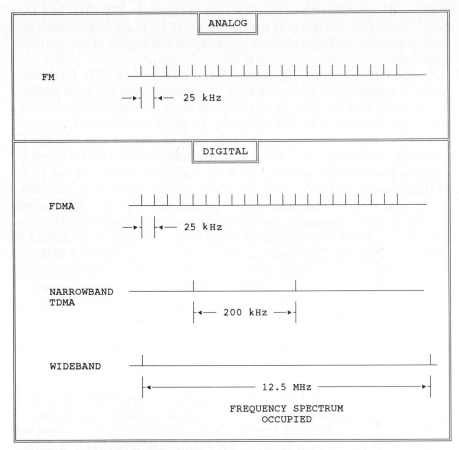

Figure 8.21 Types of cellular radio systems.

Analog FM. The first-generation cellular systems in operation were analog FM radio systems which allocated a single carrier for each call. Each carrier was frequency modulated by the caller. The carriers were typically spaced at 25 kHz intervals (i.e., carrier bandwidth). The allocated bandwidth was relatively narrow, and only a few channels (typically 12) were available.

Digital FDMA. The FDMA systems resemble analog FM, with the exception that the carrier is modulated by a digitally encoded speech signal. The bandwidth of each carrier is similar to the analog FM systems (i.e., 25 kHz).

Digital narrowband TDMA. The TDMA systems operate with several customers sharing one carrier. Each user is allocated a specific time slot for transmission and reception of short bursts or packets of information. The bandwidth of each carrier is typically 200 kHz, and the total bandwidth available is in the region of 10 to 30 MHz, which allows a reasonably large channel capacity in the region of 500 to 1000 channels.

In the United States, for example, the 823- to 849-MHz frequency band is allocated for the one-way transmission from the base station to the user, and the 868- to 894-MHz band is allocated for transmission from the user to the base station. To enable two competitive systems to operate simultaneously, only half of each of these bands is available to each operator. Each system therefore has 12.5 MHz available for transmission and 12.5 MHz for reception. Each of these 12.5-MHz bands is subdivided into 30-kHz channels for voice communication. If the cellular structure chosen is that shown in Fig. 8.5, having a total of seven frequency bands for the hexagonal base station pattern, each base station can be allocated only 12.5 MHz/7, or about 1.8 MHz of bandwidth. The resulting frequency reuse enables 1.8 MHz/30 kHz = 60 channels per cell to be used for voice communication, which is reduced to about 55 since some channels are necessary for signaling. As stated in Chap. 3, if the TDMA system is used, three users per 30-kHz channel can be accommodated without noticeable degradation, and twelve users per 30-kHz channel can be accommodated if some degradation is accepted. The final value for the number of users per cell is 55 × 3 = 165 (or 55 × 12 = 660 with degradation).

Digital wideband.   One form of digital wideband operation which has good future potential is CDMA. In these systems there is a single carrier which is modulated by the speech signals of many users. Instead of allocating each user a different time slot, each is allocated a different modulation code. Mobile users in adjacent cells all use the same frequency band. Each user contributes some interfering energy to the receivers of the fellow users, the magnitude of which depends on the processing gain (see Sec. 3.4.3). In addition to interference from users within a given cell, there is also interference from users in adjacent cells. Because the distance between adjacent users attenuates the interference considerably more than users within the same cell, the increase in interference energy by the adjacent users can be approximately 50 percent. Since this is considered to be acceptable, frequency reuse is unnecessary. Consequently, each cell can use the full available bandwidth (12.5 MHz) for CDMA operation. The total number of users per cell has been estimated to be

$$M \approx 3N/8$$

where N is the processing gain, and the signal power is assumed to be at least 10 dB higher than the thermal noise power in the receiver.

For a 12.5-MHz available bandwidth, the spreading sequence is taken to be 12.5 Mb/s, and if the voice signals are each digitized to 8.5 kHz, N = 12.5 MHz/8.5 kHz = 1470 (31.7 dB). M is therefore = 551 (or about 1900 with 2.4-kb/s voice digitization, that is, with degradation). Since about 50 percent of the interference comes from users located in adjacent cells, the number of users in each cell is approximately M/1.5 = 367 (or about 1270 with degradation).

This figure can be improved by making use of the fact that during any conversation only one voice is active, at least in any comprehensible conversation. Since a two-way link is set up for each call, this means that, statistically speaking, a two-way link is active for only 50 percent of the conversation time. This situation allows the employment of digital speech interpolation (DSI), which is a technique originally devised for satellite communications technology. A circuit senses inactivity on either direction of a two-way link, seizes that one-way channel, and gives it to a person who starts to speak by a voice activation method. This technique enables a factor of 2 enhancement of the number of users to be achieved, compared to the number of channels available. Care must be taken to ensure that such a system is operating correctly; otherwise a very annoying clipping of the first word occurs each time a responding speaker starts to talk. By including the voice activation factor, the number of users per cell could be 734 (or about 2540 with degradation). By comparing these figures with TDMA (i.e., 165 users, or about 660 with degradation), it is evident that voice activated CDMA has considerable potential for future mobile communications.

8.7 Analog Cellular Radio

Analog cellular systems were used exclusively in the early days of mobile communications. Although they are being superseded by the digital technology, a large number of systems are still in service and will probably continue to be used for several years in the future.

Analog FM. The analog FM cellular radio systems are relatively old technology (1980 to 1985) in this fast-moving industry. These are the first-generation cellular radio systems. However, it is informative to discuss briefly some aspects of these systems since they provide insight into how future systems are evolving. Analog cellular radio was initially designed for vehicle-mounted operation. By 1990, more than 50 percent of mobile stations in most networks were hand-held portables, and the demand was growing.

As far back as 1979, Bell Labs designed and installed a trial cellular mobile system called the Advanced Mobile Phone Service (AMPS). This was really the birth of cellular radio in the United States. It still is the basis of the analog systems in operation today. The AMPS system uses the hexagonal cell structure, with a base station in each cell, as illustrated in Fig. 8.22. The cells are clustered into groups of seven cells (i.e., a seven-cell repeat pattern). AMPS covers large areas with large-sized cells, and high-traffic-density areas are covered by subdividing cells. Sectorization is also used to enhance capacity. The overall control of the system is by a mobile telephone switching office (MTSO) in each metropolitan area. This digital switch connects into the regular telephone network and provides fault detection and diagnostics in addition to call processing. The mobile unit (station) was originally installed in a car, truck, or bus. The frequencies for AMPS are 870 to 890 MHz from base station to mobile station and 825 to 845 MHz from mobile to base station. Each radio channel has a pair of one-way channels separated by 45 MHz. The spacing between channels is 30 kHz. The AMPS system uses FM with 12-kHz maximum deviation. FM has a convenient capture mechanism. If a receiver detects two different signals on the same frequency, it will lock onto the stronger signal and ignore the weaker, interfering signal.

The mobile stations are microprocessor controlled. VLSI electronic

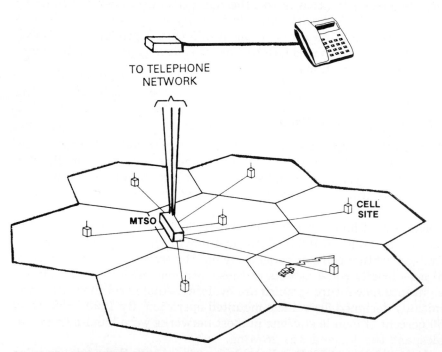

TO TELEPHONE
NETWORK

CELL
SITE

MTSO

Figure 8.22   The AMPS cellular radio network structure. (*Reproduced with permission from Ref. 329, © 1983 Artech House.*)

circuits have been crucial in reducing the size and weight of hand-held portables. The MTSO periodically monitors the carrier signal quality coming from the active mobile stations. If, during a call, a mobile station moves to the edge of a cell boundary and crosses the boundary, the signal quality from the adjacent cell gradually becomes better than the service provider, so handoff is initiated. The handoff command is a "blank and burst" message sent over the voice channel to the designated cell. A brief data burst is transmitted from the base station that is providing service to instruct the mobile station microprocessor to retune the radio to a new channel (carrier). The voice connection is momentarily blanked during the period of data transmission and base station switching. This interruption is so brief that it is hardly noticeable, and most customers are unaware of its occurrence.

All of the call setup is done by a separate channel. These are dedicated signaling channels which transmit information only in the form of binary data. These channels are monitored by all mobile stations that do not have a call in progress. When a mobile station is first switched to the ON mode or is at the end of a call, it is in the idle state. It scans the frequencies used for call setup and monitors the one providing it with the strongest signal. Each cell has its own setup channel. The mobile station periodically makes the scan to see if its change of position has made the setup channel of an adjacent cell stronger than the one it is monitoring. If a customer from, for example, a home initiates a call to the mobile station, the telephone number of the particular mobile station is transmitted over the setup channel of every base station. This is a type of paging mechanism. The mobile station identifying its own number sends a response message, and the call connection procedure starts. This is done by the MTSO which assigns an available voice channel within the cell where the mobile station is located. It then connects the home to the cell base station. Another data message is sent over the setup channel to instruct the mobile station to tune to the assigned voice channel. The mobile station activates ringing to alert the owner to pick up the handset, thereby establishing the call. When either party hangs up, another data message is sent over the setup channel to instruct the mobile station to switch off its transmitter power and revert to the idle mode. Calls initiated in the opposite direction (i.e., mobile station to fixed user) are similar. The mobile customer dials a number which is stored in a register. On pressing the Send button, the mobile station sends a message over the setup channel identifying itself and giving the number of the customer called. The MTSO then assigns an available voice channel and makes the necessary connection and ringing to the called customer. It also instructs the mobile station to tune to the assigned channel, and the call is established.

The cellular radio service has nationwide roaming capability. This is possible by cooperation between the service providers in different parts of the country. The AMPS system has been very successful, but its main disadvantage is that its total system capacity is inferior to the more advanced digital cellular radio systems.

8.8   Digital Cellular Radio

The digital cellular radio can be divided into two categories, narrowband and wideband. The narrowband systems are often considered to be the second generation of cellular radio, which is the main technology of today.

8.8.1   Digital narrowband TDMA

Although the digital narrowband TDMA systems in North America and Europe have developed along similar lines, there remain many features that are different. In an attempt to continue the international flavor of this text, the main focus of attention for digital narrowband TDMA cellular radio will be the pan-European system called GSM, to which the U.S. system called IS-54 will be compared. The acronym GSM initially stood for Groupe Speciale Mobile, the planning organization that did much of the groundwork for the TDMA cellular system. GSM now stands for global system for mobile communications. In the eyes of many, the vast complexity of this system makes it the most advanced in the world, including military systems. No doubt there are quarters in which this may be disputed, but it is undeniably highly sophisticated and uses some of the most advanced electronics technology available. It is anticipated that by 1999 there will be 10 million subscribers to the system. A description of its features serves to highlight some of the intricacies of present-day cellular radio systems.

European GSM system.   The GSM system was designed to provide good speech quality with low-cost service and low terminal cost and to be able to support hand-held portables that have a new range of services and facilities, including *international roaming*. This last feature alone is a significant technological milestone. The system became operational in July 1991. The GSM Recommendations total approximately 6000 pages, and only the main features will be discussed here. Although this system is designed to provide full ISDN capability (see Chap. 9), hand-held portables for voice communication will probably dominate the market. The maximum data rate that can presently be provided is 9.6 kb/s, which is actually less than the ISDN standard. The real difference between a mobile system and a fixed public

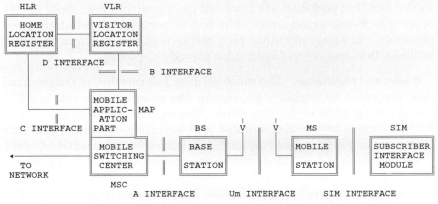

Figure 8.23 The GSM cellular radio network structure.

switched telephone network (PSTN) is the radio section. However, some network differences appear, particularly additional interfaces, and they are highlighted in Fig. 8.23. The home location register (HLR) is where all management data for the home mobiles is stored. The visitor location register (VLR) contains selected data for visiting mobile stations, including international mobile station identity and mobile station location. The HLR and VLR can be located in the same room as the mobile services switching center, and the quantity of HLRs and VLRs depends on how extensive the network becomes. As a mobile station moves into an area other than its home location, the ID number is passed on to the appropriate VLR via the nearest base station, then on to the switching center. A temporary mobile identity and roaming number is allocated to the mobile station by the network being visited. This facilitates the origination and reception of calls in that area. The HLR provides the VLR with the services that are available to a particular mobile station. Even with this very simplified view of the network, one can see that there is a high level of signaling activity over both the radio link and the fixed portion of the mobile network for call processing and mobile station management. CCITT signaling system number 7 is used for the GSM network. The mobile application part (MAP) is covered in GSM Recommendation 09.02, an extensive document of approximately 600 pages in length.

The radio link. To cater to both vehicle-mounted and hand-held portable mobile stations, the peak output power ranges from 20 W (43 dBm) for vehicles and 2 to 5 W (33 to 37 dBm) for portables. The maximum receiver sensitivity of either vehicles or base stations is -104 dBm and it is -102 dBm for the portables, giving the vehicle-mounted mobiles an operational advantage of several decibels. This is balanced by the fact that portables are intended for use within microcells in city centers or dense urban areas. The use of portables inside buildings, elevators, subways, and other poor radio reception locations is still a problem that has not yet been fully solved.

Frames and multiframes. The voice or data information is transmitted over the radio by digitally processing the voice or data into a PCM multiplexing style of format, using frames and multiframes. These TDMA signals are then modulated and superimposed on RF carriers that have a 200-kHz bandwidth. Each frame is constructed of eight time slots each of 576.923-µs duration, producing a 4.6154-ms frame, as in Fig. 8.24.

There are two types of multiframe. The 26-frame multiframe is 120 ms in duration and has 24 frames which carry traffic and two associated control channels. One channel is a slow continuous stream for supervising calls and is known as the slow associated control channel (SACCH). The other operates in the burst-stealing mode and is for power control and handoff. This is known as the fast associated control channel (FACCH). If required, a traffic channel can be taken (stolen) and used for the power control and handoff time slot.

There is also a 51-frame multiframe, where time slot zero is used for downstream broadcast control to update the mobile station with the base station identity, frequency allocation, and frequency-hopping se-

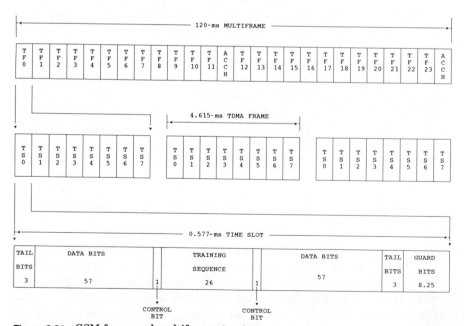

Figure 8.24 GSM frame and multiframe structure.

quence information. All other time slots in the multiframe carry voice and data traffic information. This multiframe is placed on a non-frequency-hopping RF carrier. Each base station is allocated several RF carriers, but only one of these includes the broadcast control channel.

Every frame clearly has considerable overhead incorporated. In each 156.25-bit time slot, two bursts of 57 bits are for data and the rest are overhead. Also, each time slot has a 26-bit training sequence. This specific bit sequence is already known to all mobile stations and serves to improve reception quality in a multipath fading environment. The mobile circuitry compares the known sequence with the detected sequence, and any disparity between the two allows a form of equalizer using digital filters to compensate for the errors introduced by fading. This process is invaluable for improving communication quality when the base station to mobile station distance exceeds several hundred meters, which is typically when multipath fading can become serious. A flag bit at the beginning and end of the training sequence separates it from the two data bursts. The complete time slot begins and ends with three "tail" bits, all logical zero. Finally, there is a guard time of 8.25 bits to ensure that at any base station there is no overlap of bits arriving from several mobile stations. Since 156.25 bits are transmitted in 576.923 μs, the transmission bit rate is 270.833 kb/s, of which more than one-fourth are overhead bits.

Speech coding. The initial reduction in bit rate from the 64-kb/s PCM to 32-kb/s ADPCM is inadequate for cost-effective operation of a high-capacity narrowband TDMA cellular radio network. Since the bit rate is reduced by the various techniques such as regular pulse excitation, linear predictive encoding, and other elaborate algorithms, the delay time introduced into the transmission increases. A 13-kb/s rate has been chosen for GSM, which causes an additional transmission delay of 20 to 50 ms depending on the manufacturer. Other functions are also built into the codec chips, as follows:

Voice activity detection to enable discontinuous transmission, otherwise known as voice activation. This reduces the cochannel interference and therefore allows greater frequency reuse. It also saves battery power.

Comfort noise introduction so that the person speaking does not think that connection with the called party has been lost.

Speech interpolation to smooth over errors originating in the radio path.

In addition to the inherent security offered by digital speech processing, further security can be easily applied by digital encryption, if required.

Channel coding. Channel coding is used to provide protection against error bursts and random errors. This is done by convolution coding at a rate of half-block diagonally interleaving over eight TDMA frames. Convolution coding is known to be more effective than binary block coding as a countermeasure against random errors. Also, when convolution coding is combined with block interleaving, it provides a protection against error bursts that is similar to Reed-Solomon coding (see Chap. 3). Viterbi algorithm decoding is used at the receiving end.

Modulation and spectral usage. The GSM system uses Gaussian MSK (GMSK) modulation with a modulator bandpass filter having a 3-dB bandwidth of 81.25 kHz. This is 0.3 times the bit rate and allows the signal baseband (270.833 kb/s) to be transmitted on a 200-kHz bandwidth RF carrier. The bandwidth efficiency is therefore 270.833/200 = 1.35 b/s/Hz. The total spectrum available for the network is 2×25 MHz, so 125 carriers can be used, each containing eight TDMA channels. The total channel capacity is $8 \times 125 = 1000$ channels. Other cellular radio systems can achieve a similar channel capacity, so one may ask the question, In what way is the digital system better? The GSM system is designed to operate at a carrier-to-interference ratio (C/I) of 10 to 12 dB. Since the analog systems operate at a typical C/I value of 17 to 18 dB, there is at least a 6-dB improvement in the digital system. This translates into the use of smaller cells (greater frequency reuse) and therefore higher overall capacity.

The transmission emission spectrum for the GSM system is specified to have out-of-band radiated power in adjacent channels (adjacent channel interference) to be at least 40 dB below any observed channel. The GMSK modulation assists here and takes some of the burden off the RF stage filters of the multichannel transceivers. The MSK with premodulation Gaussian low-pass filtering (i.e., GMSK) is closely related to FM. In fact, it is really a type of FM. The circuitry for GMSK can take several different forms. Recently, a technique known as *direct digital interpolation* has emerged as a high-performance contender for GSM use. It is a combination of direct digital synthesis and PLL synthesis, which uses a ROM containing a sinewave lookup table whose output addresses a fast D/A converter to directly generate the modulated carrier. This technique largely overcomes problems inherent to FM and other modulation schemes, such as residual AM, AM/PM conversion, and the presence of troublesome dc components.

Power control and handoff. Perhaps one of the most important features of any cellular radio system is handoff. In the GSM system, handoff is closely interrelated with an elaborate power control mechanism which is designed to minimize cochannel interference. The intention is to operate the mobile and base stations at the lowest possible power level, while maintaining the subjective service quality. During a call, the base station periodically sends a signaling message to the mobile station to adjust its power level. By the very nature of TDMA, the mobile and base stations must transmit pulses of power that do not cause sideband generation which would interfere with adjacent channels. The shape of the pulses of power is therefore well controlled. The range of pulse power control is 30 dB in steps of 2 dB. This adaptive power control during calls is estimated to improve the mean system C/I values by up to 2 dB.

The decision to increase or decrease the mobile and/or the base station transmit power level(s) is made by comparing measured received signal level and quality (BER) values with predetermined threshold values. The decisions to adjust transmit power are based on *incoming* received signal values and not on values obtained at the other end of the link. The question arises, By how much should a transmitter be increased or decreased if a threshold value is exceeded? Also, How often should the adjustment be updated? GSM initially has single-step incremental changes, but eventually software will be developed to hit the right level immediately. These features affect system stability. For example, if a mobile station has its transmit power level increased because its performance has fallen below threshold, that would increase the cochannel interference between itself and the mobile station in the nearest cochannel cell. Subsequently, the performance of that mobile station may require a transmitter power level increase to maintain satisfactory performance. In turn, it would affect the interference level in the original cell and the surrounding cochannel cells. This positive feedback mechanism would spiral all mobile station transmitters up to the maximum power level and defeat the intended objective of power control. A lot of work has gone into solving this problem, and although it is not completely resolved, preliminary experience has shown that such situations rarely occur. This is because of factors such as the diverse geographic distribution of users within each of the cells, the discontinuous transmission feature, the time distribution of call activity, frequency-hopping sequences, etc. Handoff decisions are based on the performance of a channel as the mobile station approaches the boundary between cells. Unfortunately, the base station does not know whether the mobile station is at the cell edge or whether a multipath fade is occurring, which would necessitate a power level increase. Clearly, the power control and handoff algorithms need to be coordinated in synchronism so that inappropriate decisions are not made. For example, a channel suffering deteriorating quality should not proceed with handoff if high quality could be restored by increasing the mobile transmitter power level. On the other hand, a mobile user approaching the edge of one cell and about to enter another should not have its power level raised just when handoff is about to take place. Since power adjustments and handoff decisions are based on the same measurements of received signal power and quality, the algorithms have to be very sophisticated. Each GSM *mobile* station monitors the broadcast control channel of as many as 16 surrounding *base* stations during the "deadtime" periods (no transmission or reception time slots). The mobile station prioritizes up to six likely candidates for handoff and sends signal strength and quality information to the base station currently providing service. The base station also makes measurements of signal strength and quality for the mobile station currently receiving service. The actual handoff decision is made by the network management system. In addition to factors already mentioned, there can be other reasons for handoff. For instance, handoff can be used to balance traffic between cells. Whereas handoff to another cell requires the mobile station to retune to a different carrier frequency, another form of handoff is possible within the same cell, using the same carrier frequency but a different time slot. This can be beneficial for interference control.

The relationship between handoff and power control can take several forms, depending on the equipment manufacturer. Two widely used algorithms are (1) the minimum acceptable performance method and (2) the power budget method. In the first method, power control takes precedence over handover. That is to say, power increases are made while they are possible and handoff is done only when a power increase will not improve the quality of a channel that has deteriorated below the threshold value. This method tends to produce poorly defined cell boundaries so that a mobile user may be well into an adjacent cell before handoff takes place. As stated previously, this situation suffers from higher cochannel interference levels than other methods. An attempt to alleviate late handoff is for the base station to make distance measurements between itself and the mobile station so that it knows how close to the cell boundary the mobile is. This is a good idea, but the accuracy of the measurement is rather poor, especially in small cells in city centers. Also, the precise shape of the cell must be memorized by the base station. This technique tends to be effective only in large rural area cells. In the second method, handoff takes precedence over power control. Handoff is made to any base station that can improve on the existing connection quality at an equal or lower power level.

Frequency hopping. Frequency agility is an essential aspect of cellular mobile radio. The mobile station has to monitor the different frequencies of all surrounding base stations in order to facilitate the handoff process as the mobile user moves to another location which, from the system point of view, is in another cell. The frequency monitoring is done during the short inactive time between transmitting and receiving information. Frequency hopping, which was originally designed for military privacy, is now featured in modern cellular radio designs to enhance performance in a multipath fading environment. Since multipath fading is a narrowband phenomenon, shifting a relatively small amount in carrier frequency can place the radio link in a nonfading condition. GSM mobile stations use slow frequency hopping at 2000 hops per second. The frequency synthesizers that achieve this agility must settle in 100 μs.

Further frequency hopping can be done by changing base station frequencies. Their synthesizers must settle in 30 μs. A minimum of five carriers are used for base station hopping, and it is their *baseband* frequencies that are changed and not the transceiver RF frequencies.

Conclusion. As one can see, there are some mandatory specifications for the GSM system which are necessary to maintain international compatibility. However, there is considerable latitude within the system framework for circuit innovations by the competing equipment manufacturers. This is essential in a system operating right at the edge of technological capability, in order to advance the state of the art.

North American IS-54 system. The TDMA system in the United States places its information on carriers spaced by 30 kHz (i.e., 30-kHz bandwidth). This is the same as the AMPS system and allows operating companies gradually to replace the analog channels with digital channels to ease base station traffic congestion. Each digital channel carries three user signals compared to eight in the GSM system. The channel transmission bit rate is 48.6 kb/s. The composition of the 40-ms frame is shown in Fig. 8.25. Each frame has six time slots of 6.67-ms duration. Each time slot carries 324 bits of information, of which 260 bits are for the 13-kb/s full-rate traffic data. The other 64 bits are overhead; 28 of these are for synchronization, and they contain a specific bit sequence that is known by all receivers to establish frame alignment. Also, as with GSM, the known sequence acts as a training pattern to initialize an adaptive equalizer. The IS-54 system has different synchronization sequences for each of the six time slots making up the frame, thereby allowing each receiver to synchronize to its own preassigned time slots. A further 12 bits in every time slot are for the SACCH (i.e., system control information). The digital verification color code (DVCC) is the equivalent of the supervisory audio tone used in the AMPS system. There are 256 different 8-bit color codes,

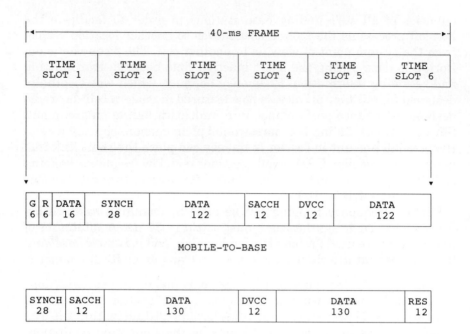

Figure 8.25   IS-54 frame structure.

which are protected by a (12,8,3) *Hamming code.* Each base station has its own preassigned color code, so any incoming interfering signals from distant cells can be ignored.

As indicated in Fig. 8.25, time slots for the mobile-to-base direction are constructed differently from the base-to-mobile direction. They essentially carry the same information but are arranged differently. Notice that the mobile-to-base direction has a 6-bit ramp time to enable its transmitter time to get up to full power and a 6-bit guard band, during which nothing is transmitted. These 12 extra bits in the base-to-mobile direction are reserved for future use.

The modulation scheme for IS-54 is $\pi/4$ differential quaternary PSK (DQPSK), otherwise known as differential 4-PSK. This technique allows a bit rate of 48.6 kb/s with a 30-kHz channel spacing, giving a bandwidth efficiency of 1.62 b/s/Hz. This value is 20 percent better than GSM. The major disadvantage with this type of linear modulation method is the power efficiency, which translates into a heavier hand-held portable, and, even more inconvenient, a shorter time between battery recharges.

The IS-54 speech coder uses the technique called vector sum excited linear prediction (VSELP) coding. This is a form of codebook excited linear prediction coder which has a source rate of 7.95 kb/s and an output bit rate of 13 kb/s.

Finally, when viewing the manufacturer's products and the technology they are working on for present and future cellular radio systems, it is probably fair to say that Europe is converging to a unified standard from several incompatible systems, whereas the United States is diverging from one continental analog system to several different types of digital systems. Unfortunately, the success or otherwise of these differing approaches will be known only in the future.

8.8.2   Future digital cellular radio

There are several possibilities for future digital cellular radio designs. The CDMA system, which has also been referred to as a form of wideband TDMA, is a serious contender. The CDMA principles will be outlined in the following.

North American digital cellular systems will increase their network channel capacity by allocating three digital voice channels for every existing analog FM channel. This can be accomplished by using TDM and an 8-kb/s coding scheme known as CDMA. It is predicted that spread-spectrum CDMA will enhance the capacity by a factor of about 20 times that of the analog systems. Theoretical calculations of CDMA system capacity are in the region of 5 to 10 million users.

As stated in Chap. 3, either frequency hopping (FH CDMA) or direct sequence (DS CDMA) provides a method of allowing multiple users to occupy the same channel (frequency band) with minimal interference. For DS CDMA, the system is asynchronous, meaning that the pulses from each user do not have to bear any phase relationship to each other. Present research is being done to maximize the number of users that can operate simultaneously while maintaining a satisfactory level of interference with each other. A major problem with DS CDMA is what is known as the *near-far* problem. This is the differing amounts of interference contributed by different mobile stations depending on their distance from the base station. Clearly, interference from mobile stations closest to the base station is the dominant source of interference unless some transmit power control mechanism is used. Furthermore, this power control must be highly sophisticated because the power level of multiple signals reaching the receiver must not vary by more than 1 or 2 dB if the maximum simultaneous user capability is to be achieved. This is no simple problem in a difficult multipath fading environment, especially if the user is moving in a vehicle on a main street lined with tall buildings in a city center.

The near-far problem is also alleviated by FH CDMA. Each user is allocated its unique, orthogonal frequency-hopping sequence. Orthogonality in this case implies that the sequences are chosen so that there will never be a situation where two users simultaneously hop onto the same frequency. Statistically, there will be a very small percentage of hop occasions when there is simultaneous frequency transmission by two (or more) users, and the probability of its occurrence increases with the number of users. Frequency hopping is classified as *fast* or *slow*, depending on the dwell time at each frequency, where multiple hops per bit is considered to be fast and multiple bits per hop is slow. In addition to interference avoidance with respect to other users, fast FH CDMA also provides some protection against multipath fading. Since this type of fading is, by definition, frequency selective, the frequency-hopping system statistically spends a smaller time on the fading frequencies than a system that is not frequency agile.

Spread-spectrum systems require the phase of the incoming signal to be determined very precisely. Phase acquisition and tracking, which is called the synchronization process in spread-spectrum technology, is of paramount importance; otherwise despreading of the desired signal is not possible (see Chap. 3).

In the United States, several serious investigations are being done to establish how soon CDMA can be implemented. It is difficult for the FCC to find an available frequency band for CDMA. However, because CDMA has the unique characteristic of being able to share a frequency band, the target for initial investigations is the sparsely used, fixed service microwave transmission band of 1.850 to 1.990 GHz. Tests are presently being made to evaluate the interference each system would have on the other, particularly when a mobile user is in the vicinity of a fixed service microwave tower.

CDMA is not the only technique receiving attention in research laboratories. For example, a scheme under investigation is packet reservation multiple access (PRMA). This is a combination of TDMA and slotted ALOHA. Without elaborating, it can be said that the initial results show very good potential for the future.

8.9   Portable Radio Telephones and PCNs

The importance of hand-held portable telephones as a strong and vital market force in the telecommunications industry cannot be overstressed. Hand-held portable radios first gained widespread recognition with the introduction of cordless extension phones. These phones allowed customers to be anywhere within their home or garden and still place or receive calls.

Because the portable phone transmitter power was only a few milliwatts, the operating range was only a hundred meters or so. The convenience of this tetherless phone facility soon prompted the idea of universal coverage for portable phones. Today's cellular radio systems go a long way toward achieving that goal, but there are still several problems to overcome. In the quest for the wireless city, the different options for accessing customers within buildings are receiving a lot of attention. This is a difficult problem because radio waves, particularly up into the microwave region, suffer severe attenuation when propagating through steel reinforced concrete. For example, recent measurements of the signal strength have shown the values taken on the first floor of buildings to be about 15 dB less than the same level outside the building. The signal strength improves as one moves further up the building. A typical measurement showed a linear 2.7-dB per floor improvement when moving from the first to the tenth floor.

This problem is becoming more important since PCNs are presently on the drawing board. There are several definitions of PCNs. Perhaps the most appealing is the following one: *the availability of all communications services anytime, anywhere, to everybody, by a single identity number and a pocketable communication terminal.* That simple statement is loaded with technological hitches. With a potential customer base in the major U.S. urban areas estimated to be up to 60 million, the stakes are high. The cost of PCN systems is not small, especially since they will probably evolve out of cellular radio systems. For example, because of the microcellular nature of PCNs, they will need more than 5 times the number of base stations of present cellular networks. Designers did not have the luxury of starting their designs from the beginning. Too much money has already been invested in cellular mobile radio to simply discard it.

Several solutions have been proposed for incorporating building interiors into the cellular mobile radio structure. One proposal is to lay a "leaky" coaxial feeder cable down the spine of a building and connect one end of the cable to an antenna on top of the building (Fig. 8.26). The outer conductor of the cable has specially designed slots so that energy can couple into or out of the cable. Each high-rise building then becomes a microcell within the cellular structure. The base stations could be placed primarily on existing street lamp poles or other overground poles.

Other methods are being considered, such as the 18-GHz microwave frequency band (i.e., 18.820 to 18.870 and 19.160 to 19.210 GHz). This frequency has limited use for long distance microwave links because of the high rainfall attenuation at this frequency. Microwaves at this high frequency do not pass through steel reinforced concrete walls but do pass through thin dielectric partitioning material used in an office environment. This implies frequency reuse for each office floor. The advancement of monolithic microwave integrated circuits (MMICs) has driven the cost of the technology to levels that are attractive for

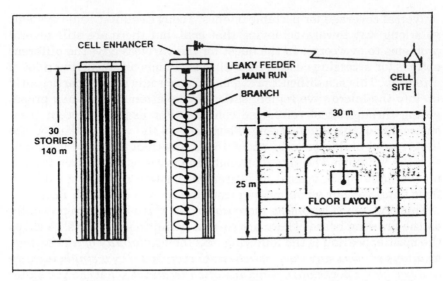

Figure 8.26 Leaky feeder cable for access to building interiors. (*Reproduced with permission from Ref. 323, © 1990 Telecommunications.*)

in-building microcell communications. When used with "smart" sectorization antennas, these low-power microwave systems offer a good solution to the last 30 to 40 m of a PCN. They are even being considered for LANs for data transmission at 15 Mb/s using 10-× 10-MHz channels in an FDMA configuration.

Radio contact to places such as elevators made completely from steel would still be a problem. Maybe wire down the center of the steel cable operating the elevator is the only solution or elevators made from nonmetallic material. However, by the time operating companies are worrying about elevators, PCNs will probably have become an overwhelming success, and the remaining problems will be only fine tuning of minute details.

The subject of PCNs is, at best, controversial, and the spread-spectrum technology often discussed in connection with PCNs is also the subject of intense debate. The future promises to be very exciting in this area of telecommunications. There are still many difficulties to be overcome. For example, in addition to the ones mentioned above, there are others, such as providing service on supersonic jet airplanes traveling at Mach-1 or even Mach-2. In these circumstances, the doppler effect adds another dimension of complication to the rather intricate problem of setting up cells to cover the vast areas of the Atlantic and Pacific Oceans. By the above definition of PCNs, it will be some years beyond the year 2000 before the objective is fully realized, although several million people should have access to PCN capability by 2000.

8.10 Data over Cellular Radio

The use of portable (laptop) computers to access an information source in a business office using the cellular network is a rapidly growing service requirement. It is understandable that data communications over cellular radio links are susceptible to relatively high BERs. As already mentioned, impairments occur which result in the received signal being below the noise level of the system. This may be caused by (1) multipath fading, (2) shadowing of the receiver from the transmitter by obstacles such as buildings and bridges, or (3) interference from other radio sources such as automobile ignition systems. Both isolated errors and error bursts can result from the above problems. When handoff occurs, breaks in the continuity of data transmission can occur for up to about 300 ms. Multipath fading or shadowing can produce fades of several seconds. In extreme cases, after a break of 5 s, the call is automatically dropped (i.e., cleared down).

Countermeasures such as FEC, interleaving, and ARQ are usually used to improve the BER. The Reed-Solomon class of FEC symbol block codes have been successfully incorporated into cellular systems to enhance the quality in the presence of error bursts. The relatively low-redundancy (72,68) code produces good improvements.

The short error burst problem can also be tackled by using "interleaving." Instead of transmitting the code blocks as complete blocks, the bits of each block are spread out in time so that there is an overlap of adjacent blocks. The depth of interleaving is denoted by the interval between each bit within the same code block.

Automatic requests for repeat are shown in Fig. 8.27. The blocks of data to be transmitted have a header at the beginning of each block. The complete block is protected by an error-detecting code which is usually a cyclic redundancy check. Each block is assigned a reference number, which is contained in the header, together with information on the reference number of the last correctly received block. For example, in Fig. 8.27, the second block is received incorrectly by the base station, which indicates the loss of block in the block C header. The mobile station then repeats block 2. Bursts of errors occupying many blocks can be corrected by this technique. The obvious disadvantage of this system is the delay in transmission caused by repeating one or several blocks. Elastic stores and control of the stores is necessary. ARQ and FEC are very effective methods of reducing all types of cellular radio link errors even at received signal strengths as low as -120 dBm.

MOBILE STATION TO BASE STATION

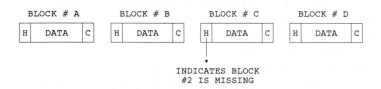

BASE STATION TO MOBILE STATION

Figure 8.27 Automatic request for repeat example.

8.11 Cellular Rural Area Networks for Developing Countries

The price of base stations and mobile station equipment has decreased over the past decade because the initial research and development costs have been recouped and because of high-volume production. The first mobile telephones were seen by many as an expensive luxury, but this viewpoint is changing as costs decrease. An outstanding feature of the cellular system is the huge area that can be covered by a single base station. It is quite acceptable for a single, high-output power base station to cover a cell radius of 30 to 45 km using an omnidirectional antenna. That is an area coverage of over 6000 km². The omnidirectional antenna allows a broadcast style of communication to the terminal station within a cell or to a repeater station in the case of additional cell creation. These digital radio multiple access systems are *point-to-multipoint* in character.

A cellular system is shown in Fig. 8.28. The subscriber terminal was initially designed to be a mobile station installed in a vehicle. In this application it is now, for example, connected to a telephone box or a medical clinic. The antennas at the terminal stations are usually the high-directivity Yagi type, which feed into the omnidirectional antenna at the repeater or base station. Each terminal usually serves about four to eight subscribers, but since the number of voice channels per RF carrier is up to about 60, this could be the number of subscribers served at one terminal if required. With as many as 10 or 12 repeaters, the total coverage radius can be extended to over 600 km. The

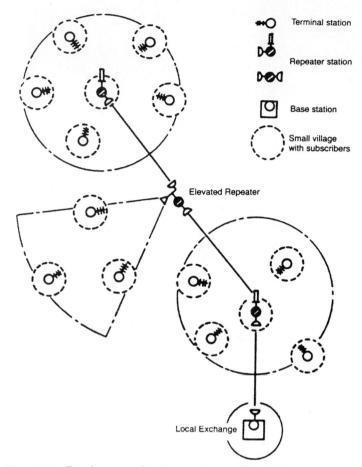

Terminal station

Repeater station

Base station

Small village
with subscribers

Elevated Repeater

Local Exchange

Figure 8.28 Rural area application (point to multipoint).

base station and repeaters operate in a microwave frequency band (typically 0.8 to 1.5 GHz or 1.5 to 2.6 GHz). The voice channels are either 64-kb/s PCM or 32-kb/s ADPCM. There are several possible modulation schemes, with 4-PSK being very popular. The choice of technology depends on the total number of subscribers, their geographical distribution, and the terrain topography.

By locating base stations at town sites along a country's backbone where multiplexers are located, penetration deep into rural areas can be achieved. A major advantage of the cellular approach is that the system can be relocated to another required place if a small community grows to a level where it is economical to introduce fixed local wire-lines. The cellular system is the fastest method of providing telephone service to remote areas.

Data Transmission and the Future Network

Technology is moving at a frantic pace. Data communications is no longer restrained by the minute bandwidth of a telecommunications voice channel (less than 4 kHz). Eventually, when optical fiber is fed all the way to the customer premises, bandwidth problems will disappear. In the meantime, bandwidth is always an important consideration. Also, data transmission discussions go beyond the bounds of the transmission media and terminal equipment, and it is necessary to have a feel for the network as a whole. In this chapter more than any other, the reader will appreciate that the industry is moving toward a digital telecommunications engineer or technician job description rather than separate switching and transmission disciplines.

As the convergence of telecommunications and data communications accelerates, the term *multimedia* is used to describe emerging applications which incorporate a combination of voice, data, and video. For example, workstation teleconferencing, image retrieval or transfer, and voice electronic mail are multimedia applications which require a wide range of bit rates. The increased acceptance of the latest generation of packet switching instead of the conventional circuit switching is central to the convergence of data communications with telecommunications. Some of the advantages of packet switching are (1) variable bit rate service capability, (2) multipoint-to-multipoint operation, (3) service integration, and (4) resource sharing. These facilities result in a lower switching cost per subscriber.

High-bit-rate local, metropolitan, and wide area networks (i.e., LANs, MANs, and WANs) are expanding their horizons by becoming broadband multimedia networks instead of just data networks. This has necessitated the development of fast packet-switching protocols. *A protocol is a set of rules that control a sequence of events which take place between equipment or layers on the same level,* as described in the appendix at the end of this chapter, "Standards." Frame relay and cell relay (see the section, "Frame relay and cell relay") are now being used in this respect to allow voice, data, and video multimedia operation. These protocols speed up packet-switching procedures by eliminating or limiting error control and flow control overheads.

When one first encounters the subject of data communications, it appears that there is a plethora of Standards. In the literature, it seems that references are made to CCITT Recommendations in every sentence. As one becomes more familiar with data communications on an engineering or technician level, the need for very stringent standards quickly becomes apparent. With so many manufacturers in the data communications business, it is absolutely essential to ensure that the equipment is fabricated so that there is compatibility with all other equipment with which it is required to interface within the network. In the early days of personal computers (1980s) it was almost impossible to interface one computer with another. Each manufacturer was trying to establish its own de facto standard and therefore corner a major share of the market. With powerful software, this problem is now slowly being overcome. The necessity for standardization is very important to the consumer, and the telecom administrations are adopting international standards to make life easier, even though many engineers may not appreciate that fact.

9.1 Standards

Standards aim to specify technical characteristics that allow compatibility and interoperability of equipment made by many different manufacturers. In addition, they recommend certain minimum levels of technical quality based on the state of the art of present-day technology.

The International Standards Organization (ISO), which was founded in 1947 and has its headquarters in Geneva, Switzerland, has become a dominant leader in the quest for global standardization of data communications. Together with the CCITT section of the ITU, a significant number of Standards have been developed which are gradually receiving global acceptance.

The ISO (with CCITT collaboration) has established a seven-layer network architecture as illustrated in Fig. 9.1. Each layer can be developed independently provided the interface with its adjacent layers meets specific requirements. To elaborate, each layer will be discussed briefly.

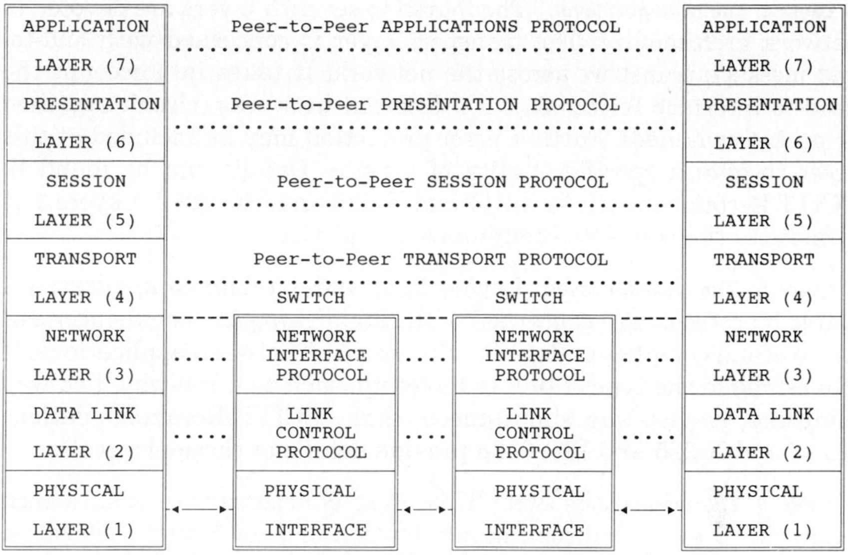

APPLICATION	Peer-to-Peer APPLICATIONS PROTOCOL	APPLICATION
LAYER (7)	....	LAYER (7)
PRESENTATION	Peer-to-Peer PRESENTATION PROTOCOL	PRESENTATION
LAYER (6)	....	LAYER (6)
SESSION	Peer-to-Peer SESSION PROTOCOL	SESSION
LAYER (5)	....	LAYER (5)
TRANSPORT	Peer-to-Peer TRANSPORT PROTOCOL	TRANSPORT
LAYER (4)	SWITCH SWITCH	LAYER (4)
NETWORK	NETWORK INTERFACE PROTOCOL .... NETWORK INTERFACE PROTOCOL	NETWORK
LAYER (3)		LAYER (3)
DATA LINK	LINK CONTROL PROTOCOL .... LINK CONTROL PROTOCOL	DATA LINK
LAYER (2)		LAYER (2)
PHYSICAL	PHYSICAL PHYSICAL	PHYSICAL
LAYER (1)	INTERFACE INTERFACE	LAYER (1)

Figure 9.1   The OSI 7-layer protocol model. (*Previously published by Peter Peregrinus Limited, Data Communications and Networks 2, by Brewster, IEE Telecommunication Series 22, 1989.*)

Layer 1: The physical level.   This layer is responsible for the electrical characteristics, modulation schemes, and general bit transmission details. For example, CCITT Recommendations V.24 and V.35 specify interfaces for analog modems, and X.21 specifies the digital interface between data terminal equipment and synchronous mode circuit-terminating equipment.

Layer 2: The link level.   This level of protocol supplements the Layer 1 raw data transfer service by including extra block formatting information to enable such features as error detection and correction, flow control, etc.

Layer 3: The network control level.   The function of this layer is to provide addressing information to guide the data through the network from the sender terminal to the receiver location; in other words, call routing. CCITT Recommendations X.21 and X.25 are the protocols involved. X.21 refers to a dedicated circuit-switched network and deals with the signaling and data transmission from terminals to switches. The X.25 protocol refers to network signaling, call routing, logical channel multiplexing, and flow control for terminals operating in the packet mode and connected to public data networks by dedicated circuits.

Layer 4: The transport level. The fourth to seventh layers are devoted to network architecture. The transport layer is concerned with end-to-end message transport across the network. It takes into account the need to interface terminals with different networks (circuit switched or packet switched). Further error protection may be included at this layer to give a specific quality of service. Details can be found in CCITT Recommendations X.214 and X.224 and ISO 8072. Layers 1 to 4 relate to the complete communication service.

Layer 5: The session level. Layers 5, 6, and 7 relate to applications. Layer 5 protocols are concerned with establishing the commencement (log-on) and completion (log-off) of a "session" between applications. It can establish the type of link to be set up, such as a one-way, two-way alternate, or two-way simultaneous link. CCITT Recommendations X.215 and X.225 and ISO 8326 provide details of these protocols.

Layer 6: The presentation level. This layer is necessary to ensure each subscriber views the incoming information in a set format, regardless of the manufacturer supplying the equipment (e.g., ASCII representation of characters). Also screen presentation of size, color, number of lines, etc., must have uniformity from one supplier to another. CCITT Recommendations X.216 and X.226 and ISO 8823 refer to these protocols.

Layer 7: The application level. Finally, Layer 7 is concerned with the interface between the network and the application. For example, the application could be a printer, a terminal, file transfer, etc. In this respect, CCITT Recommendation X.400 relates to message-handling services and ISO 8571 to file transfer and management.

These seven layers may seem somewhat academic on first reading, but they are extremely valuable. They are what is known as the Open Systems Interconnection (OSI). It must be emphasized that OSI is not a protocol and does not contain protocols. OSI defines a complete architecture which has seven layers. It defines a consistent language and boundaries establishing protocols. Systems conforming to these protocols should be "open" to each other, enabling communication. OSI is nothing new per se. It just standardizes the manner in which communication is viewed. In fact, the protocols at all seven levels of OSI can actually be applied to any type of communication regardless of whether it is data or speech. For example, one could make an analogy between the seven levels and the processing of a conventional telephone call, as follows:

1. *Physical layer.* Concerns the sounds spoken into the telephone mouthpiece and heard in the receiver earpiece.

2. *Link layer.* Involves speaking when required and listening when necessary. A repeat is requested if something is misunderstood. The person at the other end is told to slow down if he or she is talking too fast.

3. *Network layer.* Concerns dialing the number and listening for the connection to be made. If the busy signal is given, dial again. On completion of the call, disconnect by replacing the handset.

4. *Transport layer.* Involves deciding which is the most cost-effective way to make the call (which carrier to use).

5. *Session layer.* Concerns deciding if one call will suffice or whether several will be necessary. Will more than one person need to be included? If so, who will control a subsequent conference type of call? Who will resetup the call if cut off?

6. *Presentation layer.* Establishes whether or not the two parties are speaking the same language.

7. *Application layer.* Is concerned with: Am I speaking to the right person? Who will pay for the call? Is it convenient to converse now or should the call be placed again later? Does the called party have the means to make notes on the conversation?

The Standards presented above are summarized by the CCITT in its Recommendations that are published every four years, in 1988, 1992, etc. The various different series of Recommendations follow.

V Series: "Data Communication over the Telephone Network"

V.1 –V.7	General recommendations
V.10 –V.32	Interfaces and voice band modems
V.35 –V.37	Wideband modems
V.40 –V.41	Error control
V.50 –V.57	Transmission quality and maintenance
V.100–V.110	Interworking with other networks

X Series: "Data Communications Network Interfaces"

| X.1 –X.15 | Services and facilities |
| X.20 –X.32 | Interfaces (X.21 is central) |

X.40 –X.80	Transmission signaling and switching
X.92 –X.141	Network aspects
X.150	Maintenance
X.180–X.181	Administration arrangements
X.200–X.250	OSI system description techniques
X.300–X.310	Internetworking between networks
X.350–X.353	Mobile data transmission systems
X.400–X.430	Message handling systems

Sometimes two or three Recommendations will address the same issue, but in a totally different manner, and they will be called, for example:

X.21

X.21 bis (meaning second recommendation)

X.21 ter (meaning third recommendation)

G Series: "Digital Networks Transmission Systems and Multiplexing"

G.700—General aspects of digital transmission systems, terminal equipments (as in Chap. 2)
- PCM
- Higher-order multiplexers or systems

G.800—Digital networks
- Network performance
- Maintenance, etc.

G.900—Digital sections and digital line systems
- Megabit digital corrections
- FDM
- Use of coaxial and optical fiber cable

I Series: "Integrated Services Digital Network (ISDN)"

I.100–I.130	General
I.210–I.212	Service capability
I.310–I.340	Overall network aspects and functions
I.410–I.464	ISDN user-network interfaces

9.2   Data Transmission in an Analog Environment

The principles involved in data transmission are basically the same as for pulse transmission, already encountered in the earlier chapters. A brief review of the problems and how they are overcome for data transmission in an analog environment will follow.

9.2.1   Bandwidth problems

Ever since the start of data communications, perhaps the greatest impediment to progress has been the constant battle to solve the problems associated with operating with very limited bandwidth. The public telephone network was not designed to deal with isolated, individual pulses, so it is even less able to cope with high-speed bit streams. Transmitting pulses through the *analog* telephone bandwidth of 300 to 3400 Hz causes severe pulse distortion because of (1) loss of high-frequency components, (2) no dc component, and (3) amplitude and phase variation with frequency. Chapter 3 dealt with the effects of pulse transmission in a bandlimited channel which led to the Nyquist Criterion.

Ultimately, the bandwidth problem will simply go away when optical fiber is taken to every home. Since that situation is at least 10 years away, what can be done in the meantime? Packet-switching networks are springing up and special leased lines are becoming available, but the fact remains that in a mixed analog/digital environment, some temporary solutions are needed. The translation of digital signals to a condition suitable for transmission over analog voiceband telephone circuits is performed by a device called the *modem*. Although these devices will be a dying breed in the mid 1990s, they are worthy of a brief mention since they are still in use today.

9.2.2   Modems

A variety of voiceband modems has evolved, each operating at a specific bit rate. The CCITT V-series of Recommendations relates to these modems, and Table 9.1 summarizes the important details. In addition, Recommendation V.24 specifies the various data and control interchange circuit functions operating across the modem-terminal interface. Also, V.25 specifies how automatic answering or calling modems should operate. The original stand-alone box modems are now mainly replaced by chip modems which are offered as an extra item within a

TABLE 9.1   Summary of V-Series CCITT Recommendations for Modems

CCITT Rec.	Transmission rate (b/s)	Modulation type	Additional information
V21 (FDM)	300	FSK	Two-wire full duplex
V22 (FDM)	1200	PSK	Two-wire full duplex
V22bis (FDM)	2400	QAM	Two-wire full duplex
V23	1200/75	FSK	Two-wire full duplex (1200/75); half duplex (1200/1200)
V26, V26bis	2400	PSK	Half duplex
V26ter			Two-wire full duplex plus echo cancelers
V27, bis, ter	4800	PSK	Half duplex
V29	9600	QAM	Half duplex
V32			Two-wire full duplex plus echo cancelers
V33	14400	QAM	Four-wire leased

PC. This temporary situation will exist only until the full deployment of ISDN.

Modulation methods.   For data transmission beyond 10 to 20 km (in the junction or trunk circuits), line distortion necessitates translating the digital baseband signal into the usable portion of a voiceband telephone channel. This is done by using the baseband signal to modulate a carrier frequency centered within the voiceband range. FSK is the simplest method used for low-bit-rate modulation in modems up to 1200 b/s (CCITT Rec. V.21 and V.23). For high bit rates, PSK is suitable. Four-state differential PSK is used for 2400-b/s modems (V.26), and this is extended to eight-state PSK for the 4800-b/s modem (V.27). The highest bit rate modems employ the more bandwidth-efficient QAM as designated by V.29/V.32 for 9600 b/s and V.33 for 14400 b/s. Performance-enhancing techniques such as the multilevel trellis coding and Viterbi detection schemes discussed in Chap. 3 have been incorporated in recent modem designs.

Two-wire modems.   If two-wire transmission is used instead of four-wire, for full-duplex operation there must be a mechanism for organizing the two directions of transmission. FDM is an obvious choice, whereby the voiceband is split up into two halves, one for GO and the other for RETURN. This is done for 300 b/s (V.21), 1200 b/s (V.22), and 2400 b/s (V.22 bis using 16-QAM), but bandwidth constraints require other techniques for the higher bit rates. Echo cancellation is preferable to FDM for high bit rates, in which case the full voice bandwidth is used for both the transmit and receive directions. In the 9600-b/s modems (V.32), an adaptive canceler is used in each receiver to eliminate interference caused by its own transmitter.

9.3 Packet Switching

Since its introduction in the early 1970s, packet switching has received widespread acceptance. Public networks have been constructed in most developed countries and several developing countries. The internetwork CCITT X.75 protocol provides for interlinking of national networks at an international level. The CCITT X.25 Recommendation is the Standard for packet-switching architecture.

Packet switching has several advantages over conventional circuit-switched networks. Figure 9.2 highlights the difference between the two. The circuit-switched network maintains a fixed bandwidth between the transmitter and receiver for the duration of a call. Also, the circuit-switched network is bit stream transparent, meaning it is not concerned with the data content or error-checking process. This is not the case for packet switching, where bandwidth is allocated dynamically on an "as required" basis. Data is transmitted in packets, each containing a header which contains the destination of the packet and a tail, or footer, for error-checking information. Packets from different sources can coexist on the same customer-to-network physical link without interference. The simultaneous call and variable bandwidth facilities improve the efficiency of the overall network. The buffering in the system which allows terminals operating at different bit rates to interwork with each other is a significant advantage of packet switching. The obvious disadvantage is the extra dimension of complexity with respect to the switches and network-to-customer protocol.

Furthermore, in certain circumstances, packet switching has several advantages over other methods of data communication:

1. Packet switching may be more economical than using private lines if the amount of traffic between terminals does not warrant a dedicated circuit.

2. Packet switching may be more economical than dialed data when the data communication sessions are shorter than a telephone call minimum chargeable time unit.

3. Destination information is contained in each packet, so numerous messages can be sent very quickly to many different destinations. The rate depends on how fast the data terminal equipment (DTE) can transmit the packet.

4. Computers at each node allow dynamic data routing. This inherent

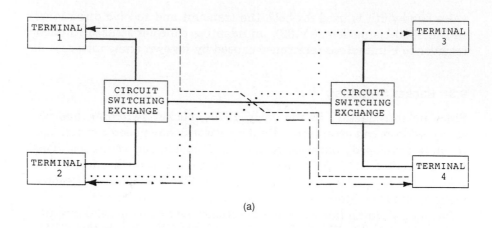

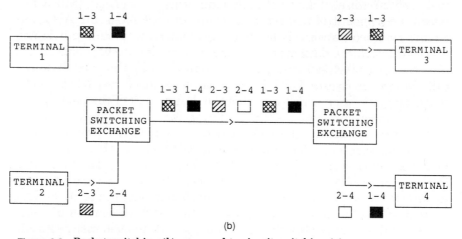

Figure 9.2 Packet switching (b) compared to circuit switching (a).

intelligence in the network picks the best possible route for a packet to take through the network at any particular time. Throughput and efficiency are therefore maximized.

5. The packet network inherent intelligence also allows graceful degradation of the network in the event of a node or path (link) failure. Automatic rerouting of the packets around the failed area causes more congestion in those areas, but the overall system is still operable.

6. Other features of this intelligence are error detection and correction, fault diagnosis, verification of message delivery, message sequence checking, reverse billing (charging), etc.

9.3.1 Packet networks

The intelligent switching nodes within packet-switching networks, in general, have the following characteristics:

1. All data messages are divided into short blocks of data, each having a specific maximum length. Within each block there is a header for addressing and sequencing. Each packet usually contains error control information.

2. The packets move between nodes very quickly and arrive at their destinations within a fraction of a second.

3. The node computers do not store data. As soon as a receiving node acknowledges to the transmitting node that it has correctly received the transmitted data, by doing an error check, the transmitting node deletes the data.

For packet switching, a data switching exchange (DSE) is required, which is a network node interlinking three or more paths. Data packets move from one DSE to another in a manner which allows packets from many sources and to many destinations to pass through the same internode path in consecutive time sequence. This, of course, is in contrast to the circuit-switched network that seizes a specific path (link) for the duration of the message transfer. Figure 9.3 shows how the

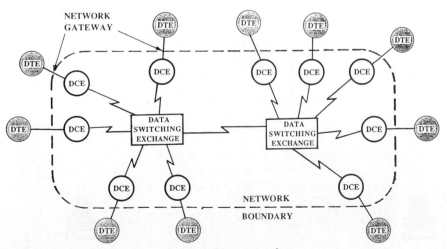

Figure 9.3 An example of a packet-switching network.

DTE, data circuit-terminating equipment (DCE), and DSE are related within a packet network. The boundary of the network is usually defined as the point where the serial interface cable is connected to the DCE. This point is also often referred to as the *network gateway*.

9.3.2 The X.25 protocol

CCITT Recommendation X.25 is recognized by the data communications fraternity as one of the most significant milestones in the evolution of networking architecture. Within the X series of Recommendations, X.25 specifies the physical, link, and network protocols for the interface between the packet switching network and the DTE/DCE at the gateway. Figure 9.4 clarifies the packet-switching network architecture by comparing it to the analogous situation for a regular telephone local loop.

The DTE is connected to the DCE, which is equivalent to the subscriber telephone connection to the central office. X.25 does not describe details of the DTE or the packet data network equipment construction, but it does specify, to a large extent, what the DTE and packet network must be able to perform. The X.25 Recommendation specifies two types of service which should be offered by a carrier: (1) Virtual call (VC) service and (2) permanent virtual circuit (PVC) service, as described in the following.

Virtual call service.   Prior to the transmission of any packets of data in a VC service, a virtual connection must first be set up between a call-

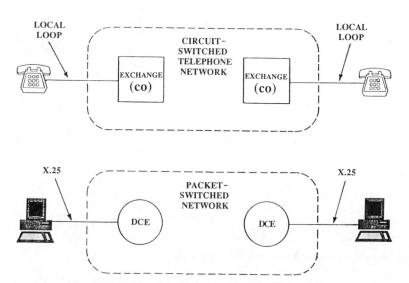

Figure 9.4 The X.25 packet-switching compared to the local loop.

ing DTE logical channel and the destination DTE logical channel. This is analogous to placing a telephone call before beginning the conversation. Special packets having specific bit sequences are used to establish and disconnect the virtual connection. Having made the virtual connection, the two DTEs can proceed with a two-way data transmission until a disconnect packet is transmitted. The term *virtual* is used because no fixed physical path exists in the network. The end points are identified as logical channels in the DTEs at each end of the connection, but the route for each packet varies depending on other packet traffic activity within the network.

Any DTE can have many active logical channel numbers simultaneously active in the same X.25 interface. Having established a virtual connection, for example between DTE logical channel X and DTE logical channel Y, the header in each data packet will ensure the data from X reaches its destination at Z via the intelligent nodes N in the network. An example of packet routing in virtual circuits is illustrated in Fig. 9.5, using logical channel numbers. The network N stores the following information for the duration of the call setup and then deletes it when the call is cleared:

16 on link XN, connected by node N to 24 on link NZ

33 on link XN, connected by node N to 8 on link NY

2 on link ZN, connected by node N to 17 on link NY

In this manner, packets originating at any terminal are routed to the desired destination.

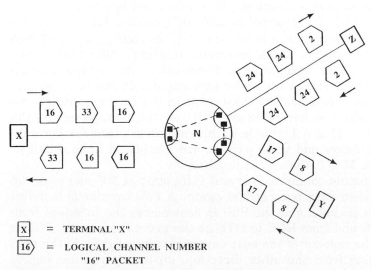

$\boxed{X}$ = TERMINAL "X"

$\boxed{16\rangle}$ = LOGICAL CHANNEL NUMBER
           "16" PACKET

Figure 9.5 Packet switching routes for virtual circuits.

Permanent virtual circuit service. A PVC service is similar to the VC in that there is no fixed, physical end-to-end connection, and the intelligent network provides the data transfer between the logical channels of any two DTEs. A PVC is analogous to a private line service. Other PVC and VC similarities are (1) the network has to deliver packets in the precise order in which they were submitted for transmission regardless of path changes due to network loading or link failure and (2) equivalence of network facilities such as error control and failure analysis.

The difference between the PVC and the VC mainly lies in the method of interconnection. A PVC is set up by a written request to the carrier providing the packet-switching network service. Once a PVC has been set up, the terminals need only transmit data packets on logical channels assigned by the carrier. Each DTE can have many logical channels active simultaneously within the X.25 interface. This means that some channels may be used as PVCs and others used as VCs at the same time. Another written request is required to break the connection. Note that there are no unique packets transmitted by the DTEs to set up or break a connection. Finally, the cost of a PVC compared to a VC is analogous to private line service versus a long distance telephone service.

Packet formation

Frame format. Figure 9.6 illustrates the X.25 DTE and DCE data packet format. At least the first three octets, as in Fig. 9.6a, are for the header. The first octet contains the logical channel group number and the general format identifier. This is followed by the second octet which gives the logical channel number information. Each X.25 gateway interface can support a maximum of 16 channel *groups* and each group can contain 256 logical channels, totaling 4096 simultaneous logical channels for each gateway. These can be distributed between VC and PVC services as set up by agreement between the customer and equipment provider. The third octet in the header contains two 3-bit data packet counters, P(S) and P(R), an M-bit, and a zero for the first bit. When M is a 1, this is an announcement that further data packets will follow and these packets are considered to be a unit or entity. When M is 0, the unit is finished.

The data packet counters, P(S) and P(R), start at 000 and count up to 111 and then start back at 000 again. A P(S) counter is included only in data packets, and the P(S) is sent across the interface from DTE to DCE and from DCE to DTE at the network gateway. P(R) is defined as the number of the next expected value of P(S), in the next packet coming from the other direction, on that particular logical

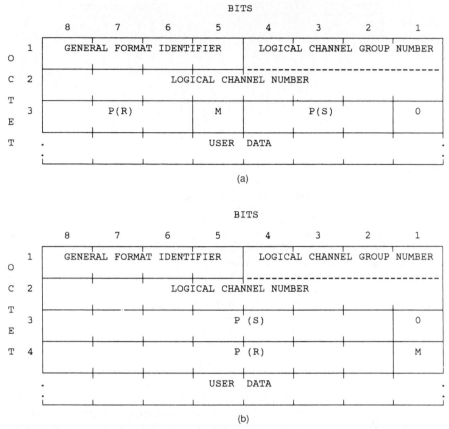

Figure 9.6 (*a*) DTE and DCE data packet format (for modulo 8); (*b*) DTE and DCE data packet format (for modulo 128). (*Previously published by the Institution of Electrical Engineers, Teacher, V., Packet Switching Data Networks, Brewster, IEE Fifth Vacation School on "Data Communications and Networks," September 1987.*)

channel. To make sense out of that, observe Fig. 9.7. This is a time sequence for a full-duplex transmission of packets from DTE to DCE and vice versa. Each block is a packet, and the values of P(S) and P(R) are shown for each packet header. P(S) is easy, since it simply increases in numerical order. P(R), on the other hand, is more complicated. It has values that are updated only after the previous P(S) has been received, free of errors.

The user data then follows the header and there are 128 octets of user data in a Standard X.25 data packet (modulo 8). Figure 9.6*b* shows, by contrast, the header composition for a modulo 128 scheme. In this case, the P(S) and P(R) counters are each 7 bits long, and the header has at least four octets instead of three.

Although most packets contain data, a few packets are reserved for

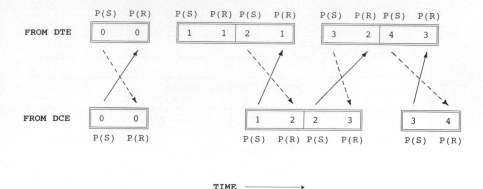

Figure 9.7   P(S) and P(R) values for a data packet sequence.

overhead functions such as control, status indication, diagnostics, etc. A data packet is identified by placing a zero for the first bit of the third octet.

Call request. The procedure for setting up and disconnecting a VC two-way simultaneous data transfer is shown in Fig. 9.8. This sequence of events resembles the various steps involved in placing and terminating a telephone call on a circuit-switched network. The contents of the packet header required to fulfill this procedure are clearly more complex than those of a typical data packet. Figure 9.9a shows the packet format for the call request/incoming call. Notice that the third octet in the header has a packet-type identifier. Figure 9.9b details the values of the third octet for the various identifiers for call request/incoming call, call accepted/call connected, clear request/clear indication, and DTE-initiated clear confirmation/network-initiated clear confirmation.

The header also contains destination addressing and VC/PVC selection information. The numerical values assigned to the logical channels at each end are never the same. The value assigned to the logical channels is chosen from a pool of numbers available when the call is set up.

X.25 summary. Figure 9.10 summarizes the three levels of the X.25 Standard. This diagram shows the various stages in transmitting and receiving data blocks. First, packets are formed which include headers, then frames are constructed at the link and then at the physical level, ready for sending to the line. On receipt of frames, the link level checks for errors, after which the data blocks are extracted from the packets. There is an enormous amount of information concerning the

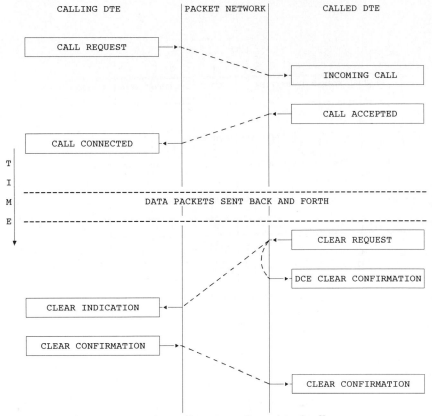

Figure 9.8 Call request/incoming call sequence for a virtual call.

X.25 Recommendation and its implementation. Only a brief outline has been presented here.

9.3.3 Network interface protocols

Figure 9.11 summarizes how the various CCITT Recommendations apply to a packet-switching network and its associated equipment. X.25 specifies how the packet mode DTE is related to the packet-switching network. X.25 also specifies the relationship between the packet assembly and disassembly [PAD (X.3)] equipment and the packet-switching network. X.28 relates the X.3 PAD to the character mode (nonpacket mode) DTE. X.29 specifies the relationship between packet mode DTEs when communicating with character mode DTEs. This protocol also specifies the interconnection of a PAD to a packet mode DTE or a PAD to another PAD.

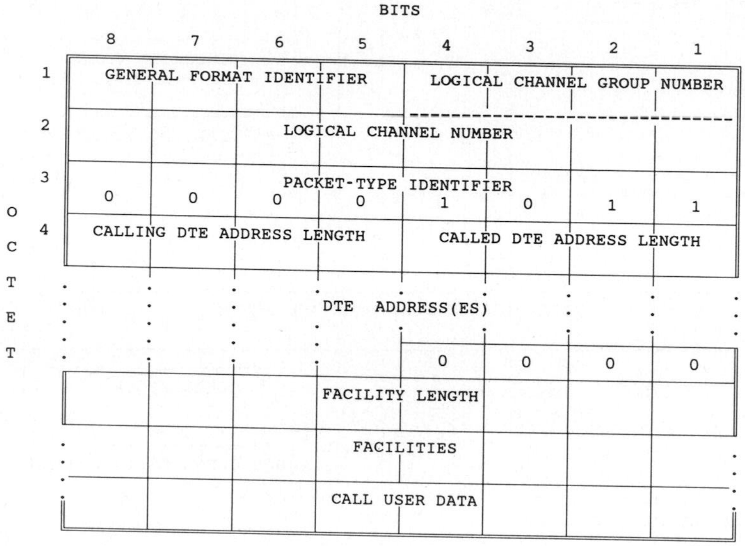

BITS

	8	7	6	5	4	3	2	1
1	GENERAL FORMAT IDENTIFIER				LOGICAL CHANNEL GROUP NUMBER			
2	LOGICAL CHANNEL NUMBER							
3	PACKET-TYPE IDENTIFIER							
	0	0	0	0	1	0	1	1
4	CALLING DTE ADDRESS LENGTH				CALLED DTE ADDRESS LENGTH			
	.	.	.	.	DTE ADDRESS(ES)	.	.	.
	.	.	.	.	0	0	0	0
					FACILITY LENGTH			
	.				FACILITIES			.
	.				CALL USER DATA			.

O C T E T

(a)

DTE TO DCE	DCE TO DTE	OCTET #3
CALL REQUEST	INCOMING CALL	0 0 0 0 1 0 1 1
CALL ACCEPTED	CALL CONNECTED	0 0 0 0 1 1 1 1
CLEAR REQUEST	CLEAR INDICATION	0 0 0 1 0 0 1 1
DTE-INITIATED CLEAR CONFIRMATION	NETWORK-INITIATED CLEAR CONFIRMATION	0 0 0 1 0 1 1 1

(b)

Figure 9.9 (a) Packet format for a call request/incoming call. (b) Identifier for call setup and clearance. (*Adapted from the Institution of Electrical Engineers, Teacher, V., Packet Switching Data Networks, Brewster, IEE Fifth Vacation School on "Data Communications and Networks," September 1987, p. 2/10.*)

9.3.4 Optical packet switching

The throughput of information in the above packet-switching networks is quite adequate for text messages. However, with the ever-increasing computing power available from workstations and PCs, it is already desirable to have the capability of sending graphics through networks. To achieve graphics transfer in a reasonable amount of time, the networks need to be able to manipulate data at the

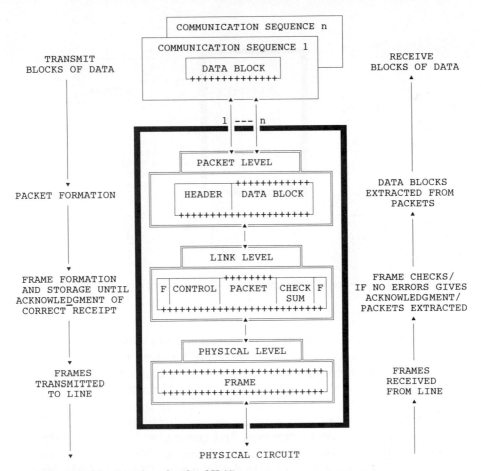

Figure 9.10 The first three levels of X.25.

multigigabits per second bit rates. Optical packet-switching promises to offer the solution to this problem.

A lot of research work is being directed toward improving multiwavelength N × N passive optical star components. Already, the potential for several hundred gigabits per second optical switches is on the foreseeable horizon. Optical packet-switching networks are still in the research stages, but it may be only a few years from now before these networks emerge onto the market with a very-high-performance capability.

9.4 Local Area Networks

The LAN was originally designed for interconnecting computers within an office. Since its inception, it has combined its original role

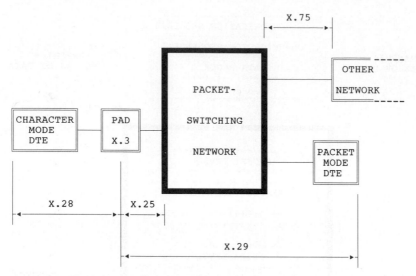

Figure 9.11   Interface protocols for a packet-switching network and its associated equipment.

with operation within the telecommunications network. The term *local* is then expanded to *wide* and *metropolitan* area networks. To understand the problems associated with linking LANs up to a telecommunications network, a description of the LAN must first be made.

The first generation of LANs have data rates in the region of 100 kb/s to 10 Mb/s and operate with wire cables over distances of a few kilometers. The most successful category of first-generation LANs is the carrier sense multiple access with collision detection (CSMA/CD) scheme. *Ethernet* is the term widely used to describe any CSMA/CD LAN. However, Ethernet is actually a proprietary Standard founded by DEC, Intel, and Xerox which has kept its name because it was initially the major player in computer networking. Ethernet is an example of a de facto Standard, although it is only slightly different from the ISO Standard 8802-3, and it allows distributed computing which encompasses file and printer sharing. The 10-Mb/s Ethernet capacity, which was initially over coaxial cable, provides good simultaneous user access to files in a file server; clearly, the more users, the slower the system. It may be necessary for each different department of an organization to have its own LAN, but with an interconnection to the other LANs via bridges. Figure 9.12 shows three different styles of LAN, each interconnected to the other two by bridges. For optimum operation, the amount of traffic flowing between LANs across the bridges should be small compared to the traffic within each individual LAN.

The second generation of LANs operate at about 50 to 150 Mb/s

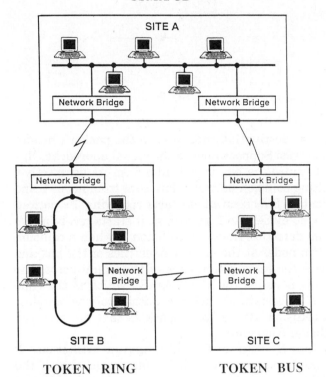

Figure 9.12   LAN bridges.

over distances up to 100 km, using optical fiber cables. This is known as the MAN and offers a wider range of services than the lower data rate LAN. The multimedia services such as voice, data, and video require even greater bit rates. The third-generation LANs are now evolving with bit rates up to and beyond 1 Gb/s. The well-established term *LAN* is now being modified to *computer networking* in many texts.

9.4.1   CSMA/CD (Ethernet)

The CSMA/CD network is characterized by a single, passive, serial bus to which all nodes are connected. Each node detects the presence or absence of data on the bus. Consider the bus of Fig. 9.13 with n nodes connected. If 1 intends to transmit to 2, the bus must first be idle, at which point the signal is transmitted in both directions from 1. On completion of its journey, the signal is absorbed by the terminations at each end. En route, any node can copy the signal into its stor-

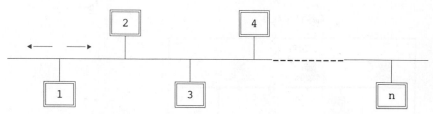

Figure 9.13 A CSMA/CD bus structure.

age buffers, provided its destination indicated in the packet's header matches the node's address. Suppose now as the signal approaches 3, 3 decides to transmit because the bus is still idle from its perspective. There will be a resulting collision between data sent from nodes 1 and 3. This will be evident to 3 as soon at it starts receiving 1's packet. Similarly, it will only be evident to 1 as soon at it starts receiving 3's packet. The maximum detection time of a collision is when a collision occurs between the two nodes at the outer extremities of the bus and could be equal to twice the propagation time of the signal across the length of the bus. This is called the *slot time* (S), which is the time taken for any node to be certain of the transmission of the smallest packet size without collision. Any smaller packets that are received are discarded as collision fragments.

The operating bit rate B, distance D, and minimum packet size P are related by the simple equation: $P = BD$. This means that if the packet size is maintained at, say 50 octets, an increase in the bit rate necessitates a reduction in the maximum length of the LAN. Calculations and experience determine these LANs to be able to transmit at 10 Mb/s over a length of about 2.5 km, constructed by several segments interconnected by repeaters. Each segment is approximately 180 to 500 m. This length is determined by the need for collision detection. A typical large-scale Ethernet configuration with five segments and several repeaters is shown in Fig. 9.14.

The collision problem associated with bus LANs is eliminated by moving to the cyclic control mechanism such as the token ring structure. Also, the installation of optical fiber cable is better suited to ring rather than to bus topologies.

9.4.2 Token rings

The token ring is a ring structure that interconnects many nodes (computer terminals). The token is a sequence of bits which circulates around the ring (Fig. 9.15). In its quiescent state, the token rotates around the ring continuously. When a user at one of the nodes wishes to transmit data, that node *captures* the token and then transmits a

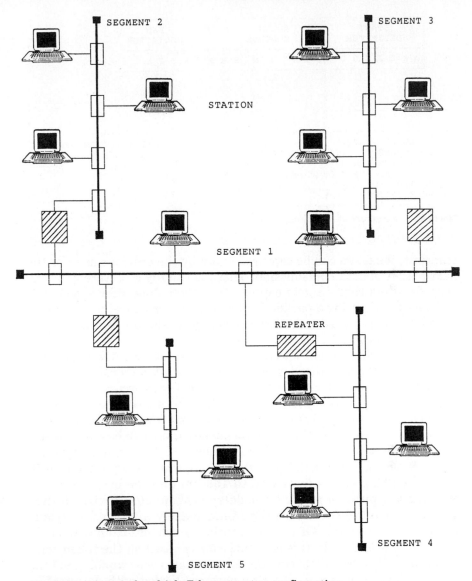

Figure 9.14 A typical multiple Ethernet system configuration.

packet of data. On capture, the token may be seized or have 1 or more bits altered to form what is often referred to as a *connector*. The size of the packet transmitted depends on the time allocated, as defined by the type of token ring. The user receiving the packet can modify the token to establish contact, after which the token and packet are usually returned to the sender. The token is then released and captured by another user who wants to transmit. If no other user wants to

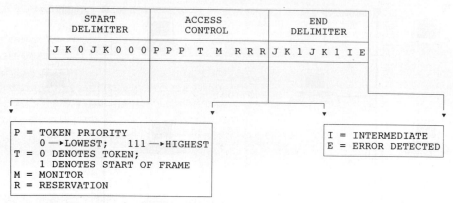

Figure 9.15   Structure of the token.

transmit, the token can be captured again by the previous user for the next packet of data transmission. There is usually a hierarchy of priority for users that want to capture the token. Just after a node has finished transmitting a packet, it places its priority level in the token, for circulation. Nodes with higher priority levels can capture the token and the original user has to wait.

The token ring suffers less congestion than other LAN structures. Unlike the Ethernet, there is no bandwidth wasted on collisions. The main disadvantage of the token ring is the delay in accessing the ring, although the priority system ensures that the more important nodes get express service. In the token ring systems where the token is released only on return of the outgoing packet, there can be a bandwidth problem for long high-bit-rate rings. When a 10-Mb/s system operates on a 1-km ring, the ring delay is about 5 μs, which is equivalent to 50 bits. If the ring size is increased to 100 km and the bit rate to 100 Mb/s, as is typical for a MAN, the delay is about 50,000 bits. In such systems, early token release (i.e., immediately after packet transmission) is essential. Of course, a priority scheme cannot work in these circumstances. In this case, priority is based on the token rotation time (TRT). In an idle ring the token rotates very rapidly, and the rotation time increases as the ring becomes loaded with data traffic. Each node has a window of rotation times during which it can transmit, and if the rotation time is too slow for its priority, it must wait for the ring to be less congested before transmitting. A node decides whether or not transmission can take place by comparing the actual token rotation time with its target token rotation time (TTRT).

An unavoidable time wastage in the token ring LAN is the time between one node releasing the token and the next node capturing it. Again, this time period increases with the length of the ring.

Although token ring technology was available in the early 1980s, it was not until 1985 that it began to receive much attention. At that time, IBM introduced its own version of a 4-Mb/s token ring based on IEEE Standard 802.5. During those early years, almost all token ring networks operated over shielded twisted pair cable.

9.4.3   10Base-T (Ethernet)

A competition has emerged between Ethernet and token ring systems. In November 1990, the IEEE plenary confirmed acceptance of Standard 802.3, which allows Ethernet 10-Mb/s data operation over *unshielded* twisted dual pair cable instead of coaxial cable. Its major attraction is the fact that it can be used to transmit data over *existing telephone cabling*. This Standard, called 10Base-T, is rapidly gaining momentum and user acceptance. Using telephone lines for LAN interconnection opens up a new dimension of versatility, not to mention reduced installation time and overall cost.

The 10Base-T Standard allows a maximum link distance of 100 m, using 24 AWG gauge wire, at an impedance within the range 85 to 110 Ω. The Standard specifies functions such as link test, minimum receive threshold, transmit voltage levels, and high-voltage isolation requirements.

A very important aspect of the 10Base-T Standard is that it can support star topologies. The star is desirable because each node is connected into the network by a separate cable, that is, between the desktop and the hub in the wiring closet. More cable is used than for other topologies, but that is more than compensated for by the improvement and simplification of centralized management and control of the network. For example, a fault can be immediately traced to a node and isolated without affecting other nodes. This degree of control cannot be achieved by bus or ring topologies. This U.S.-based Standard will no doubt penetrate the communications administrations in other parts of the world in the near future.

The competition between Ethernet and token ring systems has led to the 4-Mb/s token ring manufacturers introducing 16-Mb/s token ring systems which operate on unshielded twisted pair cables. Unfortunately, this introduction may be a little premature, because line noise, otherwise known as jitter, which is inherent in cyclic ring structures, severely affected performance of the first systems. VLSI circuits have been designed and manufactured to reduce jitter. Meanwhile the competition between Ethernet and token ring designs continues.

9.4.4   Fiber distributed data interface

The transmission rate of LANs has been increasing steadily over the past decade. The limitations of twisted pair cable (shielded or unshielded) have already been exceeded with today's data throughput requirements. The next obvious step was to move to optical fiber cable, which has enormous bandwidth capability. The FDDI was born in 1982 when the American National Standards Institute (ANSI) formulated a Standard for interconnecting computer mainframes using a 100-Mb/s token ring network.

Since then FDDI has received wide acceptance and now enjoys the distinction of a rapidly expanding market. Today's FDDI networks are usually dual counterrotating 100-Mb/s token rings, with early token release (Fig. 9.16). The two rings are called primary and secondary. This topology has evolved because it has the important advantage of isolating and bypassing faults. FDDI has now been adopted by ISO. The Standards governing FDDI operation are as follows:

1. Physical media dependent (PMD)

2. Physical protocol (PHY)

3. Medial access control (MAC)

4. Station management (SMT)

The PMD and PHY correspond to the OSI model Layer 1 (the physical link layer) which was developed by ISO. MAC is a sublayer corresponding to the lower half of Layer 2, which is the data link layer of the OSI model. SMT overlays the PMD and MAC layers.

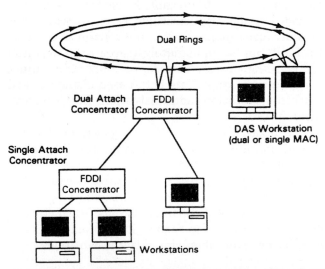

Figure 9.16 The FDDI dual token ring structure. (*Reproduced with permission from IEEE, Strohl, M. J., High Performance Distributed Computing in FDDI CPE Directions, © IEEE LTS, May 1991, Fig. 1.*)

The PMD gives details of the FDDI hardware. Each node within a ring is interconnected by a pair of multimode fibers, whose specifications are 62.5-μm core diameter and 125-μm cladding diameter, operating at a 1300-nm optical wavelength and a data frequency of 125 MHz. These characteristics allow the use of cheaper LED light sources instead of the more expensive laser. PIN diodes are used for detectors. The Standard indicates a link length limit of 2 km imposed by the chromatic dispersion of the LED.

Connection to the ring is made by either a single or dual node. Single nodes are attached to the ring via a concentrator, whereas a dual node can be attached directly to the ring. The concentrators and dual nodes require a bypassing facility so that the rings are unaffected by their presence when they are not transmitting or receiving. Figure 9.17 shows an example of an optical bypass at a dual node. The diagram indicates the situation for the bypass condition, where the equipment at the node is disconnected from the dual ring. Notice that PHY A and PHY B are connected to each other, allowing a self-test mechanism. When required, the switches are activated to allow PHY A and PHY B to be connected to the rings.

PMD also outlines the peak power, optical pulse characteristics, and data-dependent jitter. For an FDDI network, these parameters allow a guarantee of the worst-case BER to be 2.5×10^{-10} and a normal operational BER of 1×10^{-12}. Also, PMD specifications provide for a maximum fiber length of 200 km, with 100 nodes (physical connections) and a maximum distance between stations (nodes) of 2 km.

The PHY layer provides the rest of the physical layer facilities, that is, the upper half of Layer 1 of the OSI model. Data is transmitted on the ring at 125 MHz. The actual information transmission rate is limited to 100 Mb/s. A 4B/5B nonreturn-to-zero invert (NRZI) code is

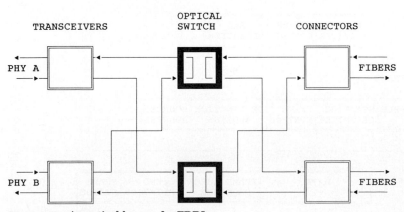

Figure 9.17 An optical bypass for FDDI.

used to perform the conversion. This code simply encodes 4 bits of information into a 5-bit pattern. NRZI is the line code in which a binary 1 is represented by a transition at the beginning of the bit interval and a binary 0 is represented by an absence of transition at the beginning of the interval. At each node, elastic buffers of at least 10 bits are used to regenerate and synchronize the clock. There is no centralized clock.

The MAC sublayer is where the token-passing protocol for FDDI networks is defined, together with packet information, addressing, and recovery mechanisms. MAC controls the data flow on the ring. The structure of the token and packet are presented in Fig. 9.18. Each frame starts with a preamble which contains idle symbols for clock synchronization, a starting delimiter, and frame control bits. The frame control is an 8-bit word for denoting whether the sequence is a token, MAC frame, management frame, type of management frame, etc. The token is completed by an end delimiter, which is one or more digits to signify the termination of the sequence. In the packet containing the data, there are some additional bits included. Following the frame control bits, there is the destination address and then the source address. Following the data is the frame check sequence, end delimiter, and, finally, the frame status. The frame status indicates any detected errors, address recognition, and frame copied. The maximum packet size is 4500 bytes.

The flow of data on the ring is controlled by the MAC. When a start delimiter is placed on the ring, the MAC monitors each packet to establish its destination. If the packet is addressed to another station, the MAC repeats the packet on the ring and observes any errors that may have occurred. A packet addressed to the MAC's station is copied

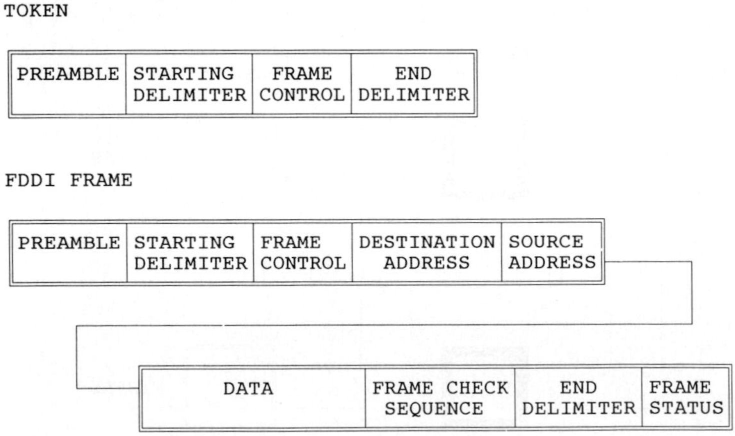

TOKEN

PREAMBLE	STARTING DELIMITER	FRAME CONTROL	END DELIMITER

FDDI FRAME

PREAMBLE	STARTING DELIMITER	FRAME CONTROL	DESTINATION ADDRESS	SOURCE ADDRESS

DATA	FRAME CHECK SEQUENCE	END DELIMITER	FRAME STATUS

Figure 9.18 Token and packet structures for FDDI.

into its storage buffers and is simultaneously repeated on the ring. If the packet was transmitted by the MAC's station itself, it is absorbed instead of being retransmitted, and only a small portion of a packet (e.g., 6 bytes) is repeated.

When the MAC receives a token and it has data to transmit, it removes the token from the ring. The MAC then goes into the transmit mode, packetizes its data, and places it on the ring. This continues until it has transmitted all of its data or the station token holding time expires. At this stage, the MAC places a token on the ring so that other stations can simultaneously access the ring.

As indicated in the token ring section, TTRT is an important parameter in determining the order of access to the ring. Each station sends claim frames with their respective TTRT bid values, and each station stores the value of the TTRT in the most recently received claim frames. Any claim frames with higher TTRT values than its own are absorbed. A station can initialize the ring and transmit data only when it has received its own claim frame after having gone around the ring and becoming the lowest TTRT bid value.

There are two modes of FDDI operation: (1) asynchronous and (2) synchronous. The normal token ring mode of operation is the asynchronous mode. Synchronous bandwidth requirements are fulfilled first, then any bandwidth remaining is dynamically allocated to asynchronous operation. Synchronous operation is used for predictable applications. When the ring is initialized, a TTRT is agreed to by all stations. If the token returns earlier than the TTRT, that station can transmit either synchronous or asynchronous data. If it returns later than the TTRT, it can send only synchronous data. Synchronous operation is managed by the SMT, which is the decision-maker in bids for bandwidth. The sum of the bandwidth allocations, the maximum packet length time, and the token circulation time must not be greater than the TTRT. There are some circumstances in which the TRT will exceed the TTRT, but the TRT will always be less than twice the TTRT.

The asynchronous mode can have either *restricted* or *nonrestricted* tokens. Nonrestricted tokens are used in normal circumstances, and bandwidth is shared equally between stations. A nonrestricted token is always used in the FDDI network initialization. The token can be restricted in cases when two or several stations want to use all of the available asynchronous bandwidth. Network management establishes which stations are making the request, and one of the stations creates a restricted token. After transmission, the station that issued the restricted token returns the ring to its normal mode of operation by issuing a nonrestricted token.

The SMT defines the beacon token process, which is used to locate ring failures. When a station realizes there is a ring failure, it enters the beacon process and starts to transmit beacon frames continuously. Beacon frames are special bit sequences, unique to each station, which are recognizable by all stations as indicating the presence of an upstream fault. If that station receives a beacon from an upstream station, it will stop transmitting its own beacon and start repeating the upstream station beacon. That process continues until eventually the only station transmitting a beacon will be the one that is immediately downstream from the failure. When the failure has been repaired, a beaconing station will receive its own beacon. This is a signal to that station to stop beaconing and return to normal by issuing a claim.

SMT specifies how the other FDDI layers are coordinated, which means that it overlaps the PMD, PHY, and MAC layers. SMT also allows for some error detection and fault isolation. The latest version of the SMT Standard (SMT 6.0) includes three aspects of management: connection management (CMT), ring management (RMT), and frame-base services and functions. CMT takes care of the PHY layer components and their interconnections at each station so that logical connections to the ring can be accomplished. Also, within a station, the CMT manages the MAC and PHY configurations. Finally, CMT monitors link quality and deals with fault detection and isolation at the PHY level. RMT manages a station's MAC layer components and also the rings to which they are attached. RMT also detects faults at the MAC layer (e.g., duplicate address identification or beacon identification problems). The frame-based services and functions perform higher-level management and control of the FDDI network such as collection of network statistics, network fault detection, isolation and correction, and overseeing the FDDI operational parameters.

The FDDI Standard specifies operating over a distance up to 2 km. A recent survey has shown that 95 percent of present FDDI network connections are within 100 m of the wiring closet. The question arises, If only 100 m of cabling are required, why not use inexpensive twisted pair cable instead of the more expensive fiber? This question is particularly relevant because the FDDI token ring LAN is experiencing strong competition from the Ethernet 10Base-T LAN. Many users are operating 4- and 16-Mb/s token ring LANs over shielded twisted pair wire, but many users have experienced problems over unshielded twisted pair wire. Although various chip circuits are being introduced to counteract jitter created in these token systems, the future of high-bit-rate token rings using unshielded twisted pair cable remains uncertain.

9.5   WANs and MANs

LANs are classified as *intra*premises communications, whereas WANs are *inter*premises communications. By the very nature of the WAN, it is implemented in cooperation with the public network operators, for example, using leased lines, dial-up PSTN connection, or X.25 packet-switching. Data rates up to 2 Mb/s are currently available with high-speed switching planned for the mid-1990s. The interconnection of LANs within a WAN is impractical in today's public network environment because of bandwidth and cost constraints. The term *MAN* has been devised for wide area interconnection of LANs within the public network. A Standard has been developed by the IEEE, known as IEEE 802.6, which does not require any change to existing computer hardware or software and allows broadband connectionless data and integrated isochronous (voice) communication services.

The MAN was originally intended for data, but voice and video traffic can be included these days. MANs provide two-way communication over a shared medium, such as optical fiber cable, and satisfy the requirements of commercial users, educational institutions, and the home. The MAN is intended to operate up to about 150 Mb/s over links up to 100 km, preferably using optical fiber cable. Although a MAN is a shared medium, privacy is ensured by authorization and address screening at the source and destination. In this manner, either accidental or intentional third-party surveillance or intervention is eliminated.

The MAN architecture is based on the dual bus topology in a form known as the distributed queue dual bus (DQDB) shown in Fig. 9.19. This simple MAP is the basis of IEEE 802.6 Standard. The DQDB structure can support both synchronous and asynchronous data, voice, or video traffic simultaneously. The DQDB MAN has two contra-directional buses which physically look like the dual ring structure, broken at the node providing the frame generators. Each node is attached to both of the buses. The logical DQDB can be redrawn as the physical ring of Fig. 9.20. The end points of the two buses are colocated in this ring configuration.

Access to the DQDB network by a user at any node is achieved in a manner which is very equitable, because requests by users are honored in their arrival order. This is often referred to as a perfect queuing structure. Figure 9.21 illustrates how this is achieved. Frame generators at each end of the two buses continuously transmit empty frames of data, in the form of packets, in opposite directions along the

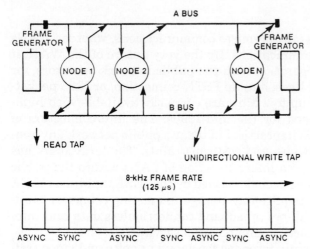

Figure 9.19   The distributed queue dual bus (DQDB) MAN. (*Reproduced with permission from Ref. 420, © 1990 Telecommunications.*)

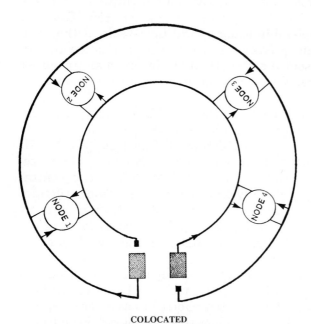

COLOCATED
FRAME GENERATORS

Figure 9.20   DQDB frame generation.

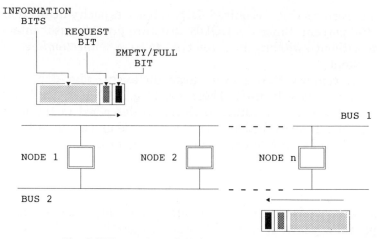

Figure 9.21   The DQDB queuing principle.

two buses. The frame generators are synchronized to the telephone network 8-kHz reference frequency so that if there is a break in either bus, the other frame generator can take over without a problem. All packets are completely absorbed at the end of their trip along a bus. All nodes can read the data passing by it or write over it as necessary, but they cannot remove data. Each packet contains a bit indicating whether the information-carrying portion of the packet is full or empty. Each packet also carries a request bit time slot. When a user at any node wants to transmit, it establishes which bus must be accessed in order to carry its information downstream to the desired destination. The user then places a request bit in the next available request bit time slot. Several packets may pass the user's node before a request time slot becomes available. In this case, each time a full request time slot passes, the node's request queue is incremented. Eventually, when a request time slot is available, it inserts a bit, indicating it has joined the queue. As more packets pass by, further user requests may be observed that were initiated prior to its own request, so the user has to wait for their completion. At any time, a node may have only one request uncompleted. A node having placed a request will eventually see empty information frames pass by, which will be filled by other nodes downstream that joined the queue at an earlier point in time. Each empty packet that passes decrements the node's queue number by 1, and when the number is 0, the node can transmit its data. If the system is idle (i.e., a condition in which no other users are transmitting or in the queue), clearly a user can transmit immediately. In these circumstances data from one node can be transmitted in packets one after the other. Comparing the DQDB to the token ring structure, the DQDB does not suffer from a time loss known as the *token slack time*. Furthermore, since the DQDB structure is not a closed ring, there are no "response bits" required. DQDB has a capacity utilization of almost 100 percent. Since the DQDB structure provides user interconnection without switches, it is referred to as a *connectionless,* or *seamless,* system.

In order to connect MANs over large distances, high-data-rate leased lines are presently used. These existing MANs are building blocks for what will be the future B-ISDN, as described in the next section. The trend is an evolution toward a national and eventually an international *broadband* telecommunications network.

9.6   ISDN

ISDN has been a term on most communication engineers' tongues over the past few years. It is revolutionary in that it is a departure from the traditional telecommunications often known as POTS (plain old telephone sets, or services). Exciting new services were envisioned. Unfortunately, after the initial optimism, it was soon realized that the 64-kb/s ISDN system was so low on bit rate (bandwidth) that only a limited number of extra services would be possible. The transmission of video signals is the obvious point that springs to mind. The 64-kb/s system supports only very low quality video signals and is now referred to as narrowband ISDN.

As disappointing as narrowband ISDN may be to many people, the important point is that the wheels of change have been set in motion. The experience gained and the problems solved during the narrowband copper-wire-based ISDN era are paving the way for the next generation of ISDN, that is, broadband optical-fiber-based ISDN, which should certainly fulfill the promise of vastly improved services. To trace the evolution of what is indeed becoming a global telecommunications revolution, the details of narrowband ISDN will be discussed first. Many books have been written on this subject alone, but it is the intention here to highlight the major aspects in as short a space as possible.

9.6.1   Narrowband ISDN

What is ISDN? During the evolution of digital networks, PCM links were initially introduced for interexchange traffic. Then digital switches were installed to form what was called an integrated digital network (IDN), supporting telephony at 64 kb/s. It was the realization that an IDN could be organized to provide data-related services in addition to telephone services that led to the ISDN. ISDN can be defined as *a network in general evolving from a telephony IDN that provides end-to-end digital connectivity to support a wide range of services, including voice and nonvoice service, to which users have access by a limited set of standard multipurpose user-network interfaces.* The ISDN standardization process has taken over a decade, and the CCITT estimates that more than 1000 person years of international meeting time has been spent on the process. End-to-end digital connectivity, which extends digital transmission to the customer, is the essence of ISDN. This digital connection from the main network to the customer is called the integrated digital access. The CCITT has defined ISDN access in terms of channels, whose main types are:

B channel	64 kb/s for carrying user information such as 64-kb/s voice encoded information or data information which is either circuit or packet switched.
D channel	Either 16 or 64 kb/s, primarily for carrying signaling information for circuit switching. It may also carry packet-switched information.

The CCITT also recommends two types of access. The first is the *basic rate access,* which contains 2B + D (16 kb/s) = 144 kb/s. Each B channel can have a different directory number if required and both channels may carry voice or data up to 64 kb/s on copper wire cable. The second type is the *primary rate access,* which has either 30 B (in 2.048-Mb/s CEPT networks) + D (64 kb/s) or 23 B (in 1.544-Mb/s North American networks) + D (64 kb/s). This access is mainly for connecting digital PBXs to the ISDN. An important aspect of ISDN is the isolation of signaling from the subscribers' data. Signaling for ISDN is specified by the CCITT to be common channel signaling. ISDN is organized in the ISO layered structure for OSI. Layer 1, the physical layer, is specified by CCITT Recommendation I.430 for the basic rate access and I.431 for the primary rate access. Layer 2, the link layer, and Layer 3, the network layer, are defined for both the basic and primary accesses by CCITT Recommendations I.441 and I.451, respectively.

ISDN call procedure. The procedure for establishing an ISDN call is indicated in Fig. 9.22. First, the customer requests a call, and the calling terminal creates a message for the call request, that is, SETUP. This is a Layer 3 process. This message is sent via the terminal Layer 2 to Layer 3 of the digital switch. The switch then makes the routing to the called customer by a Layer 3 call request (SETUP). If the call is across town, for example, this message maintains a route through the switch while the switch makes contact with the destination end switch, using the CCITT no. 7 signaling system. Once the SETUP message is received at the called terminal, an ALERT message is returned to the caller as a SETUP acknowledgment. If the called party is available, the call acceptance message (CONNECT) is sent to the

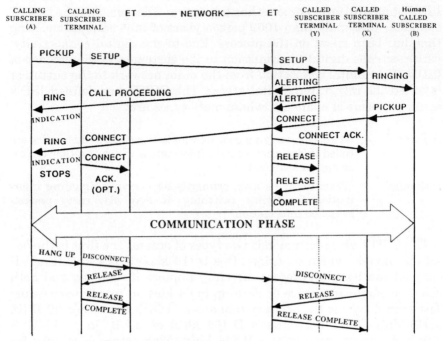

Figure 9.22   ISDN circuit-switched call procedure.

calling terminal. When the connection is completed through the network, dialogue in the form of voice or data transfer can take place. On completion of the communication phase, a CLEARDOWN can be initiated at any time by either party.

Figure 9.23 shows the basic rate user-network interface configuration as defined by the CCITT. This configuration uses the concepts of reference points and functional groupings. In this figure, NT1, NT2,

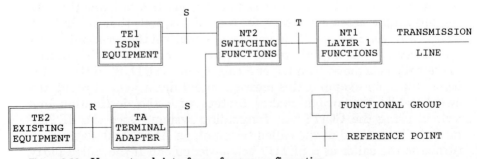

Figure 9.23   User network interface reference configuration.

TE1, and TE2 are functional groupings, whereas R, S, and T are reference points. The network termination 1 (NT1) terminates the two-wire line from the local network on its network side and supports the S interface on the customer side. The network termination 2 (NT2) distributes the access to the customer network such as PBXs and LANs. Terminal equipment (TE1) has an I-series customer network interface, whereas terminal equipment (TE2) has a non-ISDN X- or V-series interface. The *terminal adapter* (TA) is an important item which adapts the non-ISDN interface of TE2 to the ISDN interface of NT2.

The reference points R, S, and T are not specifically physical interfaces but may be points at which physical interfaces occur. The S reference point is the CCITT, ISDN user-network interface. NT2 may not be present, which makes S and T colocated. In the United States, the NT1 functions are provided by the customer, whereas in Europe the service provider owns the network termination even though it is placed on the customer premises.

User network interface. Transfer of information from Layer 2 of the terminal to the network is the responsibility of Layer 1. Conventional twisted pair cable is used from the terminal to the NT1 and from the NT1 to the local CO. If errors are introduced by the twisted pair wires, Layer 2 takes care of error detection and correction. The following characteristics are required for Layer 1 to support Layer 2.

The Layer 1 frame structures across the S interface differ depending on the direction of transmission, as indicated in Fig. 9.24. Each frame is 250 μs in duration and contains 48 bits. The transmission rate is 192 kb/s. From the terminal to the network, each frame has a group of bits which is dc-balanced by its last bit, L. The B- and D-channel groups are individually balanced because they may come from different terminals. There are bits allocated to perform the housekeeping functions of frame and multiframe alignment. From the network to the terminal, there is also a D-echo channel in each frame, in addition to the B and D channels. This D channel returns the D bits to the terminals. The complete frame is balanced using the last bit (L bit) of the frame. The M bit is used for multiframe alignment. The S bit is a spare bit.

Signaling. CCITT Rec. I.441 describes the Layer 2 procedures for a data link access, which is often known as the link access procedure for the D channel (LAP D). The Layer 2 objective is to perform an error-free exchange of I.441 frames between two physical medium connected endpoints. The call control information of Layer 3 is contained in the Layer 2 frames. I.441 frames are arranged in octet sequences as

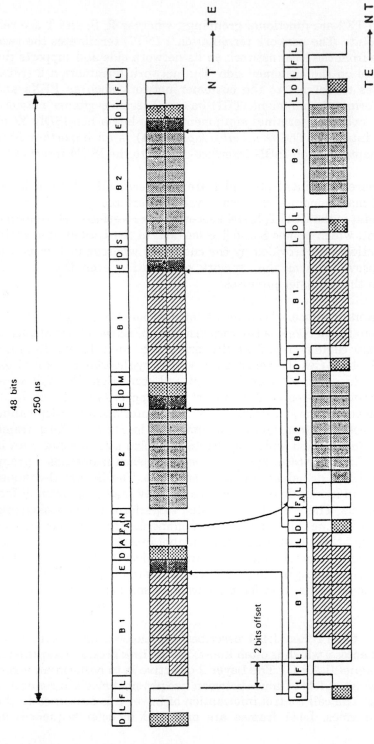

48 bits

250 µs

2 bits offset

NT ⟶ TE

TE ⟶ NT

Figure 9.24 Layer 1 basic access frame structure. D = D-channel bit, L = dc-balancing bit, F = framing bit, B1 = B-channel 1 (8 bits), E = D-echo channel bit, A = bit used for activation, F_A = frame alignment bit, N = bit set to a binary value [N = $\overline{F}_A$ (NT to TE)], B2 = B-channel 2 (8 bits), M = multiframe alignment bit, S = spare bit for future use. (*Reproduced with permission from Ref. 376, © 1987 Artech House.*)

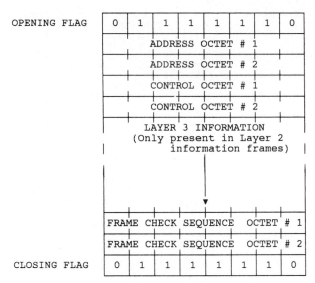

OPENING FLAG	0	1	1	1	1	1	1	0

ADDRESS OCTET # 1

ADDRESS OCTET # 2

CONTROL OCTET # 1

CONTROL OCTET # 2

LAYER 3 INFORMATION
(Only present in Layer 2
information frames)

FRAME CHECK SEQUENCE OCTET # 1

FRAME CHECK SEQUENCE OCTET # 2

CLOSING FLAG	0	1	1	1	1	1	1	0

Figure 9.25 Layer 2 frame structure.

shown in Fig. 9.25. A separate layer address is used for each LAP in layer multiplexing. The address contains a terminal endpoint identifier (TEI) and a service access point identifier (SAPI). The SAPI defines the service intended for the signaling frame, and the TEI is set for each individual terminal on installation. These two addresses identify the LAP and form a Layer 2 address. A terminal always transmits these addresses in its frames and will not accept received frames unless they contain this address. The two link control octets contain frame identification and sequence number information. The next sequence of octets is for Layer 3 information. This is followed by two octets for frame checking and error detection.

To clarify Layer 2 operation, consider the following call attempt by a terminal. For a completely new call, the caller makes a request for service, whereupon Layer 3 requests service from Layer 2. Layer 2 can offer service only when Layer 1 is ready, so a request to Layer 1 is made by Layer 2. Layer 1 initiates the start-up procedure. Subsequently, Layer 2 initiates its start-up procedure, which is called setting up a LAP to ensure correct sequencing of information frames. When the LAP has been formed, Layer 2 can carry Layer 3 information. All information frames received by the called party must be acknowledged to the calling party. If a frame is lost or corrupted, this will be detected by the called terminal within its frame timing interval, and there will be no acknowledgment sent to the calling terminal. Retransmission then takes place, and if several retransmissions occur without acknowledgment, Layer 2 presumes that there is link failure,

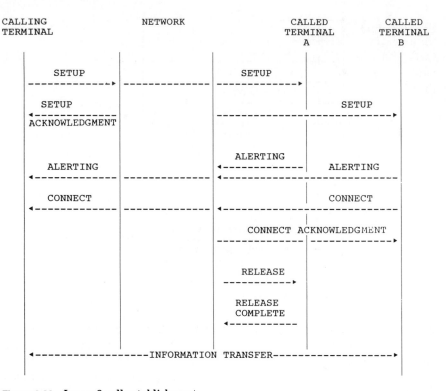

Figure 9.26 Layer 3 call establishment.

in which case it tries to reestablish the LAP. CCITT Rec. I.451 specifies the call control procedures. The Layer 3 sequence of messages required to establish and disestablish a call is shown in Fig. 9.26. The
calling terminal sends call establishing information such as calling
number, type of service required, and terminal information to ensure
compatibility. Once the call is ready for setup, the interexchange,
CCITT no. 7 signaling system is initiated. A call SETUP message is
sent from the calling to the called user via the broadcast data link.
This message can be scrutinized by all terminals at the called end to
establish if they are compatible with the calling equipment. The
ALERT message is returned to the network by all available compatible terminals. Simultaneously, an incoming call indication (ringing) is
activated by the local terminal. On answering, a terminal sends a
CONNECT message to the network and the call will be given to the
first terminal to respond with a CONNECT message. The network
connects that terminal to a B channel and returns a CONNECT AC-
KNOWLEDGMENT giving B channel information. A RELEASE message is given to all other terminals that responded to the call. When

the called terminal has received the CONNECT message, the network informs the calling terminal by also sending it a CONNECT message. Call charging starts and the communication then proceeds. This whole procedure is accomplished using a sequence of octets, designed to fulfill the protocol requirements.

In conclusion, the principal features of ISDN can be summarized as a network that has:

- User-to-user digital interconnection
- Both voice and nonvoice services
- A limited number of connection types and multipurpose interfaces
- Circuit- and packet-switched connections
- Network intelligence

Figure 9.27 summarizes the CCITT ISDN functional and corresponding protocol architecture. The CCITT ISDN functional architecture of Fig. 9.27a includes functional components such as exchange terminations (ETs), packet handlers (PHs), integrated services PBXs, (ISPBXs), NT1s, etc., which perform the switching, multiplexing, call processing operations, etc. The functional architecture can be divided into two parts, access and network. The access part of the ISDN is that portion seen and controlled by the user, often referred to as the private ISDN. The network part contains the portion of functional entities controlled by the service providers and is often referred to as the public ISDN. In each country implementing ISDN, depending on political and regulatory viewpoints, the line between user and service provider (access and network parts) can wander, even though the CCITT functional architecture does not change.

Figure 9.27b summarizes the corresponding CCITT ISDN protocol architecture for the access and network parts. This diagram indicates the various protocols used in a network containing both private and public ISDNs. As indicated by this diagram, ISDN protocols are a fairly complex business. This is not the full story, however. There are a number of additional supplementary services which have recently been standardized. Subscriber loops, which differ greatly from one country to another, have been difficult to standardize completely and modifications are still in process. For example, the basic rate access interface has several reference points. In the United States, the U-reference point has been chosen as the dividing line between the public and private ISDNs, whereas in Europe it is the S/T-reference point.

The object of all this effort is to ensure satisfactory end-to-end interworking and global feature transport.

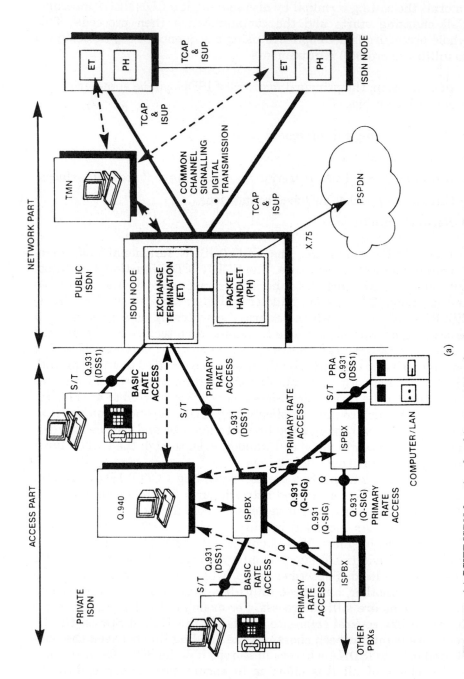

Figure 9.27 (a) CCITT ISDN functional architecture (*reproduced with permission from Ref. 414, © 1991 Telecommunications*).

(a)

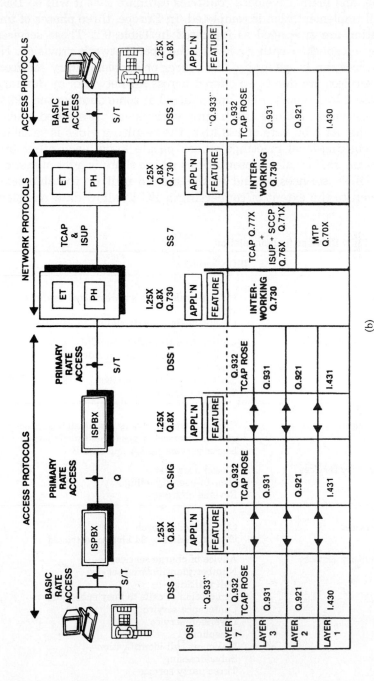

Figure 9.27 (b) CCITT ISDN protocol architecture (reproduced with permission from Ref. 414, © 1991 Telecommunications).

N-ISDN services. The time taken to establish a widespread ISDN is quite long, and many developed countries estimate that it will be 1995 before full implementation is completed. In Europe, three phases of implementation are envisaged as indicated in Table 9.2. These services have been established with a view to the open network provision. N-ISDN services can be categorized as bearer or supplementary services. Phase 1 services are due to be offered across all networks by January 1993, phase 2 by January 1994, and phase 3 at some date beyond 1994.

These services include such innovations as low-speed video conferencing. At a bit rate of 128 kb/s, the resulting video is satisfactory for working-level meetings where people know each other but not, for example, for sales presentations where there is a lot of motion activity. ISDN services extend to the mobile unit or portable telephone, even to the extent of connecting a PC to allow data transfer.

TABLE 9.2   ISDN Services Implementation

	ISDN services
Phase 1	
Bearer services	Circuit mode 64-kb/s unrestricted bearer service Circuit mode 3.1-kHz audio bearer service
Supplementary services	Calling line identification (CLI) Calling line identification restriction Direct dialing-in Multiple subscriber number Terminal portability
Phase 2	
Bearer services	Circuit mode 64-kb/s unrestricted bearer service on reserved or permanent mode packet bearer service case A and case B
Supplementary services	Closed user group User-to-user signaling Reverse charging
Phase 3	
Bearer services	Circuit mode speech Circuit mode 2 × 64 kb/s unrestricted
Supplementary services	Advice of charge services Number identification service Call waiting Completion of calls to busy subscriber Conference service Diversion service Freephone Malicious call identification Subaddressing Three-party service

Facsimile is evolving to the group 4 level, which allows for photo quality images to be transferred.

Facsimile (Fax) is growing rapidly in the United States, but the emergence of electronic mail (E-mail) is supplementing and perhaps to some extent overshadowing Fax. Videotex, teletex, and telex are declining in popularity and usage. The United States ISDN marketplace targets two groups: voice- and data- or computer-oriented and there is some overlap between the two. Eventually, the two should merge with full ISDN deployment.

In the United States, network terminators are considered to be part of the CPE, whereas in the rest of the world they are not. Regardless of the newly available bandwidth, it has been established that the CPE is critical to the success and acceptance of ISDN. Table 9.3 indicates the customer requirements for various types of CPE.

9.6.2 Broadband ISDN

B-ISDN services. The demand for high-quality video (HDTV) and high-speed data information exchange is the driving force behind the inevitable implementation of the B-ISDN. CCITT Recommendation

TABLE 9.3   CPE Customer Requirements

CPE	Customer requirements
PC with voice and data adaptor	Integrated telephone Keyboard dialing Modem replacement Multiple data channels
Stand-alone terminal adaptor to interface	Diverse terminals Host computers Data communication switches Group 2 and 3 Fax Analog telephones Special equipment
ISDN telephone	Interworkable with analog telephones High-quality voice Calling number display Voice conferencing Single-button feature activation Electronic key telephone service
ISDN telephone with adaptor	Modem replacement Data call on keypad
PC workstation	Same as PC with voice and data adaptor Integrated call management
Fax-Group 4	Transmit images on paper

I.121 classifies B-ISDN services into *interactive* or *distribution* services as follows.

Interactive services. These services consist of the following:

1. Conversational services
2. Messaging services
3. Retrieval services

Interactive implies a two-way information exchange, which can be between two subscribers or a subscriber and a service provider. Distribution services are primarily one-way information exchanges from a service provider to the subscriber.

Conversational services are the classic two-way communication with the exception that B-ISDN will include video, data, and text as well as voice. Video telephony is a major application which will probably be employed first of all in business offices. Full motion, live video will be very advantageous for users such as sales and technical personnel who need to discuss diagrams and charts, etc. As prices decrease, video telephony will become a popular service to the home. One drawback is that obscene or malicious phone calls would take on an extra dimension of reality which could create a serious social problem.

Video telephony can be expanded to provide a cheaper service to match the existing video teleconferencing by using larger camera area coverage for multiple participants and larger screens at each venue. Or, multiple video telephones could be interlinked in a conference situation by subdividing each participant's screen into several windows.

Document transfer can also be placed in the conversational category for applications such as high-resolution Fax or transfer of documents between workstation users.

Messaging services are user-to-user communications which use storage units for store-and-forward, message handling, and mailbox functions. These are not real-time conversation services. This means they make less demands on the network because both users do not have to be present simultaneously. This is analogous to the X.400 and teletex N-ISDN services.

An attractive new B-ISDN messaging service is video mail. This is analogous to the existing E- and voice mail. The main difference is that instead of allowing only text and graphics in document form, video mail would be full motion. This is the equivalent of mailing a video cassette. Electronically mailing a document with this type of messaging service capability could include a combination of text, graphs, voice, and video. This would be truly multimedia communications at the highest level.

Retrieval services allow a user to access information stored in a general public data bank. The information can be retrieved by user control, which is analogous to the N-ISDN videotex service. It is an interactive system which has a general-purpose database, for both private and business requirements, and can operate over the PSTN or an interactive metropolitan cable TV system. The N-ISDN information is in the form of pages of text (e.g., stock market information). The B-ISDN retrieval service could also include sound passages, high-resolution images, and video footage in addition to the existing text and graphics. This service could be very useful for applications such as educational remote learning. Video retrieval could be extended to full-length movie films from a video library. This would work by directly linking up the library to the user's video cassette recorder. Again, multimedia retrieval of documents containing a mixture of video, sound, text, high-resolution graphics, and even software would be possible.

Distribution services.   These are also referred to as broadcast services and may exist as follows:

1. *Without* user individual presentation control (broadcast), for example, CATV, electronic newspaper
2. *With* user individual presentation control, for example, enhanced teletext including video (cabletext)

In the first service the user has no control over the starting times and can only "tune-in" like a TV situation. In the second, the information is relayed in a sequence of frames in cyclical repetition, so the user has some degree of control over the start of the presentation. This is analogous to the narrowband teletext service. Cabletext would be a broadband enhancement of this system, using cyclical transmission of text, images, video, and sound passages.

In the United States, significant large-volume installations of B-ISDN will not begin until about 1998, so full B-ISDN may not be available until after the year 2000. B-ISDN will be achieved a few years earlier in Japan, which will probably be the first country to attain full deployment.

B-ISDN implementation.   There are three main facets of B-ISDN which must be in place before a resulting network can benefit from the synergistic combination of all three:

1. Optical FTTH
2. Synchronous digital hierarchy (SDH)
3. Asynchronous transport mode (ATM)

Optical FTTH. Central to any serious implementation of B-ISDN is the information transporting medium. It is universally accepted that optical fiber is the solution. As mentioned in Chap. 7, there are several architectures for a fiber-based B-ISDN network. While some cost saving can be made depending on which architecture is used, the fact remains that eventually there must be fiber to every home, and that is extremely expensive. Already (1992), fiber to the curb is almost reaching cost parity with copper wiring as fiber manufacturing costs decrease. That means that in most developed countries the transmission medium choice for new installations will soon automatically be fiber. In the United States, fiber will be the economical choice for all growth projects from 1992 onward and for some rehabilitation projects by 1993. It is the replacement of copper cable with fiber that incurs the very large expense.

Two possible architectures that allow steady growth and minimize costs are shown in Fig. 9.28. These two topologies have different properties and the choice depends on various factors such as service requirements, geographical situation, and reliability needs. By partitioning the subscriber loop into feeder and distribution loops, the cost is minimized. Figure 9.28a is a double star topology. High-bit-rate SDH transmission systems interlink the remote terminal (RT) and the exchange (CO) in the feeder loop. The RTs can do simple multiplexing or line concentration functions to use the available feeder bandwidth efficiently. Existing carrier system subscriber loops can be most easily upgraded using this topology.

Figure 9.28b is a ring and star topology. The feeder loop is in the form of a fiber ring and the RTs are attached to points on the ring. The ring can carry SDH variable transmission bit rates to accommodate a variety of traffic requirements. A drop and insert capability at the RTs allows variable bandwidth distribution as required around the ring. Again, traffic concentration can be done at the RTs, and some applications may require ring access protocols. The inherent route diversity of the ring, coupled with the RT bypass and loopback facilities, provides a highly reliable mechanism.

The distribution loops both use a star-connected single-mode fiber to each subscriber. Full-duplex transmission can be achieved by either using two fibers, one for each direction, or just one fiber operating at a different wavelength for each direction of transmission.

It is essential to have the flexibility to be able to deliver a wide range of bandwidths to the customer. Some subscribers may want only conventional, digital POTS. Commercial subscribers may require high capacity, warranting the use of SDH. The distribution loop, often referred to as "the last mile," will probably be the final link in the fiber chain.

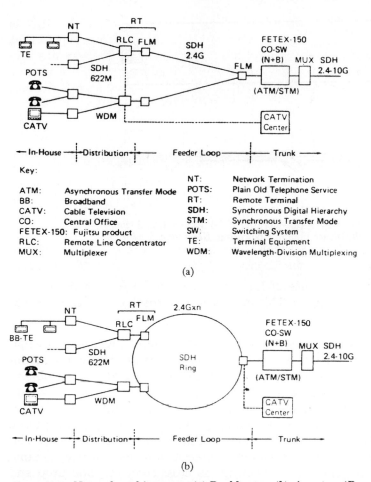

Key:

		NT:	Network Termination
ATM:	Asynchronous Transfer Mode	POTS:	Plain Old Telephone Service
BB:	Broadband	RT:	Remote Terminal
CATV:	Cable Television	SDH:	Synchronous Digital Hierarchy
CO:	Central Office	STM:	Synchronous Transfer Mode
FETEX-150:	Fujitsu product	SW:	Switching System
RLC:	Remote Line Concentrator	TE:	Terminal Equipment
MUX:	Multiplexer	WDM:	Wavelength-Division Multiplexing

(a)

(b)

Figure 9.28 Network architectures. (*a*) Double star; (*b*) ring star. (*Reproduced with permission from Ref. 394, © 1990 IEEE.*)

There is another difficulty associated with FTTH which must be overcome. That is the simple, but not trivial, question of power. Figure 9.29 is the simple picture of power supplied to the subscriber equipment by the battery in the CO. Fiber creates the problem that there are no longer any metal wires for carrying this power. There are several possible solutions: (1) use copper wires in the fiber cable up to the customer residence, (2) terminate the fiber at the curb and back-feed the optical equipment from the customer residence, and (3) use power in the customer residence to power the equipment. Solution 1 is costly and solutions 2 and 3 mean relying on commercial power, which is not always as reliable as necessary, especially in an emergency situation. Possible alternatives are solar cells or chemical fuel cells.

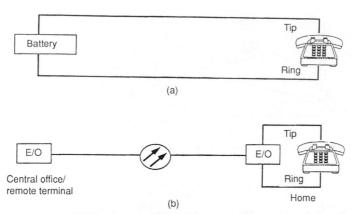

Figure 9.29 FTTH power considerations. (*a*) Conventional copper wire powering; (*b*) future photonic powering. (*Reproduced with permission from Ref. 411,* © *1991 IEEE.*)

More elegant is the use of laser power to feed very low-power consumption electronic sets. This issue is still unresolved and must be addressed very soon.

SDH/Sonet. SDH, which was explained in Chap. 2, will be a key feature in future B-ISDN networks. Not only is it primarily a synchronous multiplexing scheme, but it also has the versatility to be able to handle asynchronous data within the synchronous system. This may sound like a contradiction of terms, but it merely states that the advantages of both synchronous and asynchronous multiplexing can be realized by having both techniques available within the same system.

SDH has the advantageous characteristics of modularity, flexible network management, self-healing via cross-connects, and drop and insert capability. SDH allows ease of upgrading to higher-order bit rates. Also, cost benefits can be realized by skip multiplexing. This reduces the amount of equipment manufactured by internally multiplexing from a low bit rate to very high bit rates instead of having outputs at several specific hierarchical levels. SDH penetration into the network will be a gradual process with a possible deployment schedule as in Fig. 9.30. The CCITT has proposed a user interface at 155.52 and 622.08 Mb/s. The local exchanges will have to make a transition from wire to fiber cable. In the interim, they must be able to offer both twisted-pair wire at the basic ISDN bit rates and fiber at the B-ISDN rates. Initial SDH penetration for the trunks and feeders will use the 155.52-Mb/s (STM-1) and 622.08-Mb/s (STM-4) systems. To illustrate the versatility of SDH, the STM-1 could, for example, include 28 asynchronously multiplexed DS-1 (1.544 Mb/s) bit streams and 2 asynchronously multiplexed DS-3 (44.736 Mb/s) bit

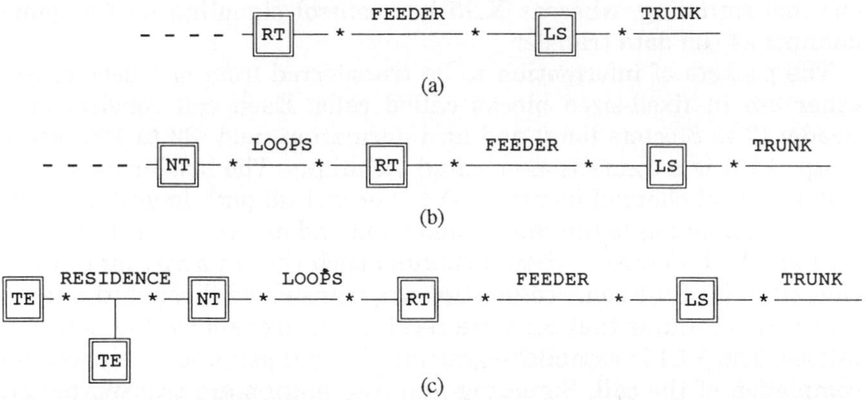

Figure 9.30 Chronological SDH penetration. (*a*) Fiber from trunks to feeder; (*b*) fiber into distribution loops; (*c*) fiber to the home. LS = local switching system, RT = remote terminal, NT = network termination, TE = terminal equipment, * = SDH interface.

streams or three synchronously multiplexed Sonet STS-1 (51.84 Mb/s) bit streams, etc. It is anticipated that the STM-16 SDH level, which supports the 2.48832-Gb/s rate, will become a popular data rate for the feeder loop. Drop and insert from this loop can be done as required. When the final phase of Fig. 9.30*c* is accomplished, subscriber premises terminals will have their own SDH interfaces. The trunk section of these networks will have bit rates at 10 Gb/s or higher. No doubt, the business community will be the first subscribers to take advantage of these high data rates. As the cost comes down with penetration, the home will be serviced by higher and higher bit rates. Already CCITT Recommendation I.121 specifies the data rate from subscriber to network need not be the same as network to subscriber. This takes into account the fact that subscribers in the home will probably want to receive more information than they transmit. Bandwidth on request at a reasonable price is the final objective.

Asynchronous transfer mode (ATM). As stated above, the asynchronous multiplexing technique will be used together with the synchronous method. This is taken one step further in the B-ISDN Standards by including packetization. In CCITT Recommendation I.121 the ATM will be used for the ISDN user network interface. The implication is that B-ISDN will be a packet-based network at the interface and probably will also have packet-based internal switching. The Recommendation states that although B-ISDN will support circuit mode applications, it will have a packet-based transport mechanism. This means that ISDN will evolve from a circuit-switched telephone network to a packet-switched B-ISDN. ATM has some similarities to and some major differences from X.25. One difference is that ATM uses common channel signaling, whereas X.25 has control signaling on the same channel as the data transfer.

The packets of information to be transferred from one user to another are in fixed-sized blocks called cells. Each cell consists of a header (3 to 8 octets long) and an information field (32 to 120 octets long). ATM is a connection-oriented technique. The header must contain a virtual channel identifier (VCI) or virtual path identifier (VPI) for cells belonging to the same connection and an error detection code. VCI and VPI identify a virtual channel such that in a virtual connection all cells of the same connection are transferred through the same route. This ensures that cells are received in the same order as transmitted. The VCI is established during the call setup and released on completion of the call. Signaling and information are transported on separate virtual channels.

There are two methods of transmission in an ATM network: (1) using ATM cells of a synchronous payload frame (Fig. 9.31a) or (2) pure ATM where there are no frames. In the frameless case, synchronization is achieved by an error-check algorithm placed in the cell header.

Even though ATM is cited extensively in the literature, there is considerable debate as to its economic viability. For example, there is strong competition for broadband services from well-established technologies such as DQDB MANs. It is essential for MANs to be compatible with B-ISDN. FDDI networks, well suited for fast transmission of data, are also very cost competitive.

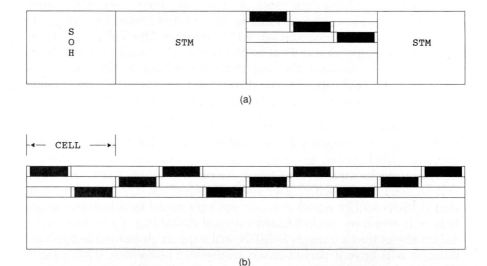

(a)

(b)

Figure 9.31 Two methods of implementing transmission in an ATM network. (a) ATM cells in a synchronous frame; SOH = section overhead, STM = synchronous transfer. (b) Frameless ATM cells.

TABLE 9.4   CCITT Defined Channel Bit Rates for N-ISDN and B-ISDN

	Narrowband (kb/s)		Broadband (Mb/s)
D	16 or 64	H21	32.768
B	64	H22	43–45
H0	384 (6 × 64 kb/s)	H4	132–138.24
H11	1536 (24 × 64 kb/s)		
H12	1920 (30 × 64 kb/s)		

ATM and other packet-switching techniques are generally accepted for noncontinuous information flow applications such as computer applications. Algorithms can be used to compensate for the variable delay inherent in packet transmission so that ATM can be used for continuous flow applications such as telephony and video. ATM has considerable flexibility in its use for all services. New services can be added without restructuring the network, or switching exchanges and services already in existence can be modified or expanded. B-ISDN channel rates have been defined by the CCITT in addition to the existing narrowband ISDN channel rates (Table 9.4).

H21 and H22 are appropriate data rates for full-motion video applicable to teleconferencing, video telephone, and video messaging. The H4 data rate is for bulk, text data transfer, enhanced video information, and facsimile. The CCITT Recommendation states that H22 and H4 should be multiples of 64 kb/s (e.g., an H22 of 43.008 Mb/s would have 672 × 64-kb/s channels). The initial CCITT Recommendation lists the following broadband services using the *new channel rates:*

- Broadband unrestricted bearer services
- High-quality broadband video telephony
- High-quality broadband video conferencing
- Existing quality TV distribution
- HDTV distribution

On a global level, data communications is eventually moving toward a single, public networking environment. This will probably use an infrastructure with a multigigabit transmission rate, optical fiber links, based on the SDH (Sonet) Standard for the physical layer. This Standard defines common timing for high-speed networks having data rates of tens of megabits per second to tens of gigabits per second with easy multiplexing and demultiplexing of low-bit-rate data into and out of a high-speed link.

Frame and cell relay.   These two terms are becoming more widespread as WANs move to higher operating rates. The main difference between the two is that frame relay uses variable-length frames to transport data, whereas cell relay uses fixed-length frames. Frame relay is available now, but cell relay will not be available in a nonproprietary form until about 1994.

Frame relay.   Frame relay is a protocol, not a service. It is a fast, simplified method of transporting packetized data. Frame relay is effectively a streamlined version of the X.25 protocol. X.25 was originally designed for analog circuits, whereas frame relay was designed for digital technology, which covers only Level 1 and part of Level 2 of the OSI scheme. There is the capability of detecting but not correcting errors. The corrections are left to the higher levels within the application itself. Frames should always follow each other along the same route because this connection type of protocol provides no sequencing information. There is only destination routing and congestion control information in the address. Since each frame is addressed individually, data from several applications can be multiplexed onto the same link. Bandwidth is not dedicated to any one application. This means that the full bandwidth of the circuit could be taken by one application for short periods of time.

The protocols relating to frame relay are the CCITT Recommendations I.122 and Q.922. They add relay and routing functions to the data link layer (Layer 2). Frame relay is a form of multiplexing that transports frames through the network as quickly as possible.

Functions such as error detection and correction, flow control, etc., are now processed on an end-to-end user device basis instead of being done by the network. Although frame relay was initially designed for the primary rate interface up to 1.92 Mb/s, it is projected that bit rates up to 34 to 45 Mb/s may be achievable. Figure 9.32 shows the construction of frames, which starts with a 1-byte flag followed by a 2-byte frame relay header. The header incorporates a data link connection identification (DLCI) for routing each frame on a hop-by-hop basis through a virtual path established at call setup or at the time of subscription. Next comes the V.120 terminal adapter header followed by the control state information. The Layer 2 messages are then transmitted, after which there is a 2-byte cyclic redundancy check, and then the end of the frame flag.

Figure 9.33a shows the conventional synchronous TDM frame structure. This is used as a comparison for frame and cell relay. Figure 9.33b shows the frame relay format. This is asynchronous multiplexing with variable frame length as opposed to the fixed frame length of the ATM cell relay asynchronous multiplexing of Fig. 9.33c. For a

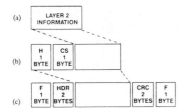

Figure 9.32 Frame relay. (*a*) From Layer 2; (*b*)
V.120 (optional); (*c*) frame relay transport.
H = V.120 terminal adapter header (optional),
CS = control state information (optional), HDR =
frame relay header, CRC = cyclic redundancy
check, F = flag. (*Reproduced with permission from
Ref. 400, © 1990 Telecommunications.*)

view of the process of a frame relay data transfer, see Fig. 9.34*a*,
which shows the data and signaling paths through the protocol architecture. For a file to be transferred from A to B, the user (application
A) initially sends a request to establish a session via the presentation
and session layers to the transport layer. Call control information is
forwarded through the ISDN D channel by the transport layer using
CCITT Rec. Q.931. As is also indicated in Fig. 9.34*a*, the signaling is
used to define a virtual path and call control necessary to set up the
data transfer. When the call has been set up, data is transferred from
A to B using the DCLI in the frame header. The amount of frame processing by the network is reduced to a minimum for frame relay,
which facilitates very fast data transfer. One problem with this technique is that in some circumstances reliability can be compromised for
speed. In other words, some data can be lost. This is because frame
relay has no error correction. This is in comparison with packet
switching, which guarantees virtually 100 percent data transfer but
takes a relatively long transfer time. Frame relay is suitable for high-
volume data transfer such as imaging and visualization (e.g., 3-D displays of engineering designs). It is less suitable for services such as
voice and video that are sensitive to time delays.

Cell relay. Communication technologies that use fixed-sized blocks of
data to transport information through a network are known as cell relay technologies. The ATM and IEEE 802.6 MAN are examples of cell
relay technologies. ATM, as described previously, is designed to operate over optical fiber (CCITT Rec. I.121). The transmission bandwidth
is organized into cells which are periodic sequences of undedicated,

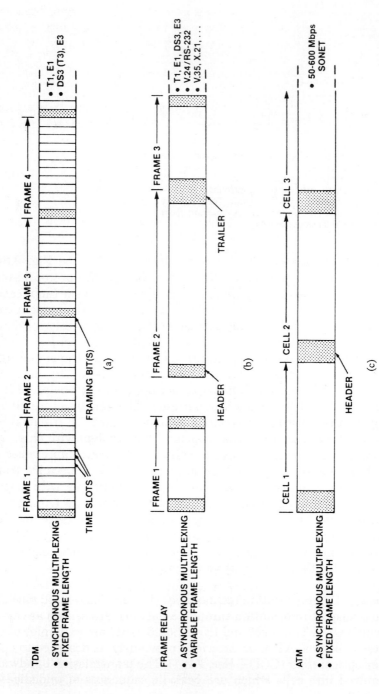

Figure 9.33 Multiplexing methods. (*Adapted with permission from Ref. 400, © 1990 Tele-communications.*)

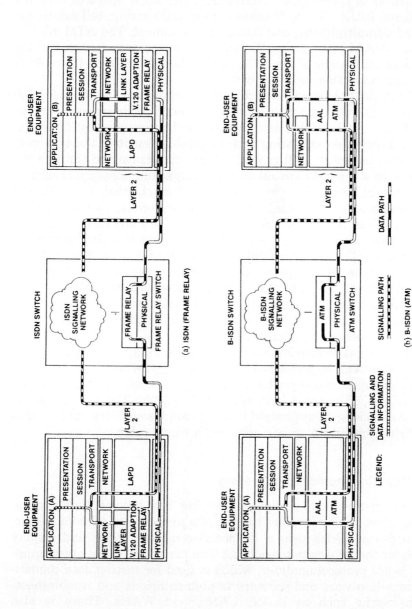

Figure 9.34 Protocol architectures. *(Reproduced with permission from Ref. 400, © 1990 Telecommunications.)*

499

fixed-sized blocks of data. This scheme is indicated in Fig. 9.33c, where each cell has a 5-byte header followed by a 48-byte block of information data. Figure 9.34b shows the signaling and data paths, as a comparison to frame relay. Notice that the ATM operates over the broadband ISDN. The ATM takes care of cell routing, cell multiplexing and demultiplexing, and header error control. The ATM adaptation layer (AAL) transforms the upper-layer message format and the fixed-size cell format of the ATM layer. The AAL also performs error detection, correction, and flow control.

ATM switches that will switch packetized data at the proposed 600 Mb/s have not yet been developed. This should occur sometime in the near future. At that time, data transfer will be initiated by, for example, application A requesting session establishment to the transport layer, etc., as before. Call control information is carried through the network on a separate virtual circuit. A signaling message is sent by the transport layer using the signaling virtual channel. The AAL and ATM layers convert the signaling message into cells, after which they are sent to the B-ISDN signaling network. Path and call parameters will probably be defined by the signaling message in a similar manner to existing ISDN for basic and primary rate services. Figure 9.34b shows the data transfer path once the call has been established. During cell transfer, cells are routed through the ATM switches by a cell VPI and routing information organized during the call setup procedure. Very fast information transfer is established by minimizing the amount of processing performed by the network.

Cell relay has some advantages over frame relay. First, the switching functions can be performed more efficiently, which translates to a lower transmission cost per bit. Also, delay-sensitive applications such as voice and video have a more acceptable quality. This is because cell relay has no long delays caused by long data bursts from one user monopolizing the circuit, as can occur with frame relay.

9.7 Data Communications Testing

Data communications equipment testing has become highly automated. Most testing involves the use of a computer or microprocessor-driven device which uses a keyboard command input and display screen for the information feedback. The main type of data test equipment is the protocol analyzer, which is used for testing data communications hardware and software implementations and troubleshooting networks such as ISDN, LANs, and WANs. There is also equipment for making measurements such as impedance, signal balance ratio, BER, pair selection, signaling, etc.

Protocol analysis. Depending on the individual manufacturer's design, a protocol analyzer can:

1. Monitor and decode protocols
2. Identify protocol implementation problems, configuration compatibilities, and communication link failure such as corrupted frames or collisions
3. Perform interactive, programmatic testing and emulate error recovery scenarios which are not easy to create
4. Evaluate performance and identify degraded network conditions before they cause network failure
5. Generate traffic to stress test products and networks
6. Perform remote circuit testing as completely as if the technician or engineer were actually at the remote site

Protocol analyzers come in a wide variety of shapes, sizes, and prices. They all invariably allow testing to be performed in conformance with several international standards including CCITT. Ideally, it would be convenient for one piece of test equipment to perform all the necessary data communications testing. At present this is not possible. One piece of test equipment may, for example, do troubleshooting of digital interfaces for, say, CCITT Recs. V.24, V.35, V.36, and X.24. Another may do troubleshooting of LANs (Ethernet and token ring) and WANs (V.24, V.35, and X.24). Another analyzer may perform only packet-switching interface troubleshooting.

Other data communications test equipment is specifically devoted to ISDN. For example, a test set may analyze the ISDN basic rate and primary rate interfaces, including interpretation of CCITT signaling system no. 7. Another may perform Layer 1 tests on basic rate access interfaces.

There is also a range of data network diagnostic and data analysis equipment, for example, for remote-controlled monitoring and switching on digital interfaces such as CCITT Recs. V.24/RS232/V.35 and X.20/X.21. Other test equipment is necessary for performance analysis such as BER tests or impedance and attenuation balancing of cables, etc.

Most of this equipment is not cheap. The point to be made here is that it is easy to order equipment which either does not do everything that is required for analyzing a specific system, or, conversely, several items may be bought that have overlapping testing functions. Both situations are unsatisfactory and considerable care should be exercised when procuring data communications test equipment.

Final comment. Data communications has come a long way in the past few years, but, to keep the progress in perspective, it is appropriate to make the following final statement on the subject. An interesting quip from a world telecommunications leader, Bell, reads: "If networking and communications today were the motor industry, this would be 1915."

Appendix: Standards

A modem must comply with

IS02110, which specifies the physical dimensions and pin assignment for the 25-way D-type connector

V.24, which lists the interface circuits and identifies their function

V.28, which specifies the electrical characteristics of the interface circuits

V.25bis, which specifies the protocol used by a terminal to communicate with the modem for the purpose of automatic dialing

V.42, which specifies the error control protocol used between a pair of modems

V.32, which defines the modulation scheme used by the modem and the means for establishing a connection

Modems have also to comply with a large number of other standards relating to electrical safety, electromagnetic radiation, flammability, and component standards.

CCITT Recommendations relating to data communications

V series Recommendations, developed by CCITT SGXVII. The V series Recommendations mainly cover:

1. Modems (e.g., V.21, V.22, V.22bis, V.23, V.32, ...)
2. Interfaces (e.g., V.24, V.35, V.230)
3. V series ISDN terminal adaptors (V.110, V.120)

In the V series Recommendations the V.24 interface and the V.22bis modem have the widest recognition.

V.10 Electrical characteristics for unbalanced double-current interchange circuits for general use with integrated circuit equipment in the field of data communications

V.11	Electrical characteristics for balanced double-current interchange circuits for general use with integrated circuit equipment in the field of data communications
V.13	Simulated carrier control
V.14	Transmission of start-stop mode characters over synchronous bearer channels
V.21	300-b/s duplex modem standardized for use in the general switched network
V.22	1200-b/s duplex modem standardized for use in the general switched network and on point-to-point two-wire leased telephone type circuits
V.22bis	2400-b/s duplex modem using the frequency division technique standardized for use in the general switched network and on point-to-point two-wire leased telephone-type circuits
V.23	600/1200-baud modem standardized for use in the general switched network
V.24	List of definitions for interchange circuits between data terminal equipment and data circuit terminating equipment
V.25bis	Automatic calling and/or answering equipment on the general switched telephone network (GSTN) using the 100 series interchange circuits
V.26	2400-b/s modem standardized for use on four-wire leased telephone-type circuits
V.26bis	2400/1200-b/s modem standardized for use in the general switched telephone network
V.26ter	2400-b/s modem using the echo cancellation technique standardized for use in the general switched network and on point-to-point two-wire leased telephone-type circuits
V.27	4800-b/s modem with manual equalizer standardized for use on leased telephone-type circuits
V.27bis	4800/2400-b/s modem with automatic equalizer standardized for use on leased telephone-type circuits
V.27ter	4800/2400-b/s modem with automatic equalizer standardized for use in the general switched network
V.28	Electrical characteristics for unbalanced double-current interchange circuits
V.29	9600-b/s modem standardized for use on point-to-point four-wire leased telephone-type circuits
V.32	9600-b/s duplex modem standardized for use on the general switched telephone network and on leased telephone-type circuits
V.33	14,400-b/s modem standardized for use on point-to-point four-wire leased telephone-type circuits
V.35	Data transmission at 48 kb/s using 60- to 108-kHz group band circuits

V.42	Error-correcting procedures for DCEs using asynchronous to synchronous conversion
V.54	Loop test devices for modems
V.110	Support of DTEs with V series-type interfaces by an ISDN
V.120	Support by an ISDN of DTE with V series-type interfaces with provision for statistical multiplexing
V.230	General data communications interface

X series Recommendations, developed by CCITT SGVII. The X series Recommendations cover several aspects of data communication networks such as interfaces, network performance, routing, and OSI. Recommendations X.21, the synchronous terminal interface, and X.25, the interface between a packet mode terminal and a packet network, are the most well recognized.

X.20	Interface between DTE and data circuit terminating equipment (DCE) for start-stop transmission services on public data networks
X.20bis	Use on public data networks of DTE which is designed for interfacing to asynchronous duplex V series modems
X.21	Interface between DTE and DCE for synchronous operation on public data networks
X.21bis	Use on public data networks of DTE which is designed for interfacing to synchronous V series modems
X.22	Multiplex DTE/DCE interface for user classes 3 to 6
X.24	List of definitions for interchange circuits between DTE and DCE on public data networks
X.25	Interface between DTE and DCE for terminal operating in the packet mode and connected to public data networks by dedicated circuits
X.26	Electrical characteristics for unbalanced double-current interchange circuits for general use with integrated circuit equipment in the field of data communications
X.27	Electrical characteristics for balanced double-current interchange circuits for general use with integrated circuit equipment in the field of data communications
X.28	DTE/DCE interface for a start-stop mode data terminal equipment accessing the PAD facility in a public data network situated in the same country
X.29	Procedures for the exchange of control information and user data between a PAD facility and a packet mode DTE or another PAD
X.30	Support of X.21- and X.21bis-based DTEs by an ISDN
X.31	Support of packet mode terminal equipment by an ISDN

X.32   Interface between a DTE and DCE for terminals operating in the
packet mode and accessing a packet-switched public data network
through a PSTN or a circuit-switched public data network

ISO data communications Standards

Perhaps the most widely recognized ISO Standard is the High-Level
Data Link Control (HDLC) procedure (ISO 3309, 7809, etc). This provides the mechanism for constructing Layer 2 data link level protocols
such as X.25 Layer 2 or V.42.

ISO2110	25 Pin DTE/DCE Interface Connector and Pin Assignments
ISO2593	34 Pin DTE/DCE Interface Connector and Pin Assignments
ISO3309	High Level Data Link Control Procedures—Frame Structure
ISO4335	High Level Data Link Control Procedures—Elements of Procedure
ISO4902	37 Pin DTE/DCE Interface Connector and Pin Assignments
ISO4903	15 Pin DTE/DCE Interface Connector and Pin Assignments
ISO7809	High Level Data Link Control Procedures—Consolidation of Classes of Procedure
ISO7776	High Level Data Link Control Procedures—Description of the X.25 LAPB Compatible DTE Data Link Procedures
ISO8208	X.25 Packet Layer Protocol for Data Terminal Equipment
ISO8348	Network Service Definition
ISO8473	Protocol for Providing the Connectionless Mode Network Service
ISO8802-2	Local Area Networks—Part 2 Logical Link Control
ISO8802-3	Local Area Networks—Part 3 CSMA/CD Access Method and Physical Layer Specifications
ISO8802-4	Local Area Networks—Part 4 Token Passing Bus Access Method and Physical Layer Specifications
ISO8802-5	Local Area Networks—Part 5 Token Ring Access Method and Physical Layer Specifications
ISO8802-7	Local Area Networks—Part 7 Slotted Ring Access Method and Physical Layer Specifications
ISO8877	Interface Connector and Contact Assignments for ISDN Basic Access Interface Located at Reference Points S and T
ISO8878	Use of X.25 to Provide the OSI Connection Oriented Network Service
ISO8880	Specification of Protocols to Provide and Support the OSI Network Service
ISO8881	Use of the Packet Layer Protocol in ISO 8802 Local Area Networks
ISO8885	High Level Data Link Control Procedures—General Purpose XID Frame Information Field and Content
ISO8886	Data Link Service Definition for Open Systems Interconnection

Bibliography

Digital Multiplexing and Baseband Composition (Chap. 2)

1. Ang, P. H., et al., Video Compression Makes Big Gains, *IEEE Spectrum*, Oct. 1991, pp. 16–19.
2. Aprille, T. J., Introducing SONET into the Local Exchange Carrier Network, *IEEE Communications Magazine*, August 1990, pp. 34–38.
3. Asatani, K., K. R. Harrison, and R. Ballart, CCITT Standardization of Network Node Interface of Synchronous Digital Hierarchy, *IEEE Communications Magazine*, August 1990, pp. 15–20.
4. Balcer, R., et al., An Overview of Emerging CCITT Recommendations for the Synchronous Digital Hierarchy: Multiplexers, Line Systems, Management, and Network Aspects, *IEEE Communications Magazine*, August 1990, pp. 21–25.
5. Ballart, R., and Y.-C. Ching, SONET: Now It's the Standard Optical Network, *IEEE Communications Magazine*, March 1989, pp. 8–14.
6. Bars, G., J. Legras, and X. Maitre, Introduction of New Technologies in the French Transmission Networks, *IEEE Communications Magazine*, August 1990, pp. 39–43.
7. Bellamy, J., *Digital Telephony*, John Wiley & Sons, 1982.
8. Bingham, J. A. C., Multicarrier Modulation for Data Transmission: An Idea Whose Time Has Come, *IEEE Communications Magazine*, May 1990, pp. 5–14.
9. Blakemore, P., Performance Analysis for Digital Networks, *Telecommunications*, August 1990, pp. 31–34.
10. Boehm, R. J., Progress in Standardization of SONET, *IEEE Lightwave Communications Systems Magazine*, May 1990, pp. 8–16.
11. Boehm, R. J., SONET: The Next Phase, *Telecommunications*, June 1989, pp. 37–40.
12. Bright, J., High Order Multiplexers, *Telecommunications*, May 1990, pp. 65–68.
13. Bylanksi, P., and T. W. Chong, Advances in Speech Coding for Communications, *GEC Journal of Research*, vol. 2, no. 1, 1984, pp. 16–22.
14. CCITT Recommendation G.721: 32 kbit/s Adaptive Differential Pulse Code Modulation (ADPCM), ITU, Geneva, 1986.
15. CCITT Recommendations G.732 and G.733: Characteristics of Primary PCM Multiplex Equipment Operating at 2048 and 1544 kbit/s, ITU, Geneva, 1986.
16. CCITT Recommendation G.732: Characteristics of Primary PCM Multiplex Equipment Operating at 2048 kbit/s, ITU, Geneva, 1986.
17. CCITT Recommendation G.733: Characteristics of Primary PCM Multiplexing Equipment Operating at 1544 kbit/s, ITU, Geneva, 1986.
18. Chen, T. C., et al., Multiple Block-size Transform Video Coding using a Subband Structure, *IEEE Trans. on Circuits and Systems for Video Tech.*, vol. 1, no. 1, March 1991, pp. 59–71.
19. den Hollander, C. J., Third-Generation Higher-Order Digital Multiplex Equipment, *Philips Telecommunication Review*, vol. 43, no. 1, April 1985, pp. 17–30.

20. Dinn, N. F., A. G. Weygand, and D. M. Garvey, Digital Interconnection of Dissimilar Digital Networks, *IEEE Communications Magazine*, vol. 24, no. 4, April 1986, pp. 12–17.
21. Farndale, L. L., CCITT G.821 and ISDN Error Performance Testing, *Telecommunications*, October 1986, pp. 60–66.
22. Finnie, G., Which Way Next for the CCITT?, *Telecommunications*, November 1988, pp. 77–80.
23. Fleury, B., Asynchronous High Speed Digital Multiplexing, *IEEE Communications Magazine*, vol. 24, no. 8, August 1986, pp. 17–25.
24. Gaggioni, H. P., The Evolution of Video Technologies, *IEEE Communications Magazine*, vol. 25, no. 11, November 1987, pp. 20–36.
25. Henderson, A., Into the Synchronous Era, *Telecommunications*, December 1988, pp. 29–36.
26. Kasai, H., T. Murase, and H. Ueda, Synchronous Digital Transmission Systems Based on CCITT SDH Standard, *IEEE Communications Magazine*, August 1990, pp. 50–59.
27. King, J., Creative Tension in Video Compression, *Telecommunications*, May 1990, pp. 99–103.
28. Kishimoto, R., and I. Yamashita, HDTV Communication Systems in Broadband Communication Networks, *IEEE Communications Magazine*, August 1991, pp. 28–35.
29. Mazzei, U., et al., Evolution of the Italian Telecommunication Network Towards SDH, *IEEE Communications Magazine*, August 1990, pp. 44–49.
30. McDowall, R., Interface Testing for Digital Transmission, *Communications International*, September 1989, pp. 75–79.
31. Mermelstein, P., G.722, A New CCITT Coding Standard, *Communications Magazine*, vol. 26, no. 1, January 1988, pp. 8–15.
32. Miki, T., and C. A. Siller, Jr., An International Perspective on Evolution to a Synchronous Digital Network, *IEEE Communications Magazine*, August 1990, pp. 7–10.
33. Nickelson, R. L., HDTV Standards—Understanding the Issues, *Telecommunication Journal*, vol. 57, V/1990, pp. 302–312.
34. Omura, J. K., Novel Applications of Cryptography in Digital Communications, *IEEE Communications Magazine*, May 1990, pp. 21–29.
35. Owen, F. F. E., *PCM and Digital Transmission Systems*, Texas Instruments Electronics Series, McGraw-Hill, 1982.
36. Rutkowski, A. M., An Overview of the Forums for Standards and Regulations for Digital Networks, *Telecommunications*, October 1986, pp. 68–80.
37. Sandesara, N. B., G. R. Ritchie, and B. Engel-Smith, Plans and Considerations for SONET Deployment, *IEEE Communications Magazine*, August 1990, pp. 26–33.
38. Schumann, A. H. G., Measuring Jitter with a Spectrum Analyzer, *Wandel & Goltermann News and Information* (Bits 52), pp. 10–12.
39. Shafi, M., L. Davey, and W. Smith, The Impact of Synchronous Digital Hierarchy on Digital Microwave Radio: A View from Australasia, *IEEE Communications Magazine*, May 1990, pp. 16–20.
40. Shafi, M., and B. Mortimer, The Evolution of SDH: A View From Telecom New Zealand, *IEEE Communications Magazine*, August 1990, pp. 60–66.
41. Sosnosky, J., and T.-H. Wu, SONET Ring Applications for Survivable Fiber Loop Networks, *IEEE Communications Magazine*, June 1991, pp. 51–58.
42. Steenberg, N., Higher-Order Network Testing, *Telecommunications*, August 1990, pp. 39–41.
43. Stewart, A., The Trouble with Sonet, *Communications International*, October 1989, pp. 59–60.
44. Troutman, R., T1 Circuits—In-Service Monitoring and Out-of-Service Testing, *Telecommunications*, July 1986, pp. 42–48.

Modulation (Chap. 3)

45. Amoroso, F., Pulse and Spectrum Manipulation in the Minimum Frequency Shift Keying Format, *IEEE Trans.*, COM-24, March 1976, p. 381.
46. Calhoun, G., *Digital Cellular Radio*, Artech House, 1988.
47. Daido, Y., et al., 256 QAM Modem for High Capacity Digital Radio Systems, *IEEE 3d Global Telecommunications Conference*, November 1984.
48. Davarian, F., and J. T. Sumida, A Multipurpose Digital Modulation, *IEEE Communication Magazine*, February 1989, p. 36.
49. Hamming, R. W., Error Detecting and Correcting Codes, *Bell System Technical Journal*, vol. 35, 1956, pp. 203–234.
50. Harris, R. M., Coherent Frequency Exchange Keying (CFEK), *Royal Aerospace Est.*, Report 8808, February 1988.
51. Bandwidth and Power Efficient Coded Modulation, *IEEE Journal of Selected Area Communications*, August and December 1989.
52. Ivanek, F. (ed.), *Terrestrial Digital Microwave Communications*, Artech House, 1989.
53. Kavehrad, M., Convolutional Coding for High Speed Microwave Radio Communications, *AT&T Technical Journal*, vol. 64, no. 7, September 1985, pp. 1625–1637.
54. Killen, H. B., *Digital Communications with Fiber Optics and Satellite Applications*, Prentice-Hall, 1988.
55. Lee, W. C. Y., Spectrum Efficiency in Cellular, *IEEE Trans.*, VT-38, May 1989, p. 69.
56. McKay, R. G., et al., Trellis-Coded Modulation on Digital Microwave Radio Systems—Simulations for Multipath Fading Channels, *Conference Proceedings GLOBECOM 88*, November 1988.
57. Meyers, R. A., and P. H. Waters, Synthesiser Review for Pan-European Digital Cellular Radio, *IEEE Colloquium on VLSI for Mobile Telecomms Systems*, March 14, 1990.
58. Murota, K., and K. Hirade, GMSK Modulation for Digital Mobile Radio Telephones, *IEEE Trans.*, COM-29, no. 7, July 1981, p. 1044.
59. Noguchi, T., Y. Daido, and J. A. Nossek, Modulation Techniques for Microwave Digital Radio, *IEEE Communications Magazine*, vol. 24, no. 11, September 1986, pp. 21–30.
60. Pasupathy, S., Minimum Shift Keying: A Spectrally Efficient Modulation, *IEEE Communications Magazine*, July 1979, p. 14.
61. Thapar, H. K., Real Time Application of Trellis Coding to High Speed Voice Band Data Transmission, *IEEE Journal on Selected Areas in Communication*, vol. SAC-2, no. 5, September 1984, pp. 648–658.
62. Ungerboeck, G., Channel Coding with Multilevel/Phase Signals, *IEEE Trans. Info. Theory*, IT-28, January 1982, p. 55.
63. Ungerboeck, G., Trellis Coded Modulation with Redundant Signal Sets, Part 1—Introduction, *IEEE Communications Magazine*, vol. 25, no. 2, February 1987, pp. 5–11.
64. Viterbi, A. J., et al., A Pragmatic Approach to Trellis-Coded Modulation, *IEEE Communications Magazine*, July 1989, p. 11.

Digital Microwave Radio (Chaps. 4 and 5)

65. Balaban, P., E. A. Sweedyk, and G. S. Axeling, Angle Diversity with Two Antennas: Model and Experimental Results, *Conference Proceedings, ICC '87*, Seattle, Washington, June 1987.

66. Barnett, W. T., Multipath Propagation at 4, 6, and 11 GHz, *Bell System Technical Journal*, vol. 51, no. 2, February 1972, pp. 321–361.

67. Bean, B. R., J. D. Horn, and A. M. Ozanich, Jr., *Climatic Charts and Data of the Radio Refractive Index for the United States and the World*, National Bureau of Standards Monograph 22, U.S. Government Printing Office, Washington, D.C., November 25, 1960.

68. Brodhage, H., and W. Hormuth, *Planning and Engineering of Radio Relay Links*, Siemens *Aktiengesellschaft*, 8th ed., 1977.

69. Campbell, J. C., and R. P. Coutts, Outage Prediction of Digital Radio Systems, *Electronics Letters*, December 1982.

70. CCIR Recommendation 556: Hypothetical Reference Digital Path for Radio Relay Systems Which May Form Part of an Integrated Services Digital Network with a Capacity Above the Second Hierarchical Level, ITU, Dusseldorf, 1990.

71. CCIR Report 338-5, Propagation Data and Prediction Methods Required for Line-of-Sight Radio Relay Systems, Study Group 5, XVIth Plenary Assembly, Dubrovnik, Yugoslavia, ITU, Geneva, 1986.

72. CCIR, XVIth Plenary Assembly, Dubrovnik, Yugoslavia, 1986, vol. IX, Report 784-2, Effects of Propagation on the Design and Operation of Line-of-Sight Radio-Relay Systems, ITU, Geneva, 1986.

73. Chamberlain, J. K., et al., Receiver Techniques for Microwave Digital Radio, *IEEE Communications Magazine*, vol. 24, no. 11, November 1986, pp. 43–54.

74. Conner, W. A., Direct RF Modulation 256 QAM Microwave System, *Globecom '88, Conference Record*, Miami, pp. 52.7.1–52.7.6.

75. Dudek, M. T., et al., The Performance of an Adaptive Equaliser in an Experimental 16 QAM Digital Radio Relay System, *Third International Conference on Telecommunications Transmission, Conference Record 246*, London, March 1985, pp. 52–56.

76. *Engineering Considerations for Microwave Communications Systems*, GTE Lenkurt Incorporated, 1975.

77. Feher, K., *Digital Communications: Microwave Applications*, Prentice-Hall, 1981.

78. Freeman, R. L., *Radio System Design for Telecommunications (1–100 GHz)*, John Wiley & Sons, 1987.

79. Gardina, M. F., and A. Vigants, Measured Multipath Dispersion of Amplitude and Delay at 6 GHz in a 30 MHz Bandwidth, *Conference Record, 1984 Int. Conf. on Commun.*, vol. 3, paper 46.1, pp. 1433–1436.

80. Giger, A. J., and W. T. Barnett, Effects of Multipath Propagation on Digital Radio, *1980 International Zurich Seminar on Digital Communications*, Zurich, Switzerland, March 4–6, 1980.

81. Grafinger, W., J. A. Nossek, and G. Sebald, Design and Realization of a High-Speed Multilevel QAM Digital Radio Modem with Time-Domain Equalization, *ICC '85, Conference Record*, Chicago, pp. 971–976.

82. Greenstein, L. J., Analysis/Simulation Study of Cross Polarization Cancellation in Dual Polarization Digital Radio, *AT&T Technical Journal*, vol. 64, no. 10, December 1985, pp. 2261–2280.

83. Greenstein, L. J., and M. Shafi, Outage Calculation Methods for Microwave Digital Radio, *IEEE Communications Magazine*, vol. 25, no. 2, February 1987.

84. Ivanek, F. (ed.), *Terrestrial Digital Microwave Communications*, Artech House, 1989.

85. Kavehrad, M., and J. Salz, Cross-Polarization Cancellation and Equalization in Digital Transmission over Dually Polarized Multipath Fading Channels, *AT&T Bell Lab. Tech. J.*, vol. 64, no. 10, December 1985, pp. 2211–2245.

86. Kurokawa, H., Otsuki, Y., and T. Matsumoto, *Microwave System Design*, Giken Company, Ltd., 1966.

87. Lin, E. H., A. J. Giger, and G. D. Alley, Angle Diversity on Line-of-Sight Microwave Paths Using Dual-Beam Dish Antennas, *Conference Record, 1987 Int. Conf. on Commun.*, vol. 2, paper 23.5, pp. 831–841.

88. Lin, S. H., Impact of Microwave Depolarization During Multipath Fading on Digital Radio Performance, *Bell System Technical Journal*, vol. 56, no. 5, May–June 1977, pp. 645–674.

89. Liniger, M., and D. Vergeres, Field Test Results for a 16-QAM and a 64-QAM Digital Radio, Compared with the Prediction Based on Sweep Measurements, *Conference Record, 1986 Int. Conf. on Commun.,* paper 15.2, vol. 1, pp. 457–461.

90. Lundgren, C. W., and W. D. Rummler, Digital Radio Outage due to Selective Fading—Observation versus Prediction from Laboratory Simulation, *Bell System Technical Journal,* vol. 58, no. 5, May–June 1979, pp. 1073–1100.

91. Mahle, C., and H. C. Huang, MMICs in Communications, *IEEE Communications Magazine,* vol. 23, no. 9, September 1985, pp. 8–16.

92. Matsue, H., et al., Digitalized Cross-Polarization Interference Canceller for Multilevel Digital Radio, *IEEE Journal on Selected Areas in Communications,* vol. SAC-5, no. 3, April 1987, pp. 493–501.

93. Meyers, M. H., and V. K. Prabhu, Future Trends in Microwave Digital Radio: A View from North America, *IEEE Communications Magazine,* vol. 24, no. 2, February 1986, pp. 46–49.

94. Morita, K., et al., A Method for Estimating Cross Polarization Discrimination Ratio during Multipath Fading, *Trans. of the IECE of Japan,* vol. E 62, no. 11, November 1979, pp. 810–811.

95. Panter, P. F., *Communication Systems Design,* McGraw-Hill, 1972.

96. Prabhu, V. K., and L. J. Greenstein, Analysis of Multipath Outage with Applications to 90 Mb/s PSK Systems at 6 GHz and 11 GHz, *Conference Record, ICC '78,* Toronto, June 1978, paper 47.2.

97. *Recommendations and Reports of the CCIR,* XVIth Plenary Assembly, Dubrovnik, 1986, vol. IX—Part 1: Fixed Service Using Radio Systems, ITU, Geneva, 1986.

98. *Recommendations and Reports of the CCIR,* XVIth Plenary Assembly, Dubrovnik, 1986, vol. I—Spectrum Utilization and Monitoring, ITU, Geneva, 1986.

99. *Recommendations and Reports of the CCIR,* XVIth Plenary Assembly, Dubrovnik, 1986, vol. V: Propagation in Non-Ionized Media, Recommendation 341-2, The Concept of Transmission Loss for Radio Links, ITU, Geneva, 1986.

100. *Recommendations and Reports of the CCIR,* XVIth Plenary Assembly, Dubrovnik, 1986, vol. V: Propagation in Non-Ionized Media, Recommendation 369-3, Reference Atmosphere for Refraction, ITU, Geneva, 1986.

101. *Recommendations and Reports of the CCIR,* XVIth Plenary Assembly, Dubrovnik, 1986, vol. IX—Part 1: Fixed Service Using Radio-Relay Systems, Report 376-5, Diversity Techniques for Radio-Relay Systems, ITU, Geneva, 1986.

102. *Recommendations and Reports of the CCIR,* XVIth Plenary Assembly, Dubrovnik, 1986, vol. V: Propagation in Non-Ionized Media, Recommendation 453-1, The Formula for the Radio Refractive Index, ITU, Geneva, 1986.

103. *Recommendations and Reports of the CCIR,* XVIth Plenary Assembly, Dubrovnik, 1986, vol. V: Propagation in Non-Ionized Media, Recommendation 525-1, Calculation of Free-Space Attenuation, ITU, Geneva, 1986.

104. *Recommendations and Reports of the CCIR,* XVIth Plenary Assembly, Dubrovnik, 1986, vol. V: Propagation in Non-Ionized Media, Report 715-2, Propagation by Diffraction, ITU, Geneva, 1986.

105. *Recommendations and Reports of the CCIR,* XVIth Plenary Assembly, Dubrovnik, 1986, vol. V: Propagation in Non-Ionized Media, Report 722-2, Cross-Polarization Due to the Atmosphere, ITU, Geneva, 1986.

106. Rummler, W. D., A New Selective Fading Model: Application to Propagation Data, *Bell System Technical Journal,* vol. 58, no. 5, May–June 1979, pp. 1037–1071.

107. Rummler, W. D., A Comparison of Calculated and Observed Performance of Digital Radio in the Presence of Interference, *IEEE Transactions on Communications,* vol. COM-30, no. 7, July 1982, pp. 1693–1700.

108. Rummler, W. D., R. P. Coutts, and M. Liniger, Multipath Fading Channel Models for Microwave Digital Radio, *IEEE Communications Magazine,* vol. 24, no. 11, November 1986, pp. 30–42.

109. Rummler, W. D., Characterizing the Effects of Multipath Dispersion on Digital Radios, in *Conference Record, GLOBECOM '88,* vol. 3, paper 52.5, pp. 1727–1733.

110. Rummler, W. D., A Statistical Model of Multipath Fading on a Space Diversity Radio Channel, *Bell System Technical Journal,* vol. 61, no. 9, part 1, November 1982, pp. 2185–2221.

111. Rummler, W. D., Modeling the Diversity Performance of Digital Radios with Maximum Power Combiners, *Conference Proceedings, ICC '84,* paper 22.6, vol. 2, pp. 657–660.

112. Ryu, T., J. Uchibori, and Y. Yoshida, A Stepped Square 256 QAM for Digital Radio System, *Conference Proceedings ICC 86,* pp. 1477–1481.

113. Saito, Y., et al., 256 QAM Modem for High Capacity Digital Radio System, *IEEE Transaction on Communication,* vol. COM-34, August 1986, pp. 799–805.

114. Sebald, G., B. Lankl, and J. A. Nossek, Advanced Adaptive Equalisation of Multilevel QAM Digital Radio Systems, *ICC 86 Conference Record,* pp. 46.5.1–46.5.5.

115. Sebald, G., J. A., Lankl, and J. A. Nossek, Advanced Time- and Frequency-Domain Adaptive Equalization in Multilevel QAM Digital Radio Systems, *IEEE Journal on Selected Areas in Communication,* vol. SAC-5, no. 3, April 1987, pp. 448–456.

116. Shafi, M., and D. P. Taylor, Influence of Terrain Induced Reflections on the Performance of High Capacity Digital Radio Systems, *Conference Proceedings, ICC '86,* Toronto, June 1986, paper no. 5.2.

117. Taylor, D. P., and P. R. Hartmann, Telecommunications by Microwave Digital Radio, *IEEE Communications Magazine,* vol. 24, no. 8, August 1986, pp. 11–16.

118. Vigants, A., Space-Diversity Engineering, *Bell System Technical Journal,* vol. 54, no. 1, January 1975, pp. 103–142.

119. Wu, K. T., Measured Statistics on Multipath Dispersion of Cross Polarization Interference, *Conference Record, 1984 Int. Conf. on Commun.,* vol. 3, paper 46.3.

120. Yeh, Y. S., and L. J. Greenstein, A New Approach to Space Diversity Combining in Microwave Digital Radio, *AT&T Technical Journal,* April 1985.

Fiber Optics (Chaps. 6 and 7)

Introduction

121. Henry, P. S., Introduction to Lightwave Transmission, *IEEE Communications Magazine,* vol. 23, no. 5, May 1985, pp. 12–16.

Fibers

122. Bachmann, P. K., et al., Loss Reduction in Fluorine-Doped SM- and High N.A.-PCVD Fibers, *IEEE Journal of Lightwave Technology,* vol. LT-4, no. 7, July 1986, pp. 813–816.

123. Bhagavatula, V. A., et al., Segmented-Core Fiber for Long-Haul and Local-Area-Network Applications, *IEEE Journal of Lightwave Technology,* vol. 6, no. 10, October 1988, pp. 1466–1468.

124. Hatano, S., et al., Optical Coupling Between Coated Fibers in a Compact Fiber Ribbon, *IEEE Journal of Lightwave Technology,* vol. LT-4, no. 3, March 1986, pp. 335–340.

125. Jay, J. A., and E. M. Hopiavuori, Dispersion-Shifted Fiber Hits Its Stride, *Photonics Spectra,* September 1990, pp. 153–158.

126. Kiang, Y. C., and T. E. Klieber, Macrobending Effects on Fiber Numerical Aperture, *IEEE Journal of Lightwave Technology,* vol. LT-5, no. 5, May 1987, pp. 709–711.

127. Kimura, T., Factors Affecting Fiber-Optic Transmission Quality, *IEEE Journal of Lightwave Technology,* vol. 6, no. 5, May 1988, pp. 611–619.

128. Ogai, M., et al., Development and Performance of Fully Fluorine-Doped Single-Mode Fibers, *IEEE Journal of Lightwave Technology,* vol. 6, no. 10, October 1988, pp. 1455–1461.

129. Sunak, H. R. D., and A. L. Deus, Dispersion Characteristics of Bimodal Heavy Metal Fluoride Fibers at 1.3 μm, *IEEE Photon. Technol. Lett.,* vol. 2, no. 9, September 1990, pp. 659–664.

130. Sunak, H. R. D., and S. P. Bastien, Characteristics of Dispersion-Flattened Depressed-Cladding Single-Mode Fluoride Fibers from 1.55 to 3.1 μm, *IEEE Photon. Technol. Lett.,* vol. 1, no. 8, August 1989, pp. 244–247.
131. Surat, C., Fiber Cable: Market Trends and Technology, *Photonics Spectra,* December 1990, pp. 95–98.
132. Titchmarsh, J., Optical Fiber Stresses, *Communications International,* April 1989, pp. 66–68.
133. Tokiwa, H., and Y. Mimura, Ultra-Loss Fluoride-Glass Single-Mode Fiber Design, *IEEE Journal of Lightwave Technology,* vol. LT-4, no. 8, August 1986, pp. 1260–1266.

Splicing

134. Fujise, M., Y. Iwamoto, and S. Takei, Self-Core-Alignment Arc-Fusion Splicer Based on a Simple Local Monitoring Method, *IEEE Journal of Lightwave Technology,* vol. LT-4, no. 8, August 1986, pp. 1211–1218.
135. Khoe, G.-D., J. A. Luijendik, and L. J. C. Vroomen, Arc-Welded Monomode Fiber Splices Made with the Aid of Local Injection and Detection of Blue Light, *IEEE Journal of Lightwave Technology,* vol. LT-4, no. 8, August 1986, pp. 1219–1222.
136. Miller, C. M., Mechanical Optical Fiber Splices, *IEEE Journal of Lightwave Technology,* vol. LT-4, no. 8, August 1986, pp. 1228–1231.
137. Sankawa, I., et al., Fault Location Techniques for In-Service Branched Optical Fiber Networks, *IEEE Photon. Technol. Lett.,* vol. 2, no. 10, October 1990, pp. 766–768.
138. So, V. C. Y., et al., Splice Loss Measurement Using Local Launch and Detect, *IEEE Journal of Lightwave Technology,* vol. LT-5, no. 12, December 1987, pp. 1663–1666.
139. Tachikura, M., N. Kashima, and M. Hirai, Fusion Mass-Splicing for Optical Fibers, *Review of the ECL,* NTT, Jpn., vol. 33, no. 6, 1985, pp. 953–959.
140. Tamaki, Y., et al., Field-Installable Plastic Multifiber Connector, *IEEE Journal of Lightwave Technology,* vol. LT-4, no. 8, August 1986, pp. 1248–1254.

Lasers

141. Adams, A. C., Multiple Quantum Well Spatial Light Modulators, *Photonics Spectra,* May 1990, pp. 191–194.
142. Arima, I., et al., 1.5 μm GaInAsP/InP Distributed Reflector (DR) Lasers with SCH Structure, *IEEE Photon. Technol. Lett.,* vol. 2, no. 6, June 1990, pp. 385–387.
143. Begley, D. L., and B. Boscha, A Bright Future for Laser Diodes, *Photonics Spectra,* June 1990, pp. 165–172.
144. Fletcher, P. W., K. Ibbs, and C. Seaton, New Developments in Ultrafast Lasers, *Photonics Spectra,* July 1990, pp. 111–120.
145. Goodwin, A. R., et al., The Design and Realization of a High Reliability Semiconductor Laser for Single-Mode Fiber-Optical Communication Links, *IEEE Journal of Lightwave Technology,* vol. 6, no. 9, September 1988, pp. 1424–1434.
146. Hafskjaer, L., and A. S. Sudbo, Attenuation and Bit-Rate Limitations in LED/Single-Mode Fiber Transmission Systems, *IEEE Journal of Lightwave Technology,* vol. 6, no. 12, December 1988, pp. 1793–1797.
147. Ishida, O., H. Toba, and Y. Tohmori, 0.04 Hz Relative Optical-Frequency Stability in a 1.5 μm Distributed-Bragg-Reflector (DBR) Laser, *IEEE Photon. Technol. Lett.,* vol. 1, no. 12, December 1989, pp. 452–454.
148. Kitamura, M., et al., High Power and Narrow Spectral Linewidth 1.5 μm MQW-DFB-LDs with Low FM Dip Frequency, *IEEE Photon. Technol. Lett.,* vol. 2, no. 11, November 1990, pp. 778–780.
149. Kitamura, M., et al., 250 kHz Spectral Linewidth Operation of 1.5 μm Multiple Quantum Well DFB-LDs, *IEEE Photon. Technol. Lett.,* vol. 2, no. 5, May 1990, pp. 310–311.

150. Lee, T. P., and C.-E. Zah, Wavelength-Tunable and Single-Frequency Semiconductor Lasers for Photonic Communications Networks, *IEEE Communications Magazine,* October 1989, pp. 42–52.
151. Lin, M. S., et al., Nearly Dispersion-Penalty-Free Transmission Using Blue-Shifted 1.55-μm Distributed Feedback Lasers, *IEEE Photon. Technol. Lett.,* vol. 2, no. 10, October 1990, pp. 741–742.
152. Lo, Y. H., et al., Multigigabit/s 1.5 μmλ/4-Shifted DFB OEIC Transmitter and its Use in Transmission Experiments, *IEEE Photon. Technol. Lett.,* vol. 2, no. 9, September 1990, pp. 673–674.
153. Morgan, R. A., VCSEL: A New Twist in Semiconductor Lasers, *Photonics Spectra,* December 1990, pp. 89–92.
154. Okai, M., et al., Corrugation-Pitch-Modulated MQW-DFB Laser with Narrow Spectral Linewidth (170 kHz), *IEEE Photon. Technol. Lett.,* vol. 2, no. 8, August 1990, pp. 529–530.
155. Slater, N., A New Breed of Diode Laser, *Photonics Spectra,* April 1990, pp. 181–186.
156. Takeshita, T., M. Okayasu, and S. Uehara, High-Power Operation in 0.98-μm Strained-Layer InGaAs-GaAs Single-Quantum-Well Ridge Waveguide Lasers, *IEEE Photon. Technol. Lett.,* vol. 2, no. 12, December 1990, pp. 849–851.
157. Wang, S. J., et al., Dynamic and CW Linewidth Measurements of 1.55-μm InGaAs-InGaAsP Multiquantum Well Distributed Feedback Lasers, *IEEE Photon. Technol. Lett.,* vol. 2, no. 11, November 1990, pp. 775–777.
158. Woodward, S. L., T. L. Koch, and U. Koren, The Onset of Coherence Collapse in DBR Lasers, *IEEE Photon. Technol. Lett.,* vol. 2, no. 6, June 1990, pp. 391–394.
159. Yano, M., et al., Extremely Low-Noise Facet-Reflectivity-Controlled InGaAsP Distributed-Feedback Lasers, *IEEE Journal of Lightwave Technology,* vol. LT-4, no. 10, October 1986, pp. 1454–1459.
160. Yasaka, H., et al., Optical Frequency Spacing Tunable Four-Channel Integrated 1.55 μm Multielectrode Distributed-Feedback Laser Array, *IEEE Photon. Technol. Lett.,* vol. 1, no. 4, April 1989, pp. 75–76.

Modulators

161. Gnauck, A. H., 16 Gbit/s Transmission Experiments Using a Directly Modulated 1.3 μm DFB Laser, *IEEE Photon. Technol. Lett.,* vol. 1, no. 10, October 1989, pp. 337–339.
162. Goto, M., et al., A 10-Gb/s Optical Transmitter Module with a Monolithically Integrated Electroabsorption Modulator with a DFB Laser, *IEEE Photon. Technol. Lett.,* vol. 2, no. 12, December 1990, pp. 896–898.
163. Hui, R., Optical PSK Modulation Using Injection-Locked DFB Semiconductor Lasers, *IEEE Photon. Technol. Lett.,* vol. 2, no. 10, October 1990, pp. 743–746.
164. Imai, M., et al., Wide-Frequency Fiber-Optic Phase Modulator Using Piezoelectric Polymer Coating, *IEEE Photon. Technol. Lett.,* vol. 2, no. 10, October 1990, pp. 727–729.
165. Jackel, J. L., and J. J. Johnson, Nonsymmetric Mach-Zehnder Interferometers Used as Low-Drive-Voltage Modulators, *IEEE Journal of Lightwave Technology,* vol. 6, no. 8, August 1988, pp. 1348–1351.
166. Kobrinski, H., et al., Simultaneous Fast Wavelength Switching and Intensity Modulation Using a Tunable DBR Laser, *IEEE Photon. Technol. Lett.,* vol. 2, no. 2, February 1990, pp. 139–142.
167. Korotky, S., et al., 4-Gb/s Transmission Experiment over 117 km of Optical Fiber Using a Ti:LiNbO$_3$ External Modulator, *IEEE Journal of Lightwave Technology,* vol. LT-3, no. 5, October 1985, pp. 1027–1030.
168. Mak, G., et al., High-Speed Bulk InGaAsP-InP Electroabsorption Modulators with Bandwidth in Excess of 20 GHz, *IEEE Photon. Technol. Lett.,* vol. 2, no. 10, October 1990, pp. 730–733.
169. Murphy, E. J., et al., Simultaneous Single-Fiber Transmission of Video and Bidirectional Voice/Data Using LiNbO$_3$ Guided-Wave Devices, *IEEE Journal of Lightwave Technology,* vol. 6, no. 6, June 1988, pp. 937–944.

170. Soref, R. A., D. L. McDaniel, and B. R. Bennett, Guided-Wave Intensity Modulators Using Amplitude-and-Phase Perturbations, *IEEE Journal of Lightwave Technology*, vol. 6, no. 3, March 1988, pp. 437–444.
171. Suzuki, M., et al., Electrical and Optical Interactions between Integrated InGaAsP/InP DFB Lasers and Electroabsorption Modulators, *IEEE Journal of Lightwave Technology*, vol. 6, no. 6, June 1988, pp. 779–784.
172. Vodhanel, R. S., 5 Gbit/s Direct Optical DPSK Modulation of a 1530-nm DFB Laser, *IEEE Photon. Technol. Lett.*, vol. 1, no. 8, August 1989, pp. 218–220.
173. Wang, S. Y., and S. H. Lin, High Speed III-V Electrooptic Waveguide Modulators at λ = 1.3 μm, *IEEE Journal of Lightwave Technology*, vol. 6, no. 6, June 1988, pp. 758–770.
174. Wood, T. H., Multiple Quantum Well (MQW) Waveguide Modulators, *IEEE Journal of Lightwave Technology*, vol. 6, no. 6, June 1988, pp. 743–756.

Polarizers

175. Okoshi, T., Recent Advances in Coherent Optical Fiber Communication Systems, *IEEE Journal of Lightwave Technology*, vol. LT-5, no. 1, January 1987, pp. 44–51.
176. Sanford, N. A., J. M. Connors, and W. A. Dyes, Simplified Z-Propagating DC Bias Stable TE-TM Mode Converter Fabricated in Y-Cut Lithium Niobate, *IEEE Journal of Lightwave Technology*, vol. 6, no. 6, June 1988, pp. 898–902.
177. Tsubokawa, M., T. Higashi, and Y. Sasaki, Measurement of Mode Couplings and Extinction Ratios in Polarization-Maintaining Fibers, *IEEE Journal of Lightwave Technology*, vol. 7, no. 1, January 1989, pp. 45–50.
178. Zervas, M. N., Surface Plasmon-Polariton Fiber-Optic Polarizers Using Thin Chromium Films, *IEEE Photon. Technol. Lett.*, vol. 2, no. 8, August 1990, pp. 597–599.

Photodetectors

179. Crawford, D. L., et al., High Speed InGaAs-InP p-i-n Photodiodes Fabricated on a Semi-Insulating Substrate, *IEEE Photon. Technol. Lett.*, vol. 2, no. 9, September 1990, pp. 647–652.
180. Kuwatsuka, H., et al., An $Al_xGa_{1-x}Sb$ Avalanche Photodiode with a Gain Bandwidth Product of 90 GHz, *IEEE Photon. Technol. Lett.*, vol. 2, no. 1, January 1990, pp. 54–55.
181. Shiba, T., et al., New Approach to the Frequency Response Analysis of an InGaAs Avalanche Photodiode, *IEEE Journal of Lightwave Technology*, vol. 6, no. 10, October 1988, pp. 1502–1506.
182. Tarof, L. E., Planar InP-InGaAs Avalanche Photodetectors with n-Multiplication Layer Exhibiting a Very High Gain-Bandwidth Product, *IEEE Photon. Technol. Lett.*, vol. 2, no. 9, September 1990, pp. 643–646.
183. Wei, C. J., et al., Lateral High-Speed Metal-Semiconductor-Metal Photodiodes on High-Resistivity InGaAs, *IEEE Electron Device Lett.*, vol. 11, no. 8, August 1990, pp. 334–335.

Amplifiers

184. Aoki, Y., Properties of Fiber Raman Amplifiers and Their Applicability to Digital Optical Communication Systems, *IEEE Journal of Lightwave Technology*, vol. 6, no. 7, July 1988, pp. 1225–1239.
185. Aspell, J., and N. S. Bergano, Erbium Doped Fiber Amplifiers for Future Undersea Transmission Systems, *IEEE Lightwave Communication Systems*, November 1990, pp. 63–66.
186. Becker, P. C., et al., Erbium-Doped Fiber Amplifier Pumped in the 950-1000 nm Region, *IEEE Photon. Technol. Lett.*, vol. 2, no. 1, January 1990, pp. 35–37.
187. Becker, P. C., et al., High-Gain and High-Efficiency Diode Laser Pumped Fiber Amplifier at 1.56 μm, *IEEE Photon. Technol. Lett.*, vol. 1, no. 9, September 1989, pp. 267–269.

188. Choy, M. M., et al., A High Gain, High-Output Saturation Power Erbium-Doped Fiber Amplifier Pumped at 532 nm, *IEEE Photon. Technol. Lett.,* vol. 2, no. 1, January 1990, pp. 38–40.

189. Desurvire, E., et al., Study of Spectral Dependence of Gain Saturation and Effect of Inhomogeneous Broadening in Erbium-Doped Aluminosilicate Fiber Amplifiers, *IEEE Photon. Technol. Lett.,* vol. 2, no. 9, September 1990, pp. 653–655.

190. Edagawa, N., et al., 12 300 ps/nm, 2.4 Gb/s Nonregenerative Optical Fiber Transmission Experiment and Effect of Transmitter Phase Noise, *IEEE Photon. Technol. Lett.,* vol. 2, no. 4, April 1990, pp. 274–276.

191. Eisenstein, G., Semiconductor Optical Amplifiers, *IEEE Circuits and Devices Magazine,* July 1989, pp. 25–30.

192. Gabla, P. M., S. Gauchard, and I. Neubauer, 279 km, 591 Mb/s Direct Detection Transmission Experiment Using Four In-Line Semiconductor Optical Amplifiers, *IEEE Photon. Technol. Lett.,* vol. 2, no. 8, August 1990, pp. 594–596.

193. Giles, C. R., and D. DiGiovanni, Spectral Dependence of Gain and Noise in Erbium-Doped Fiber Amplifiers, *IEEE Photon. Technol. Lett.,* vol. 2, no. 11, November 1990, pp. 797–800.

194. Henmi, N., et al., Rayleigh Scattering Influence on Performance of 10 Gb/s Optical Receiver with Er-Doped Optical Fiber Preamplifier, *IEEE Photon. Technol. Lett.,* vol. 2, no. 4, April 1990, pp. 277–278.

195. Iqbal, M. Z., et al., An 11 Gbit/s, 151 km Transmission Experiment Employing a 1480 nm Pumped Erbium-Doped In-Line Fiber Amplifier, *IEEE Photon. Technol. Lett.,* vol. 1, no. 10, October 1989, pp. 334–336.

196. Kagi, N., A. Oyobe, and K. Nakamura, Efficient Optical Amplifier Using a Low-Concentration Erbium-Doped Fiber, *IEEE Photon. Technol. Lett.,* vol. 2, no. 8, August 1990, pp. 559–561.

197. Lassen, H. E., P. B. Hansen, and K. E. Stubkjaer, Crosstalk in 1.5-μm InGaAsP Optical Amplifiers, *IEEE Journal of Lightwave Technology,* vol. 6, no. 10, October 1988, pp. 1559–1565.

198. Nakagawa, K., and S. Shimada, Optical Amplifiers in Future Optical Communication Systems, *IEEE Lightwave Communication Systems,* November 1990, pp. 57–62.

199. Nakashima, T., et al., Theoretical Limit of Repeater Spacing in an Optical Transmission Line Utilizing Raman Amplification, *IEEE Journal of Lightwave Technology,* vol. LT-4, no. 8, August 1986, pp. 1267–1272.

200. Noe, R., et al., Optical Amplifier with 27 dB Dynamic Range in a Coherent Transmission System, *IEEE Photon. Technol. Lett.,* vol. 2, no. 2, February 1990, pp. 120–121.

201. Olsson, N. A., Semiconductor Optical Amp Characteristics, *Communications International,* September 1988, pp. 81–88.

202. O'Mahony, M. J., Semiconductor Laser Optical Amplifiers for Use in Future Fiber Systems, *IEEE Photon. Technol. Lett.,* vol. 6, no. 4, April 1988, pp. 531–544.

203. Pedersen, B., et al., Detailed Theoretical and Experimental Investigation of High-Gain Erbium-Doped Fiber Amplifier, *IEEE Photon. Technol. Lett.,* vol. 2, no. 12, December 1990, pp. 863–865.

204. Ryu, S., et al., Over 1000 km FSK Heterodyne Transmission System Experiment Using Erbium-Doped Optical Fiber Amplifiers and Conventional Single-Mode Fibers, *IEEE Photon. Technol. Lett.,* vol. 2, no. 6, June 1990, pp. 428–430.

205. Sankawa, I., et al., An Optical Fiber Amplifier for Wide-Band Wavelength Range Around 1.65 μm, *IEEE Photon. Technol. Lett.,* vol. 2, no. 6, June 1990, pp. 422–424.

206. Schlager, J. B., P. D. Hale, and D. L. Franzen, Subpicosecond Pulse Compression and Raman Generation Using a Mode-Locked Erbium-Doped Fiber Laser-Amplifier, *IEEE Photon. Technol. Lett.,* vol. 2, no. 8, August 1990, pp. 562–564.

207. Shimizu, M., et al., Concentration Effect on Optical Amplification Characteristics of Er-Doped Silica Single-Mode Fibers, *IEEE Photon. Technol. Lett.,* vol. 2, no. 1, January 1990, pp. 43–45.

208. Sugawa, T., T. Komukai, and Y. Miyajima, Optical Amplification in Er^{3+}-Doped Single-Mode Fluoride Fiber, *IEEE Photon. Technol. Lett.,* vol. 2, no. 7, July 1990, pp. 475–476.

209. Taga, H., et al., Power Penalty Due to Optical Back Reflection in Semiconductor Optical Amplifier Repeater Systems, *IEEE Photon. Technol. Lett.*, vol. 2, no. 4, April 1990, pp. 279–281.

210. Taga, H., et al., 5 Gbit/s, 233 km Optical Fiber Transmission Experiment Employing Five Semiconductor Laser Amplifiers, *IEEE Photon. Technol. Lett.*, vol. 1, no. 10, October 1989, pp. 332–333.

211. Takada, A., K. Iwatsuki, and M. Saruwatari, Picosecond Laser Diode Pulse Amplification up to 12 W by Laser Diode Pumped Erbium-Doped Fiber, *IEEE Photon. Technol. Lett.*, vol. 2, no. 2, February 1990, pp. 122–124.

212. Tomofuji, H., et al., Cumulative Waveform Distortion in Cascaded Optical Amplifier Repeaters for Multigigabit IM/DD Systems, *IEEE Photon. Technol. Lett.*, vol. 2, no. 10, October 1990, pp. 756–758.

213. Whitley, T. J., et al., Laser Diode Pumped Er^{3+}-Doped Fiber Amplifier in a 565 Mbit/s DPSK Coherent Transmission Experiment, *IEEE Photon. Technol. Lett.*, vol. 1, no. 12, December 1989, pp. 425–427.

214. Willner, A. E., et al., Use of LD-Pumped Erbium-Doped Fiber Preamplifiers with Optimal Noise Filtering in a FDMA-FSK 1 Gb/s Star Network, *IEEE Photon. Technol. Lett.*, vol. 2, no. 9, September 1990, pp. 669–672.

215. Yamada, M., et al., Er^{3+}-Doped Fiber Amplifier Pumped by 0.98 μm Laser Diodes, *IEEE Photon. Technol. Lett.*, vol. 1, no. 12, December 1989, pp. 422–424.

Solitons

216. Bell, T. E., Light That Acts Like Natural Bits, *IEEE Spectrum*, August 1990, pp. 56–57.

217. Iwatsuki, K., et al., 5 Gb/s Optical Soliton Transmission Experiment Using Raman Amplification for Fiber-Loss Compensation, *IEEE Photon. Technol. Lett.*, vol. 2, no. 7, July 1990, pp. 507–509.

218. Iwatsuki, K., et al., 20 Gb/s Optical Soliton Data Transmission Over 70 km Using Distributed Fiber Amplifiers, *IEEE Photon. Technol. Lett.*, vol. 2, no. 12, December 1990, pp. 905–907.

219. Iwatsuki, K., S. Nishi, and K. Nakagawa, 3.6 Gb/s All Laser-Diode Optical Soliton Transmission, *IEEE Photon. Technol. Lett.*, vol. 2, no. 5, May 1990, pp. 355–357.

220. Nakazawa, M., K. Suzuki, and Y. Kimura, 3.2-5 Gb/s, 100 km Error-Free Soliton Transmissions with Erbium Amplifiers and Repeaters, *IEEE Photon. Technol. Lett.*, vol. 2, no. 3, March 1990, pp. 216–218.

221. Olsson, N. A., et al., 4 Gb/s Soliton Data Transmission Over 136 km Using Erbium Doped Fiber Amplifiers, *IEEE Photon. Technol. Lett.*, vol. 2, no. 5, May 1990, pp. 358–359.

Filters

222. Habbab, I. M. I., S. L. Woodward, and L. J. Cimini, Jr., DBR-Based Tunable Optical Filter, *IEEE Photon. Technol. Lett.*, vol. 2, no. 5, May 1990, pp. 337–339.

223. Kobrinski, H., and K.-W. Cheung, Wavelength-Tunable Optical Filters: Applications and Technologies, *IEEE Communications Magazine*, October 1989, pp. 53–63.

224. Numai, T., 1.5 μm Optical Filter Using a Two-Section Fabry-Perot Laser Diode with Wide Tuning Range and High Constant Gain, *IEEE Photon. Technol. Lett.*, vol. 2, no. 6, June 1990, pp. 401–403.

225. Salehi, J. A., R. C. Menendez, and C. A. Brackett, A Low-Pass Digital Optical Filter for Optical Fiber Communications, *IEEE Journal of Lightwave Technology*, vol. 6, no. 12, December 1988, pp. 1841–1847.

Couplers

226. Abebe, M., C. A. Villarruel, and W. K. Burns, Reproducible Fabrication Method for Polarization Preserving Single-Mode Fiber Couplers, *IEEE Journal of Lightwave Technology*, vol. 6, no. 7, July 1988, pp. 1191–1198.

227. Berenbrock, G. A., and B. Schlemmer, Active Controlled Fiber Optical 90° Hybrid for Coherent Communications, *IEEE Photon. Technol. Lett.,* vol. 1, no. 4, April 1989, pp. 86–87.
228. Bricheno, T., Fused Taper Couplers Make the Most of Singlemode, *Communications International,* April 1985, p. 72.
229. Cheng, H. C., and R. V. Ramaswamy, Determination of the Coupling Length in Directional Couplers from Spectral Response, *IEEE Photon. Technol. Lett.,* vol. 2, no. 11, November 1990, pp. 823–825.
230. Gillham, F., and H. A. Roberts, Fiber Optic Couplers for Multiplexing, *Photonics Spectra,* April 1984, pp. 45–52.
231. Gillham, F., and H. A. Roberts, Fiber Optic Couplers for Multiplexers: Part II, *Photonics Spectra,* May 1984, pp. 45–50.
232. Hawkins, R. T. II, and J. H. Goll, Method for Calculating Coupling Length of Ti: $LiNbO_3$ Waveguide Directional Couplers, *IEEE Journal of Lightwave Technology,* vol. 6, no. 6, June 1988, pp. 887–891.
233. Murphy, K. W., Building Blocks Offer Coupler Flexibility, *Photonics Spectra,* August 1990, pp. 133–138.
234. Saleh, A. A. M., and H. Kogelnik, Reflective Single-Mode Fiber-Optic Passive Star Couplers, *IEEE Journal of Lightwave Technology,* vol. 6, no. 3, March 1988, pp. 392–398.
235. Severin, P. J., A. P. Severijns, and C. H. van Bommel, Passive Components for Multimode Fiber-Optic Networks, *IEEE Journal of Lightwave Technology,* vol. LT-4, no. 5, May 1986, pp. 490–495.
236. Toussaint, H.-N., and G. Winzer, Technology and Applications of Couplers for Optical Communications, *Siemens Telcom Report 6,* 1983, Special Issue "Optical Communications," pp. 102–107.

Switches and switching

237. Erman, M., et al., Mach-Zehnder Modulators and Optical Switches on III-V Semiconductors, *IEEE Journal of Lightwave Technology,* vol. 6, no. 6, June 1988, pp. 837–846.
238. Hinton, H. S., Photonic Time-Division Switching Systems, *IEEE Circuits and Devices Magazine,* July 1989, pp. 39–43.
239. Johnson, M., Fiber Optic Switching: Sensor- and Test-Network Applications, *Photonics Spectra,* May 1990, pp. 119–124.
240. Midwinter, J. E., Photonic Switching Technology: Component Characteristics versus Network Requirements, *IEEE Journal of Lightwave Technology,* vol. 6, no. 10, October 1988, pp. 1512–1519.
241. Murdocca, M., V. Gupta, and M. Majidi, Logic and Interconnects in Optical Computers, *Photonics Spectra,* December 1990, pp. 129–134.
242. Personick, S. D., Photonic Switching: Technology and Applications, *IEEE Communications Magazine,* vol. 25, no. 5, May 1987, pp. 5–8.
243. Shimosaka, N., et al., Photonic Wavelength-Division and Time-Division Hybrid Switching System Utilizing Coherent Optical Detection, *IEEE Photon. Technol. Lett.,* vol. 2, no. 4, April 1990, pp. 301–303.
244. Silvernail, L. P., Optical Computing: Does Its Promise Justify the Present Hype?, *Photonics Spectra,* September 1990, pp. 127–129.
245. Smith, P. W., On the Role of Photonic Switching in Future Communications Systems, *IEEE Circuits and Devices Magazine,* May 1987, pp. 9–14.
246. Zucker, J. E., et al., InGaAs-InAlAs Quantum Well Intersecting Waveguide Switch Operating at 1.55 μm, *IEEE Photon. Technol. Lett.,* vol. 2, no. 11, November 1990, pp. 804–806.

Multiplexing

247. Carter, A. C., Wavelength Multiplexing for Enhanced Fibre-Optic Performance, *Telecommunications,* October 1986, pp. 30–36.

248. Elrefaie, A., M. W. Maeda, and R. Guru, Impact of Laser Linewidth on Optical Channel Spacing Requirements for Multichannel FSK and ASK Systems, *IEEE Photon. Technol. Lett.*, vol. 1, no. 4, April 1989, pp. 88–90.

249. Hillerich, B., et al., Wavelength Division Multiplexing in Fibre-Optical Systems, *Telecommunications*, July 1985, pp. 73–78.

250. Lin, Y.-K. M., D. R. Spears, and M. Yin, Fiber-Based Local Access Network Architectures, *IEEE Communications Magazine*, October 1989, pp. 64–73.

251. Maeda, M. W., et al., Electronically Tunable Liquid-Crystal-Etalon Filter for High-Density WDM Systems, *IEEE Photon. Technol. Lett.*, vol. 2, no. 11, November 1990, pp. 820–822.

252. Meriem, T. B., Wavelength Division Multiplexing Applied to Local and Trunk Optical Fibre Networks, *Telecommunication Journal*, vol. 52-VII/1985, pp. 408–416.

253. McMahon, D. H., and W. A. Dyes, Self-Heterodyne Multiterminal System Concepts for Frequency Division Multiplexed Fiber-Optic Communication, *IEEE Journal of Lightwave Technology*, vol. 6, no. 7, July 1988, pp. 1162–1170.

254. Nosu, K., Advanced Coherent Lightwave Technologies, *IEEE Communications Magazine*, vol. 26, no. 2, February 1988, pp. 15–30.

255. Nosu, K., and K. Iwashita, A Consideration of Factors Affecting Future Coherent Lightwave Communication Systems, *IEEE Journal of Lightwave Technology*, vol. 6, no. 5, May 1988, pp. 686–694.

256. Oguchi, K., Y. Hakamada, and J. Minowa, Optical Design and Performance of Wavelength-Division-Multiplexed Optical Repeater for Fiber-Optic Passive Star Networks Connection, *IEEE Journal of Lightwave Technology*, vol. LT-4, no. 6, June 1986, pp. 665–670.

257. Parry, D., and S. Hughes, Multiplexing Fibre Optics, *Communications International*, January 1986, pp. 40–43.

258. Rottmann, F., et al., Integrated-Optic Wavelength Multiplexers on Lithium Niobate Based on Two-Mode Interference, *IEEE Journal of Lightwave Technology*, vol. 6, no. 6, June 1988, pp. 946–952.

259. Trischitta, P. R., and D. T. S. Chen, Repeaterless Undersea Lightwave Systems, *IEEE Communications Magazine*, March 1989, pp. 16–21.

260. Wagner, S. S., and H. Kobrinski, WDM Applications in Broadband Telecommunication Networks, *IEEE Communications Magazine*, March 1989, pp. 22–30.

261. Way, W. I., et al., Simultaneous Distribution of Multichannel Analog and Digital Video Channels to Multiple Terminals Using High-Density WDM and a Broad-Band In-Line Erbium-Doped Fiber Amplifier, *IEEE Photon. Technol. Lett.*, vol. 2, no. 9, September 1990, pp. 665–668.

LANs

262. Edwards, T., Fibre Optics for the Local Loop, *Communications International*, January 1991, pp. 36–40.

263. Finnie, G., Lighting Up the Local Loop, *Telecommunications*, January 1989, pp. 31–40.

264. Foschini, G. J., and G. Vannucci, Using Spread-Spectrum in a High-Capacity Fiber-Optic Local Network, *IEEE Journal of Lightwave Technology*, vol. 6, no. 3, March 1988, pp. 370–378.

265. Henry, P. S., High-Capacity Lightwave Local Area Networks, *IEEE Communications Magazine*, October 1989, pp. 20–26.

266. Liew, S. C., and K. W. Lu, New Architectures for Diversity in Fiber Loop Networks, *IEEE Communications Magazine*, December 1989, pp. 31–37.

267. Moustakas, S., The Standardization of IEEE 802.3 Compatible Fiber Optic CSMA/CD Local Area Networks: Physical Topologies, *IEEE Communications Magazine*, vol. 25, no. 2, February 1987, pp. 22–29.

268. Prisco, J. J., and R. J. Hoss, Fiber Optic Regional Area Networks, *IEEE Communications Magazine*, vol. 23, no. 11, November 1985, pp. 26–39.

269. Rawson, E. G., The Fibernet II Ethernet-Compatible Fiber-Optic LAN, *IEEE Journal of Lightwave Technology*, vol. LT-3, no. 3, June 1985, pp. 496–501.

270. Shibagaki, T., H. Ibe, and T. Ozeki, Video Transmission Characteristics in WDM Star Networks, *IEEE Journal of Lightwave Technology*, vol. LT-3, no. 3, June 1985, pp. 490–495.
271. Shimada, S., K. Hashimoto, and K. Okada, Fiber-Optic Subscriber Loop Systems for Integrated Services—The Strategy for Introducing Fibers into the Subscriber Network, *IEEE Journal of Lightwave Technology*, vol. LT-5, no. 12, December 1987, pp. 1667–1675.
272. Suh, S. Y., S. W. Granlund, and S. S. Hegde, Fiber-Optic Local Area Network Topology, *IEEE Communications Magazine*, vol. 24, no. 8, August 1986, pp. 26–32.
273. Yamashita, K., et al., 100-Mbit/s LED-p-i-n Transmitter and Receiver Modules for High-Speed Local Area Networks, *IEEE Journal of Lightwave Technology*, vol. LT-3, no. 3, June 1985, pp. 560–564.

Receivers (heterodyne/homodyne)

274. Atlas, D. A., and L. G. Kazovsky, An Optical PSK Homodyne Transmission Experiment Using 1320 nm Diode-Pumped Nd: YAG Lasers, *IEEE Photon. Technol. Lett.*, vol. 2, no. 5, May 1990, pp. 367–370.
275. Glance, B., Performance of Homodyne Detection of Binary PSK Optical Signals, *IEEE Journal of Lightwave Technology*, vol. LT-4, no. 2, February 1986, pp. 228–234.
276. Jacobsen, G., J.-X. Kan, and I. Garrett, Tuned Front-End Design for Heterodyne Optical Receivers, *IEEE Journal of Lightwave Technology*, vol. 7, no. 1, January 1989, pp. 105–114.
277. Kawanishi, S., A. Takada, and M. Saruwatari, Wide-Band Frequency-Response Measurement of Optical Receivers Using Optical Heterodyne Detection, *IEEE Journal of Lightwave Technology*, vol. 7, no. 1, January 1989, pp. 92–98.
278. Kazovsky, L. G., et al., Wide-Linewidth Phase Diversity Homodyne Receivers, *IEEE Journal of Lightwave Technology*, vol. 6, no. 10, October 1988, pp. 1527–1536.
279. Kazovsky, L. G., P. Meissner, and E. Patzak, ASK Multiport Optical Homodyne Receivers, *IEEE Journal of Lightwave Technology*, vol. LT-5, no. 6, June 1987, pp. 770–790.
280. Kazovsky, L. G., and D. A. Atlas, 560 Mb/s Optical PSK Synchronous Heterodyne Experiment, *IEEE Photon. Technol. Lett.*, vol. 2, no. 6, June 1990, pp. 431–434.
281. Norimatsu, S., K. Iwashita, and K. Sato, PSK Optical Homodyne Detection Using External Cavity Laser Diodes in Costas Loop, *IEEE Photon. Technol. Lett.*, vol. 2, no. 5, May 1990, pp. 374–376.
282. Salz, J., Modulation and Detection for Coherent Lightwave Communications, *IEEE Communications Magazine*, vol. 24, no. 6, June 1986, pp. 38–49.
283. Vodhanel, R. S., B. Enning, and A. F. Elrefaie, Bipolar Optical FSK Transmission Experiments at 150 Mbit/s and 1 Gbit/s, *IEEE Journal of Lightwave Technology*, vol. 6, no. 10, October 1988, pp. 1549–1552.
284. Yamamoto, S., et al., 1.55-μm Fiber-Optic Transmission Experiments for Long-Span Submarine Cable System Design, *IEEE Journal of Lightwave Technology*, vol. 6, no. 3, March 1988, pp. 380–391.

Coherent optical communications

285. Basch, E. E., and T. G. Brown, Introduction to Coherent Optical Fiber Transmission, *IEEE Communications Magazine*, vol. 23, no. 5, May 1985, pp. 23–30.
286. Chandrasekhar, S., et al., An InP/InGaAs p-i-n/HBT Monolithic Transimpedance Photoreceiver, *IEEE Photon. Technol. Lett.*, vol. 2, no. 7, July 1990, pp. 505–506.
287. Cline, T. W., et al., A Field Demonstration of 1.7 Gb/s Coherent Lightwave Regenerators, *IEEE Photon. Technol. Lett.*, vol. 2, no. 6, June 1990, pp. 425–427.
288. Day, T., A. D. Farinas, and R. L. Byer, Demonstration of a Low Bandwidth 1.06 μm Optical Phase-Locked Loop for Coherent Homodyne Communication, *IEEE Photon. Technol. Lett.*, vol. 2, no. 4, April 1990, pp. 294–296.

289. Deri, R. J., et al., Low-Loss Monolithic Integration of Balanced Twin-Photodetectors with a 3 dB Waveguide Coupler for Coherent Lightwave Receivers, *IEEE Photon. Technol. Lett.*, vol. 2, no. 8, August 1990, pp. 581–584.

290. Garrett, I., and G. Jacobsen, Theoretical Analysis of Heterodyne Optical Receivers for Transmission Systems Using (Semiconductor) Lasers with Nonnegligible Linewidth, *IEEE Journal of Lightwave Technology*, vol. LT-4, no. 3, March 1986, pp. 323–334.

291. Glance, B., Polarization Independent Coherent Optical Receiver, *IEEE Journal of Lightwave Technology*, vol. LT-5, no. 2, February 1987, pp. 274–276.

292. Gnauck, A. H., et al., 4-Gb/s Heterodyne Transmission Experiments Using ASK, FSK, and DPSK Modulation, *IEEE Photon. Technol. Lett.*, vol. 2, no. 12, December 1990, pp. 908–910.

293. Green, P. E., and R. Ramaswami, Direct Detection Lightwave Systems: Why Pay More?, *IEEE Lightwave Communications Systems*, November 1990, pp. 36–49.

294. Gross, R., et al., Coherent Transmission of 60 FM-SCM Video Channels, *IEEE Photon. Technol. Lett.*, vol. 2, no. 4, April 1990, pp. 288–290.

295. Hou, A. S., R. S. Tucker, and G. Eisenstein, Pulse Compression of an Actively Modelocked Diode Laser Using Linear Dispersion in Fiber, *IEEE Photon. Technol. Lett.*, vol. 2, no. 5, May 1990, pp. 322–324.

296. Kahn, J. M., BPSK Homodyne Detection Experiment Using Balanced Optical Phase-Locked Loop with Quantized Feedback, *IEEE Photon. Technol. Lett.*, vol. 2, no. 11, November 1990, pp. 840–843.

297. Kahn, J. M., et al., 4-Gb/s PSK Homodyne Transmission System Using Phase-Locked Semiconductor Lasers, *IEEE Photon. Technol. Lett.*, vol. 2, no. 4, April 1990, pp. 285–287.

298. Kahn, J. M., 1 Gbit/s PSK Homodyne Transmission System Using Phase-Locked Semiconductor Lasers, *IEEE Photon. Technol. Lett.*, vol. 1, no. 10, October 1989, pp. 340–342.

299. Kao, C. K., *Optical Fiber Systems: Technology, Design, and Applications*, McGraw-Hill, 1986.

300. Kavehrad, M., and B. S. Glance, Polarization-Insensitive Frequency Shift Keying Optical Heterodyne Receiver Using Discriminator Demodulation, *IEEE Journal of Lightwave Technology*, vol. 6, no. 9, September 1988, pp. 1386–1394.

301. Kazovsky, L. G., Multichannel Coherent Optical Communications Systems, *IEEE Journal of Lightwave Technology*, vol. LT-5, no. 8, August 1987, pp. 1095–1102.

302. Kazovsky, L. G., Balanced Phase-Locked Loops for Optical Homodyne Receivers: Performance Analysis, Design Considerations, and Laser Linewidth Requirements, *IEEE Journal of Lightwave Technology*, vol. LT-4, no. 2, February 1986, pp. 182–195.

303. Kazovsky, L. G., D. A. Atlas, and R. W. Smith, Optical Phase-Locked PSK Heterodyne Experiment at 4 Gb/s, *IEEE Photon. Technol. Lett.*, vol. 2, no. 8, August 1990, pp. 588–590.

304. Kazovsky, L. G., Performance Analysis and Laser Linewidth Requirements for Optical PSK Heterodyne Communications Systems, *IEEE Journal of Lightwave Technology*, vol. LT-4, no. 4, April 1986, pp. 415–425.

305. Kazovsky, L. G., and J. L. Gimlett, Sensitivity Penalty in Multichannel Coherent Optical Communications, *IEEE Journal of Lightwave Technology*, vol. 6, no. 9, September 1988, pp. 1353–1365.

306. Linke, R. A., Optical Heterodyne Communications Systems, *IEEE Communications Magazine*, October 1989, pp. 36–41.

307. Linke, R. A., and P. S. Henry, Coherent Optical Detection: A Thousand Calls on One Circuit, *IEEE Spectrum*, February 1987, pp. 52–57.

308. Mears, C. L., and T. E. Batchman, An Evaluation of the Quantum Efficiency of Optical Heterodyne Detectors, *IEEE Journal of Lightwave Technology*, vol. LT-5, no. 6, June 1987, pp. 827–836.

309. O'Byrne, V., A Method for Reducing the Channel Spacing in a Coherent Optical Heterodyne System, *IEEE Photonics Technol. Lett.*, vol. 2, no. 7, July 1990, pp. 513–514.

310. Sandbank, C. P. (ed.), *Optical Fibre Communication Systems,* John Wiley & Sons, 1980.
311. Suematsu, Y., and K.-I. Iga, *Introduction to Optical Fiber Communications,* John Wiley & Sons, 1982.
312. Sun, L., and P. Ye, Optical Homodyne Receiver Based on an Improved Balance Phase-Locked Loop with the Data-to-Phaselock Crosstalk Suppression, *IEEE Photon. Technol. Lett.,* vol. 2, no. 9, September 1990, pp. 678–679.
313. Tajima, K., Bandwidth of a Single-Mode Optical Fiber in PSK Coherent Optical Transmission Systems, *IEEE Journal of Lightwave Technology,* vol. 6, no. 2, February 1988, pp. 322–328.
314. Wagner, R. E., and R. A. Linke, Heterodyne Lightwave Systems: Moving Towards Commercial Use, *IEEE Lightwave Communications Systems,* November 1990, pp. 28–35.
315. Walker, G. R., R. C. Steele, and N. G. Walker, Performance Limitation of Amplified Coherent Optical Transmission Systems Due to Interfering Echoes, *IEEE Photon. Technol. Lett.,* vol. 2, no. 5, May 1990, pp. 377–379.
316. Way, W. I., Fiber-Optic Transmissions of Microwave 8-Phase-PSK and 16-ary Quadrature-Amplitude-Modulated Signals at the 1.3-μm Wavelength Region, *IEEE Journal of Lightwave Technology,* vol. 6, no. 2, February 1988, pp. 273–280.
317. Yamazaki, S., et al., 2.5 Gb/s CPFSK Coherent Multichannel Transmission Experiment for High Capacity Trunk Line System, *IEEE Photon. Technol. Lett.,* vol. 2, no. 12, December 1990, pp. 914–916.

Mobile Radio (Chap. 8)

318. Agnew, C. E., Efficient Spectrum Allocation for Personal Communications Services, *IEEE Communications Magazine,* vol. 29, no. 2, February 1991, pp. 52–54.
319. Ahola, K., and H. Makila, Introducing Hand Portables onto Cellular Networks, *Communications International,* June 1986, pp. 67–71.
320. Beddoes, E., and M. Pinches, Cellular Radio Telephony—the Racal-VODAFONE Network in Great Britain, *Ericsson Review,* no. 3, 1987, pp. 18–28.
321. Bosworth, R., The Mobile Potential, *Communications International,* November 1990, pp. 37–42.
322. Bultitude, R. J. C., Measurement, Characterization and Modeling of Indoor 800/900 MHz Radio Channels for Digital Communications, *IEEE Communications Magazine,* vol. 25, no. 6, June 1987, pp. 5–12.
323. Burke, T., In-Building Cellular Systems, *Telecommunications,* September 1990, pp. 73–77.
324. Chien, E. S. K., D. J. Goodman, and J. E. Russell, Sr., Cellular Access Digital Network (CADN): Wireless Access to Networks of the Future, *IEEE Communications Magazine,* vol. 25, no. 6, June 1987, pp. 22–31.
325. Clarke, G., Mobile: The Shape of the Future?, *Communications International,* November 1990, pp. 46–54.
326. Cox, D. C., Portable Digital Radio Communications—An Approach to Tetherless Access, *IEEE Communications Magazine,* July 1989, pp. 30–40.
327. Delisle, G. Y., et al., Propagation Loss Prediction: A Comparative Study with Application to the Mobile Radio Channel, *IEEE Trans. Veh. Technology,* vol. VT-34, no. 2, May 1985, pp. 86–96.
328. Devasirvathem, D. M. J., Multipath Time Delay Spread in the Digital Portable Radio Environment, *IEEE Communications Magazine,* vol. 25, no. 6, June 1987, pp. 13–21.
329. Ehrlich, N., Advanced Mobile Phone Service Using Cellular Technology, *Microwave Journal,* August 1983, pp. 119–126.
330. Evans, R., Transmission Trunked Mobile Radio, *Communications International,* May 1986, pp. 75–77.
331. Freeburg, T. A., Enabling Technologies for Wireless In-Building Network Communications—Four Technical Challenges, Four Solutions, *IEEE Communications Magazine,* April 1991, pp. 58–64.
332. Ginn, S., Personal Communication Services: Expanding the Freedom to Communicate, *IEEE Communications Magazine,* vol. 29, no. 2, February 1991, pp. 30–33.

333. Goodman, D. J., Trends in Cellular and Cordless Communications, *IEEE Communications Magazine*, June 1991, pp. 31–40.
334. Graham-Rack, N., A New Generation for Trunked PMR, *Telecommunications*, March 1991, pp. 68–70.
335. Jarvis, R. A., Data on Cellular, *Communications International*, August 1985, pp. 33–36.
336. Jismalm, G., and N. Rydbeck, Ericsson Telephones for Cellular Systems, *Ericsson Review*, no. 3, 1987, pp. 29–38.
337. Kavehrad, M., and P. J. McLane, Spread Spectrum for Indoor Digital Radio, *IEEE Communications Magazine*, vol. 25, no. 6, June 1987, pp. 32–40.
338. Kawasaki, R., Cordless Telephone Using Multi-Channel Access Technique, *JTR*, April 1984, pp. 101–105.
339. Kuramoto, M., and M. Shinji, Second Generation Mobile Radio Telephone System in Japan, *IEEE Communications Magazine*, vol. 24, no. 2, February 1986, pp. 16–20.
340. Lecours, M., et al., Statistical Modeling of a Mobile Radio Channel, *IEEE Veh. Technol. Conf.*, Dallas, May 1986.
341. Lee, W. C. Y., In Cellular Telephone, Complexity Works, *IEEE Circuits and Devices*, January 1991, pp. 26–32.
342. Lee, W. C. Y., *Mobile Communications Engineering*, McGraw-Hill, 1982.
343. Lejdal, J.-O., and H. Lindqvist, Cellular Network Planning Is Maximizing System Economy, *Ericsson Review*, no. 3, 1987, pp. 10–17.
344. Lindell, F., J. Swerup, and J. Uddenfeldt, Digital Cellular Radio for the Future, *Ericsson Review*, no. 3, 1987, pp. 48–56.
345. Lynch, R. J., PCN: Son of Cellular? The Challenges of Providing PCN Service, *IEEE Communications Magazine*, vol. 29, no. 2, February 1991, pp. 56–57.
346. Macario, R. C. V. (ed.), *Personal and Mobile Radio Systems*, *IEE Telecommunications* Series 25, Peter Peregrinus, 1991.
347. Mathews, P., Integrating Frequency Synthesis for Mobile Radio, *Communications International*, May 1985, p. 66.
348. Mikulski, J. J., DynaT*A*C Cellular Portable Radiotelephone System Experience in the U.S. and the U.K., *IEEE Communications Magazine*, vol. 24, no. 2, February 1986, pp. 40–46.
349. Niyonizeye, G., M. Lecours, and T. H. Huynh, Mutual Interferences in a Binary FH-FSK Spread Spectrum for Mobile Radio, *IEEE Trans. Veh. Technology*, vol. VT-34, no. 1, February 1985, pp. 28–34.
350. Obuchowski, J., Wireless Communications and Spectrum Conservation: Sending a Signal to Conserve, *IEEE Communications Magazine*, vol. 29, no. 2, February 1991, pp. 26–29.
351. Potter, R., Personal Communications for the Mass Market, *Telecommunications*, September 1990, pp. 79–80.
352. Preller, H. G., and W. Koch, MATS-E, an Advanced 900 MHz Cellular Radio Telephone System: Description, Performance, Evaluation, and Field Measurements, *IEEE Communications Magazine*, vol. 24, no. 2, February 1986, pp. 28–39.
353. Reljonen, P., GSM Base Station Development, *Telecommunications*, September 1990, pp. 85–92.
354. Ross, I. M., Wireless Network Directions, *IEEE Communications Magazine*, vol. 29, no. 2, February 1991, pp. 40–42.
355. Schilling, D. L., et al., Spread Spectrum for Commercial Communications, *IEEE Communications Magazine*, April 1991, pp. 66–79.
356. Seki, S., N. Kanmuri, and A. Sasaki, Detachable Unit Service in 800 MHz-Band Cellular Radiotelephone System, *IEEE Communications Magazine*, vol. 24, no. 2, February 1986, pp. 47–52.
357. Singer, R. M., and D. A. Irwin, Personal Communications Services: The Next Technological Revolution, *IEEE Communications Magazine*, vol. 29, no. 2, February 1991, pp. 62–66.
358. Smith, D., Spread Spectrum for Wireless Phone Systems: The Subtle Interplay between Technology and Regulation, *IEEE Communications Magazine*, vol. 29, no. 2, February 1991, pp. 44–46.
359. Soderholm, G., J. Widmark, and E. Ornulf, Ericsson Cellular Mobile Telephone Systems, *Ericsson Review*, no. B, 1987, pp. 2–9.

360. Stewart, A., Wireless Telephone for New York, *Communications International,* November 1990, pp. 10–12.
361. Symington, I., A RACE for Freedom, *Communications International,* October 1990, pp. 49–51.
362. Taylor, J. T., and J. K. Omura, Spread Spectrum Technology: A Solution to the Personal Communications Services Frequency Allocation Dilemma, *IEEE Communications Magazine,* vol. 29, no. 2, February 1991, pp. 48–51.
363. Whitehead, J. F., Cellular System Design: An Emerging Engineering Discipline, *IEEE Communications Magazine,* vol. 24, no. 2, February 1986, pp. 8–14.
364. Van der Hoek, H., From Cordless PABX to PCN, *Telecommunications,* March 1991, pp. 49–52.
365. Vincent, G., The Cordless Office—and Beyond, *Telecommunications,* September 1990, pp. 49–52.

Data Communications and the Future Network (Chap. 9)

366. Abate, J. E., et al., AT&T's New Approach to the Synchronization of Telecommunication Networks, *IEEE Communications Magazine,* April 1989, pp. 35–45.
367. Anderton, C., Implementing 10BaseT, *Telecommunications,* March 1991, pp. 61–64.
368. Banks, S., Just What is FDDI?, *Communications MEA,* vol. 1, no. 4, pp. 14–16.
369. Barrett, M., The Challenges of Fiber in the Loop, *IEEE Lightwave Communication Systems Magazine,* August 1990, pp. 12–16.
370. Boyer, G. R., A Perspective on Fiber in the Loop Systems, *IEEE Lightwave Communication Systems Magazine,* August 1990, pp. 6–11.
371. Brewster, R. L. (ed.), *Data Communications and Networks 2,* IEE Telecommunications Series 22, Peter Peregrinus, 1989.
372. Byrne, W. R., et al., Evolution of Metropolitan Area Networks to Broadband ISDN, *IEEE Communications Magazine,* January 1991, pp. 69–82.
373. Chalmers, J., User-Friendly, Affordable, and Available..., *CEI,* June 1990, pp. 12–20.
374. Colombo, M., and P. Ferrari, Multifunction Packet Terminal for Voice and Data Integrated PABXs, *Contributions,* no. 1, 1990, pp. 9–14.
375. DeWitt, R. G., ISDN Symposia: A Historical Overview, *IEEE Communications Magazine,* April 1990, pp. 10–11.
376. Dicenet, G., *Design and Prospects for the ISDN,* Artech House, 1987.
377. Dougall, C. J., Broadband Network Evolution in Telecom Australia, *IEEE Communications Magazine,* April 1990, pp. 52–54.
378. Eigen, D. J., Narrowband and Broadband ISDN CPE Directions, *IEEE Communications Magazine,* April 1990, pp. 39–46.
379. Frame, M., Broadband Service Needs, *IEEE Communications Magazine,* April 1990, pp. 59–62.
380. Friend, G. E., et al., *Understanding Data Communications, The Texas Instruments Learning Center,* Howard W. Sams, 1984.
381. Galbraith, M., Facsimile Broadcasts for Japan, *Communications International,* November 1990, p. 23.
382. Gasbarrone, G., Future Developments in Packet-Switching Services, *Telecommunications,* October 1990, pp. 51–56.
383. Goodman, M. S., Multiwavelength Networks and New Approaches to Packet Switching, *IEEE Communications Magazine,* October 1989, pp. 27–35.
384. Hills, T., Broadband Strategy, *Communications International,* October 1990, pp. 25–32.
385. Kahl, P., ISDN Implementation Strategy of the Deutsche Bundespost Telekom, *IEEE Communications Magazine,* April 1990, pp. 47–51.
386. Karppala, A., Broadband ISDN: Future Challenges and Benefits, *Discovery,* vol. 19, February 1990, pp. 30–32.

387. Large, D., Tapped Fiber vs. Fiber-Reinforced Coaxial CATV Systems, *IEEE Lightwave Communication Systems Magazine,* February 1990, pp. 12–18.
388. Lazzari, A., Packet Switching in an ISDN: Implementation Considerations, *Contributions,* no. 1, 1990, pp. 2–7.
389. Liou, M. L., Visual Telephony as an ISDN Application, *IEEE Communications Magazine,* February 1990, pp. 30–38.
390. Marcoux, B., Designing International Networks with Data Compression Multiplexers, *Telecommunications,* June 1990, pp. 63–68.
391. McDonald, J., ISDN Symposium—Future Directions, *IEEE Communications Magazine,* April 1990, pp. 13–14.
392. Morgan, D., M. Lach, and R. Bushnell, ISDN as an Enabler for Enterprise Integration, *IEEE Communications Magazine,* April 1990, pp. 23–27.
393. Morreale, P. A., and G. M. Campbell, Metropolitan-Area Networks, *IEEE Spectrum,* May 1990, pp. 40–42.
394. Murano, K., et al., Technologies Towards Broadband ISDN, *IEEE Communications Magazine,* April 1990, pp. 66–70.
395. Olshansky, R., A Migration Path to BISDN, *IEEE Lightwave Communication Systems Magazine,* August 1990, pp. 30–33.
396. Pahlavan, K. and J. L. Holsinger, Voice-Band Data Communications Modems—A Historical Review: 1919–1988, *IEEE Communications Magazine,* vol. 26, no. 1, January 1988, pp. 16–24.
397. Radley, P., Optical Technology: The Key to Advanced Services, *Telecommunications,* June 1990, pp. 33–36.
398. Rando, R., and R. C. Purkey, Power System Design for Fiber to the Home Systems, *IEEE Lightwave Communication Systems Magazine,* August 1990, pp. 17–19.
399. Raychaudhuri, D., Data Communications, *IEEE Spectrum,* January 1991, pp. 48–49.
400. Roy, D., Frame Relay Technology: Complement or Substitute?, *Telecommunications,* October 1990, pp. 39–48.
401. Roy, R., ISDN Applications at Tenneco Gas, *IEEE Communications Magazine,* April 1990, pp. 28–30.
402. Rudvin, S., ISDN Video Phone Now a Reality, *Discovery,* vol. 19, February 1990, pp. 6–8.
403. Rzeszewski, T., A Two-Layer Fiber Network for Broadband Integrated Services, *IEEE Lightwave Communication Systems Magazine,* February 1990, pp. 77–80.
404. Semilof, M., More Work to Do on Token-Ring, *Communications Week International,* March 18, 1991, pp. 23–24.
405. Sexton, M. J., and F. P. Kelly, Protected Networks, *Communications International,* November 1990, pp. 57–60.
406. Sharma, P., The Application of ONP to ISDN, *Telecommunications,* March 1991, pp. 45–46.
407. Shumate, P. W., Jr., Cost Projections for Fiber in the Loop, *IEEE Lightwave Communication Systems Magazine,* February 1990, pp. 73–76.
408. Smouts, M., and G. Adams, Options for Packet Switching over ISDN, *Telecommunications,* October 1990, pp. 61–70.
409. Snelling, R. K., J. Chernak, and K. W. Kaplan, Future Fiber Access Needs and Systems, *IEEE Communications Magazine,* April 1990, pp. 63–65.
410. Spencer, J. L., Reliability Challenges for Deployment of Fiber in the Loop, *IEEE Lightwave Communication Systems Magazine,* August 1990, pp. 20–24.
411. Snelling, R. K., Bringing Fiber to the Home, *IEEE Circuits and Devices,* January 1991, pp. 22–25.
412. Stallings, W., CCITT Standards Foreshadow Broadband ISDN, *Telecommunications,* May 1990, pp. 89–96.
413. Tat, T. N., D. Fisher, and A. Berglund, An Evolution Path to ATM, *Telecommunications,* July 1990, pp. 41–46.
414. Thomas, M. W., ISDN: Some Current Standards Difficulties, *Telecommunications,* March 1991, pp. 33–43.
415. Toda, I., Migration to Broadband ISDN, *IEEE Communications Magazine,* April 1990, pp. 55–58.

416. Turton, P., The Economics of ISDN in the UK, *Telecommunications,* March 1991, pp. 27–30.
417. Wagner, S. S., and R. C. Menendez, Evolutionary Architectures and Techniques for Video Distribution on Fiber, *IEEE Communications Magazine,* December 1989, pp. 17–24.
418. White, M., Networking Europe, *Communications International,* January 1991, pp. 28–33.
419. Yokoi, T., and K. Kodaira, Grade of Service in the ISDN Era, *IEEE Communications Magazine,* April 1989, pp. 46–50.
420. Zitsen, W., Metropolitan Area Networks: Taking LANs into the Public Network, *Telecommunications,* June 1990, pp. 53–60.

Index

Absolute delay equalization, 266
Absorption, 151
Acousto-optic filter, 345–346
Adaptive differential PCM (ADPCM), 28, 133, 429, 441
Adaptive equalizer, 204
Additive white Gaussian noise (AWGN), 110
Administrative unit (AU), 77–78
Advanced mobile phone service (AMPS), 424–426, 433
Aliasing distortion, 20
Alternate digit inversion (ADI), 43
Alternate mark inversion (AMI), 42–43
Amplifiers:
  doped fiber, 336, 380–381
  high electron mobility transfer (HEMT), 307, 333, 348
  semiconductor optical, 335
  traveling wave tube (TWT), 238, 288
Amplitude modulation (AM), 89
Amplitude shift keying (ASK), 358, 363
Analog baseband, composition of, 283
Analog cellular radio, 423–426
Analog color TV, 82–83
  chrominance, 82
  luminance, 82
Analog microwave radio (AMR), 10
  baseband:
    composition (illus.), 285
    group, 283
    mastergroup, 283
    supergroup, 283
    supermastergroup, 283
  comparison with DMR, 282–293
    single-sideband amplitude modulation (SSB-AM), 282
  discriminator, 288
  FM modulator, 287
    Carson's rule, 287

Analog microwave radio (AMR) (*Cont.*):
  noise:
    CCIR Recommendation 395-2, 289
    noise power ratio (NPR), 290
      measurement (illus.), 291
  output spectrum, 290
  performance comparison with DMR (illus.), 183
  preemphasis and deemphasis, 287
  transceiver (illus.), 286
  TWT amplifiers, 288
Antennas, 138–149, 240–241
  beamwidth, 141–143
  front-to-back ratio, 140
  gain, 138–141
  high performance, 146
  horn (illus.), 147
  isotropic (illus.), 139
  mobile radio base station, 417–419
  mobile unit, 419–420
  noise, 144–146
  parabolic (illus.), 138
  periscope, 190–191
  polarization, 143–144
  radiation pattern for (illus.), 141
  towers, 146–149
Asymptotic coding gain (ACG), 112–113
Asynchronous and synchronous interfaces, comparison of, 76–77
Asynchronous digital hierarchy (illus.), 47
Asynchronous higher order digital multiplexing, 47–57
  bit-by-bit, 48
  DS1-C, 58
  DS2, 60
  DS3, 60
  DS4, 63
  fifth order, 56
  fourth order, 55
  hierarchy of, 47
  second order, 48
  third order, 54
  word-by-word, 48

Asynchronous transfer mode (ATM),
    493–495
  cell relay, 497–500
Atmospheric effects, 151–157
  absorption, 151
  ducting, 157
  effective earth radius, 153
  k-factor, 153–157
  refraction, 151–157
Automatic gain control (AGC) circuit,
    110, 241, 248–241, 262, 283
Automatic protection switching (APS),
    69
Automatic request for repeat (ARQ), 110,
    120–121, 439
Availability, 169–178
Avalanche photodiodes (APD), 307,
    331–332

Bandwidth efficiency, 89–105
Baseband (BB), 6
Baseband adaptive transversal equaliz-
    ers, 211–220
Baseband composition, 17
Binary N zero substitution (BNZS), 45
Bit error ratio (BER) 9, 110–113,
    118–122, 124–125, 172–178,
    194–196, 266–272, 277–279
  eye pattern (illus.), 270
  improvement, 388–389
  residual, 175
Block codes, 111–113
Branching networks, 238
Broadband ISDN, 15, 87, 385, 487–500

Cable measurements (optical), 390–391
Carrier-to-noise ratio (C/N), 101,
    104–105, 201, 204–206, 253, 268, 277
  versus availability (illus.), 202
Carrier sense multiple access with
    collision detection (CSMA/CD),
    463–464
CCITT Recommendations for data
    communications, 502–504
Cell relay, 497–500
Cell sectorization, 405–406
Cellular mobile radio (illus.), 14
Cellular radio:
  analog, 423–426
  antennas, 416–420
    base station, 417–419
    mobile unit, 419–420
  data over, 439

Cellular radio (*Cont.*):
  digital, 426–436
    European GSM system, 426–433
      channel coding, 430
      frequency hopping, 432–433
      modulation and spectral usage,
        430
      power control and handoff,
        430–432
      speech coding, 429
    narrowband TDMA, 426
    North American IS-54 system,
      433–435
  field strength predictions, 410–413
  irregular terrain, effects of, 414–416
  propagation problems, 409–416
  spread spectrum techniques in,
    124–132
Cellular rural area networks, for
    developing countries, 440–441
Cellular structures 14, 400–407
Central office (CO), 3, 382–383, 385, 479,
    491
Chirping, 338–339, 358, 379
CMI to NRZ decoding, 226
Cochannel interference (illus.), 193
Code division multiple access (CDMA),
    133, 422
  cellular radio, for, 435–436
  direct sequence (DS CDMA), 435
  frequency hopping (FH CDMA),
    435–436
Code interleaving, 118–119
Coded mark inversion (CMI), 45, 226
Coding gain, 112
Coherent demodulation 11
Coherent detection, 307
Coherent optical transmission systems,
    359–368
  balanced PLL, 366–368
  heterodyne, 360
  homodyne, 360–361
  nonlinear PLL, 365–366
Combining techniques, 204–206
Companding, 25–27
  A-law (illus.), 26
  μ-law, 26–27
Concatenated codes, 119–120
Concatenation, 72–73
Cone of acceptance, 296–297
Conference of European Posts and
    Telecommunications (CEPT), 85
  frequency spectrum limits (illus.), 104

Constellation analysis, 274–282
Constellation diagram (illus.), 101
Container, 78
Convolution codes, 113–118
  Viterbi algorithm, 117–119
Cosine roll-off filter (illus.), 103
Costas loop, 106–109
  2-PSK analog (illus.), 107
  4-PSK digital demodulator (illus.), 108
Couplers, optical, 341–343
Cross-polarization discrimination (XPD),
  143
Cross-polarization interference canceler
  (XPIC), 255
Customer, 3
Customer premises (CP), 382
Cycloid curve, 287

Data circuit-terminating equipment
  (DCE), 454, 457
Data communications, 14
  CCITT Recommendations relating to,
    502–504
  ISO standards, 505
  testing, 500–502
    protocol analysis, 501
Data modems, 449–451
  standards, 502–505
Data switching exchange (DSE), 453
Data terminal equipment (DTE), 451,
  454–458
Data transmission:
  analog environment, in, 449–451
    bandwidth problems, 449
    modems, 449–451
  ISDN, 476–500
    broadband, 487–500
      distribution services, 489
      implementation, 489–495
      interactive services, 488
      CCITT functional architecture
        (illus.), 484
      CCITT protocol architecture (illus.),
        485
      circuit-switched call procedure
        (illus.), 478
      narrowband, 476–487
        services, 486–487
      primary rate access, 477
      protocol architectures (illus.), 499
      signaling, 479–483
        layer-3 call establishment (illus.),
          482

Data transmission, ISDN, signaling
    (Cont.):
      link access procedure (LAP), 479
      service access point identifier
        (SAPI), 481
      terminal adapter (TA), 474
      user network interface reference
        configuration (illus.), 478
  local area networks, 461–472
    10Base-T (Ethernet), 467
    carrier sense multiple access with
      collision detection (CSMA/CD),
      462–464
      slot time, 464
    Ethernet, 462–464
    fiber distributed data interface
      (FDDI), 467–473
      American National Standards
        Institute (ANSI), 468
      FDDI dual token ring structure
        (illus.), 468
      optical bypass for FDDI (illus.),
        469
      token and packet structures for
        FDDI (illus.), 470
      TRT, 471
      TTRT, 471
    LAN bridges (illus.), 463
    token rings, 464–467
      structure of token (illus.), 466
      target token rotation time (TTRT),
        466, 471
      token rotation time (TRT), 466, 471
  metropolitan area networks (MAN),
    473–476
    distributed queue dual bus (DQDB),
      473–476, (illus.) 474
    DQDB queuing principle (illus.), 475
    token slack time, 475
  multimedia telecommunications, 443
  packet formation, 456–459
    call request, 458
    DTE and DCE data packet format
      (illus.), 457
    format for call request/incoming call
      (illus.), 460
    frame format, 456–458
    X.25:
      summary, 458
      first three levels (illus.), 461
  packet switching, 451–461, (illus.)
    452
    network, 453–454, (illus.) 453

Data transmission, packet switching
(*Cont.*):
network interface protocols, 459,
(illus.) 462
optical, 460–451
packet assembly and disassembly
(PAD), 459
X.25 protocol, 454–459
permanent virtual circuit (PVC),
454, 456, 458
virtual call (VC), 454, 456, 458
protocol, 444
standards, 444–449
CCITT Recommendations, 447–448
G series: digital networks
transmission systems and
multiplexing, 448
I series: integrated services digital
network (ISDN), 448
V series: data communication over
the telephone network, 447
X series: data communications
network interface, 447
International Standards Organization (ISO), 444–447
OSI 7-layer protocol model (illus.),
445
wide area networks (WAN), 473–476
Decision value, 21
Decoding, differential, 244
Demodulation, 105–110
Costas loop, 106–109
2-PSK analog (illus.), 107
4-PSK digital demodulator (illus.),
108
decision-directed method, 109–110
phase lock loop (PLL) (illus.), 106
subbaseband (SBB) signal, 108
Descrambling, 245–246
Differential decoding, 244
Differential encoding, 12, 231–236
Differential pulse code modulation
(DPCM), 83
Diffraction, 163–166
Digital cellular radio, 426–436
European GSM system, 426–433
narrowband TDMA, 426
Digital crossconnect (DCC), 66
Digital microwave radio (DMR), 10,
220–261
12 + 1 frequency diversity system
(illus.), 224
16-QAM (illus.), 221

Digital microwave radio (DMR) (*Cont.*):
64- and 256-QAM, 253–256
FEC in, 253
RF power amplifier, 237–238
140-Mb/s transmitter (illus.), 225
baseband adaptive transversal
equalizers, 211–220, (illus.) 219
postcursor and precursor equalization (illus.), 218
combining techniques, 204–206
adaptive equalizer, 204
baseband combiner (illus.), 205
diversity protection switching, 201–202
availability versus C/N (illus.), 202
error analysis, 266–271
hot-standby protection, 202–204
frequency diversity, 202
IF adaptive equalizers, 206–211
cycloid curve, 207
multipath propagation (illus.), 207
performance (illus.), 212
link analysis, 182–186
hop calculations in, 182–186
fade margin, 184
link availability, 184–185
noise, 191–195
calculations, 194–195
interference, 191–194
passive repeaters, 186–191
far or near field, 186–191
passive reflectors (illus.), 187
low-capacity, 256–259
rainfall, 256
attenuation, 257
CCIR Report 563 on, 256
noise calculations, 194–195
performance and measurements,
261–282
absolute delay equalization, 266
BER, 266–267
constellation analysis, 274–282
amplitude imbalance, 279
phase-lock error, 278
plot for 16-QAM (illus.), 277
sinusoidal interfering tone, 278
error analysis, 266–271
CCITT Recommendation G.821,
267
errored seconds, 267
eye pattern (illus.), 270
IF-IF group delay, 265–266
microwave link analyzer (MLA),
265–266

Digital microwave radio (DMR), performance and measurements (*Cont.*):
  IF output spectrum, 265
  jitter analysis, 271–274, (illus.) 271
    carrier phase (illus.), 276
    input port and wander tolerance (illus.), 274–275
    maximum output, 272, (illus.) 273
    maximum tolerable input, 272
    timing phase (illus.), 276
    transfer function, 272
    256-QAM system, 272
  pressurization, 264–265
  pulse position noise (illus.), 271
  transmitter distortion level, 263–264
  waveguide return loss, 264–265
performance comparison with AMR (illus.), 183
performance objectives, 171–178
  BER, 172–177
    residual, 175
  CCIR Report 1052-1 (1990) on error performance (illus.), 174
  errored seconds (ES), 173, 177–178
    severely (SES), 177
  high-grade circuits, 173–175
  hypothetical reference digital path (HRDP), 171, 173
  hypothetical reference digital section (HRD), 175–176
  hypothetical reference connection (HRX), 171–178
  local-grade circuits, 176–177
  medium-grade circuits, 175–176
regenerative repeater (illus.), 223
RF channel arrangement:
  18-GHz band (illus.), 258
  CCIR Recommendation 595, 257
service channel methods, 251–253
surface acoustic wave (SAW) filter, 92
system configuration, 10
transmitter/receiver, 18-GHz band transceiver (illus.), 260
transmitter components:
  antennas, 240–241
    140-Mb/s receiver (illus.), 243
    parabolic-reflector (illus.), 241
    shell (Gregorian) type (illus.), 242
  branching networks, 238
    channel branching filters (illus.), 239
  CMI to NRZ decoding, 226

Digital microwave radio (DMR), transmitter components (*Cont.*):
  descrambling, 245–246
    circuit (illus.), 247
  differential decoding, 244
  differential encoding, 231–236
    16-QAM modulator (illus.), 233
  IF in-phase combiner, 248–251, (illus.) 250
  NRZ to CMI encoding, 246–248
    CMI encoder circuit (illus.), 249
    dc wander, 246, (illus.) 248
  parallel-to-serial conversion, 244–245
  RF amplifier predistorter, 236
    amplifier, 238
    pre-distorters, 238
    traveling wave tube (TWT)
  scrambling, 226–231
    circuit (illus.), 230
  serial-to-parallel conversion, 231
  waveguide run, 240
Digital multiplex hierarchy, 47
Digital multiplexing, 17
  hierarchy, 58
Digital speech interpolation (DSI), 422
Digital TV transmission system, 84–86
Digitization, 8–10
  TV signals, 81–82
Direct RF modulation, 11
Direct sequence spread spectrum (DSSS), 126–127, 129–132
  pseudorandom noise generation, 127
Discrete cosine transform (DCT), 84
Dispersion, optical fiber, 300–303
  chromatic, 301
  modal, 301–302
  polarization, 388
  waveguide, 303
Distortion, aliasing, 20
Distributed feedback (DFB) laser
  structures, 346, 358
  amplifier, 346
  diode (LD), 328, 330, 363
  semiconductor filters, 346
Distributed queue dual bus (DQDB), 473–476
Diversity, 178–182
  angle, 181
  frequency, 181–182
    hitless switching, 182
    improvement factor, 182
  multipath fading, 178

Diversity (*Cont.*):
    space, 179–181
        improvement factor, 180
        Vigants equation, 180–181
Diversity protection switching,
        201–206
    combining techniques, 204–206
    hot standby, 202–204
        frequency diversity, 202
        space diversity, 201
DMR noise calculations, 194–195
Double sideband suppressed carrier
        (DSB-SC), 100, 231
Ducting, 157

E_b/N_o, 122, 195
Effective earth radius, 153
Elastic memory, 53
Electro-optic filter, 346
Encoding/decoding, 27–28
    differential, 12, 231–236
    iterative encoder, 27–28
Error analysis, 266–271
Error control, 110–124
    additive white Gaussian noise
        (AWGN), 110
    automatic request for repeat (ARQ),
        110, 120–121
    bit error rate (BER), 110–113,
        118–122, 124–125
    block codes, 111–113
        asymptotic coding gain (ACG),
            112–113, 122
        block encoder (illus.), 111
        coding gain, 112, 122
        Reed-Solomon (RS), 112–113,
            118–120, 439
    code interleaving, 118–119
    concatenated codes, 119–120
    convolution codes, 113–118
        very large-scale integration (VLSI)
            circuits, 117
        Viterbi algorithm, 117–119
    forward correction, 111–120, 253
    Hamming distance, 113, 117
    trellis-coded modulation (TCM),
        110–111, 114–117, 121–124
        free distance, 123
        set partitioning, 122
        16-QAM, of (illus.), 123
Errored seconds, 267
Ethernet:
    10Base-T, 467
    CSMA/CD, 463–464

Eye pattern, 269–271

Fade margin, 184–185
Fading, 166–169
    flat, 166–167
    frequency selective, 167–169
Fast frequency shift keyed (FFSK)
        modulation, 94
Federal Communications Commission
        (FCC), 407
    FCC frequency spectrum limits (illus.),
        103
Fiber, optical:
    attenuation, 297–300
    bending, 304–306
    cable, 7, 310–323
    characteristics of, 295–306
    distributed data interface (FDDI), 386,
        467–473
    home, to the (FTTH), 15, 370, 382,
        490–492
Field, near/far, 186–191
Field strength predictions, 410–413
Filters:
    characteristics (illus.), 91
    cosine roll-off (illus.), 103
    optical, 344–347
        acousto-optic, 345–346
        electro-optic, 346
    pulse transmission through, 90–92
    surface acoustic wave (SAW), 92
    tunable, 346–347
        free spectral range (FSR),
            346
Fixed location mapping, 70
FM modulator, 287
Forward error correction (FEC), 86,
        111–120, 253, 388–389
Frame:
    relay, 496–497
    structure:
        3.153-Mb/s (illus.), 59
        6.132-Mb/s (illus.), 61
        8-Mb/s (illus.), 52
        34-Mb/s (illus.), 55
        44.736-Mb/s (illus.), 62
        274.176-Mb/s (illus.), 64
        140-Mb/s (illus.), 56
        565-Mb/s (illus.), 57
Free space loss, 149–150
Free space propagation, 149–150
Frequency diversity, 202, 224
Frequency division multiplexing (FDM),
        17, 356, 370–372, 450

Frequency division multiple access
    (FDMA), 132, 421
  single channel per carrier (SCPC), 132
Frequency hopping, 432–433
Frequency-hopping spread spectrum
    (FHSS), 126–127, 129–130,
  pseudorandom noise generation, 127
Frequency modulation (FM), 89
Frequency reuse, 14
Frequency shift keyed (FSK) modulation,
    90, 93–94
Fresnel zones, 159–163
Front-to-back ratio, 140

Gallium-arsenide field-effect transistor
    (GaAs FET), 132, 238
Group delay, 265–266

Hamming distance, 113, 117
Handoff, 402, 430–432
High-density bipolar 3 (HDB3), 30,
    43–45
Heterodyne, optical, 360
High-definition television (HDTV), 81,
    83, 86–87
High electron mobility transfer (HEMT)
    amplifier, 307, 333, 348
High performance antennas, 146
Hitless switching, 182
Home location register (HLR), 427
Homodyne, optical, 360–361
Hop calculations, 182–186
Hot-standby protection, 202
Hypothetical reference connection (HRX),
    171–178
Hypothetical reference digital path
    (HRDP), 171, 173–176

IF adaptive equalizers, 206–-211
IF-IF group delay, 265–266
IF in-phase combiner, 248–251
IF output spectrum, 265
Integrated digital network (IDN), 476
Integrated services digital network
    (ISDN) 9, 476–500
  asynchronous transfer mode (ATM),
    493–495
  broadband (BISDN), 15, 87, 487–500
  cell relay, 497–500
  frame relay, 496–497
  narrowband, 476–487
Interference, 191–194
  extra-system, 194
  interchannel, 192

Interference (Cont.):
  interhop, 192
  intersymbol (ISI), 90, 300, 330
  intra-system, 191
International Radio Consultative
    Committee (CCIR), 16
International standards, 16
International Standards Organization
    (ISO), 444–447, 468, 477
International Telecommunication Union
    (ITU), 16, 400
International Telegraph and Telephone
    Consultative Committee (CCITT), 16
Intersymbol interference (ISI) 90, 300,
    330
Iridium, 5
Isolators, optical, 343–344

Jitter, 9
  analysis of, 271–274, (illus.) 271
Jointing, optical fiber cables, 311–321
Junction (interexchange) optical routes,
    374–378
Justification, 50

k-factor, 153–157

Laser diode (LD), 306, 323–330, 335–336,
    376–379
  distributed feedback (DFB LD), 328,
    330, 363
Light detectors, 330–333
Light emitting diode (LED), 306,
    323–330, 355, 469
Line codes, 39–46
  alternate digit inversion (ADI), 43
  alternate mark inversion (AMI), 42–43
  binary N zero substitution (BNZS),
    45
  coded mark inversion (CMI), 45
  comparison of (illus.), 46
  HDB3, 43–44
  non–return-to-zero (NRZ), 25
  return-to-zero (RZ), 25
Line terminal equipment (LTE),
    374–375, 378, 391–397
Line terminal measurements (optical),
    391–397
Line termination (LT), 66
Link analysis, microwave, 182–186
Link design, optical, 306–310
Link security, 7
Lithium niobate, 334, 339–340, 348–349
  integrated circuits, 349–350

Local area network (LAN), 3, 381–387,
    461–472
  10Base-T, 467
  carrier sense multiple access with
      collision detection (CSMA/CD),
      463–464
  fiber distributed data interface (FDDI),
      467–473
  medial access control (MAC), 468,
      470–472
  token rings, 464–467
Local loop 3, 382–385
Local oscillator (LO), 97, 255, 262
Long-haul optical links, 378–380

Mach-Zender interferometer, 340
Measurements:
  DMR, 261–282
  optical fiber cable, 390–391
  optical line terminal, 391–397
Metropolitan area networks (MAN), 385,
    473–476
  distributed queue dual bus (DQDB),
      473–476
Microwave frequency bands, 137
Microwave integrated circuit (MIC),
    237
Microwave link:
  analyzer (MLA), 265
  antennas, 138–149
    beamwidth, 141–143
    boresight, 140, 159
    front-to-back ratio, 140
    gain, 138–141
    horn (illus.), 147
    isotropic (illus.), 139
    noise, 144–146
      radome, 146
      temperature (illus.), 144
    parabolic (illus.), 138, 145
    polarization, 143
      cross-polarization discrimination
          (XPD), 143
      waveguide (illus.), 143
    radiation pattern (illus.), 141
    towers, 146–149
  atmospheric effects, 151–157
    absorption, 151
    ducting, 157
    rain attenuation (illus.), 152
    refraction, 151–157
    k-factor, 153–157
  backbone, 135

Microwave link (Cont.):
  fading, 166–169
    flat, 166–167
      standard atmosphere, 166
      subrefractive atmosphere, 167
      substandard atmosphere, 166
    frequency selective, 167–169
      multipath (illus.), 168
  free space:
    loss, 149, (illus.) 150
    propagation through, 149–150
  hop, (illus.) 10, 135, (illus.) 136
  terrain effects on, 157–166
    diffraction, 163–166
      grazing incidence, 164
      knife-edge, 164
      path clearance (illus.), 164
    duct (illus.), 159
    earth profile template (illus.), 158
    Fresnel zones, 159–163, (illus.) 160
      radius, 161, (illus.) 162
    reflections, 157–159
      ground (illus.), 159
Microwave system (analog/digital)
    comparison (illus.), 11–12
Minimum shift keying (MSK) (illus.), 94
Mobile application part (MAP), 472
Mobile radio antennas, 416–420
  base station, 417–419
  unit, 419–420
Mobile radio communications:
  analog cellular, 423–426
    advanced mobile phone service
        (AMPS), 424–426
    mobile telephone switching office
        (MTSO), 424
  analog FM, 89
  antennas, 416–420
    base station, 417–419
      bent or folded dipole, 417
      corner reflector, 419
      dipole radiation pattern (illus.),
          417
      ground plane, 417–419
      stacked, 419
    mobile unit, 419–420
      whip, 420
  cellular radio, data over, 439
    automatic request for repeat (illus.),
        440
  cellular rural area networks for
      developing countries, 440–441
    digital radio multiple access, 440

Mobile radio communications (*Cont.*):
  cellular structures, 400–407, (illus.) 401
    frequency reuse (illus.), 404
    handoff, 402
    handover, 402
    honeycomb, 400
    microcell, 405
    mobile location, 407
    real (illus.), 402
    sectorization, 405–406, (illus.) 405
  digital cellular, 426–436
    European GSM system, 426–433
      cellular radio network structure
        (illus.), 427
      channel coding, 430
      frame and multiframe structure
        (illus.), 428
      frequency hopping, 432–433
      home location register (HLR), 427
      mobile application part (MAP),
        427
      modulation and spectral usage,
        430
        direct digital interpolation, 430
      power control and handoff,
        430–432
      public switched telephone network
        (PSTN), 427
      speech coding, 429
      visitor location register (VLR), 427
    future design possibilities for,
      CDMA, 435–436
    narrowband TDMA, 426
    North American IS-54 system,
      433–435
      Hamming code, 434
      vector sum excited linear predic-
        tion (VSELP) coding, 434–435
  frequency allocations, 407–409
    personal communications network
      (PCN), 409
  personal communications network,
    436–439
  portable radio telephones, 436–439
  propagation problems, 409–416
    effects of irregular terrain, 414–416
      correction factor, 414–416
    field strength predictions, 410–413
      attenuation against frequency in
        urban areas (illus.), 411
      attenuation relative to free space,
        410
      correction factor, 412, 413

Mobile radio communications, propaga-
    tion problems, field strength
    predictions (*Cont.*):
      field strength of moving mobile
        station (illus.), 410
  system types (illus.), 421
    CDMA, 422–423
      digital speech interpolation (DSI),
        423
    FDMA and TDMA, 421
Mobile radio systems 13, 420–423
Mobile telephone switching office
    (MTSO), 424–425
Modulation:
  amplitude (AM), 89
  amplitude shift keyed (ASK), 358, 363
  bandwidth efficiency, 89–105
  constellation diagram (illus.), 101
  differential pulse code (DPCM), 83
  differential quaternary PSK (DQPSK
    or 4-PSK), 434
  double sideband suppressed carrier
    (DSB-SC), 100, 231
  fast FSK (FFSK), 94
  frequency (FM), 89, 338–339
  frequency shift keyed (FSK), 90, 93–94
  Gaussian minimum shift keyed
    (GMSK), 94–95, 430
  minimum shift keyed (MSK), 94
    2-phase (illus.), 95
    4-phase (illus.), 96
    8-phase (illus.), 96
  Nyquist filter, 92–93
  phase shift keyed (PSK), 90, 95–101,
    (illus.) 95, 338–339
  pulse amplitude (PAM), 21, 90, 92
  pulse code (PCM), 17–39
  quadrature amplitude (QAM), 90,
    97–105
    modem (illus.), 97
  single-sideband amplitude (SSB-AM),
    282–283
  trellis-coded, 110–111, 114–117, 121–124
Modulators, external optical, 339–341
Monolithic microwave integrated circuit
    (MMIC), 437–438
Monolithic optical integrated circuits,
    multiple-quantum-well (MQW), 349
Multiframe alignment, 37–38
Multimode fiber, 299
Multipath fading, 167–169, 206–211
Multiple quantum well (MQW), 349, 351,
    353

Multiplexers:
  asynchronous digital, 48
  synchronous digital, 48
Multiplexing:
  frequency division, 17
  time division, 9, 18

N x N star coupler, 343, 387, 461
Narrowband (ISDN), 476–487
Nesting, 74–76
  signals (illus.), 75
Network interface protocols, 459
Network node interface (NNI), 65–66,
    (illus.) 66
  universal (illus.), 76
Noise, 191–195
Noise power ratio (NPR), 290
Non–return to zero (NRZ), 39–41,
    338–339
Non–return-to-zero invert (NRZI),
    469–470
Nonuniform quantizing, 25–27
North American IS-54 system, 433–435
NRZ-to-CMI encoding, 246–248
  dc wander, 246
Numerical aperture, 296–298, 323
Nyquist:
  bandwidth, 90
  filter, 90, 92, 94
  Theorem, 90, 92

Open Systems Interconnection (OSI),
    446, 468, 477
Optical fibers:
  amplifiers, 335–338
    doped fiber, 336–338
      erbium (Er) (illus.), 337
    semiconductor, 335–336
      Fabry-Perot, 336
      traveling wave, 336
  cables, 310–323, (illus.) 312–313
    buffering (illus.), 310
    dispersion-shifted, 310
    home, to the (FTTH), 490–492
    ribbon-matrix, 311
    splicing, or jointing, 311–321
      cladding leakage light, 316
      fusion, 314–318
      loss due to lateral misalignment
        (illus.), 324
      loss due to numerical aperture
        differential (illus.), 323
      mechanical, 318, (illus.) 319–320

Optical fibers, cables, splicing, or jointing
    (Cont.):
      prefusion (illus.), 316
      preparation (illus.), 315
      splice organizer, 318
      submarine, 311, (illus.) 315
  characteristics of, 295–306
    attenuation, 297–300
    bending, 304–306
      loss (illus.), 305
      microbending (illus.), 304, 305
    dispersion, 300–303
      intersymbol interference, 300
      intramodal or chromatic, 301
      material, 301–302
      modal, 301–302
      waveguide, 303
    numerical aperture, 296
      cone of acceptance angle, 296
      light-acceptance cone (illus.), 297
      splice loss due to differential of
        (illus.), 323
      waveguide (illus.), 298
    polarization, 303–304
      modal birefringence, 304
  communication systems (illus.), 356
  configurations (illus.), 357
  couplers, 341–343
    N x N star, 343
    proximity (illus.), 341
    two-mode interference (TMI), 342
    Y-junction (illus.), 341
    zero-gap (illus.), 341
  filters, 344–347
    frequency tunable, 344–347, (illus.)
        345
    acousto-optic, 345–346
    electro-optic, 346
  isolators, 343–344
  light detectors, 330–333
    avalanche photodiode (APD), 331–332
    PIN diode, 332–333
    receiver sensitivity, 332
  light sources, 323–330
    laser diode (LD), 326–330
      distributed feedback, 326–330
      external cavity semiconductor
        (illus.), 329
      Fabry-Perot, 326–330
      linewidth, 328
    light-emitting diode (LED), 325–326
  line terminal equipment (LTE),
      374–375, 378

Optical fibers (*Cont.*):
 link design, 306–310
  coherent detection, 307, 359
  power budget, 308
 modulators, 338–341
  direct, 338–339
   frequency chirping in, 338
  external, 339–341
   frequency chirping in, 339
   lithium niobate phase (illus.),
    339
   Mach-Zender interferometer, 340
 multimode, 299
  graded index (illus.), 300
  step index (illus.), 300
 photonic switches, 347–348
  stored program controlled (SPC),
   347
 polarization controllers, 333–335
  lithium niobate, 334, (illus.) 334
 single-mode, 299, (illus.) 300
 solid-state circuit integration, 348–353
  lithium niobate, 348–350
  multiple-quantum-well (MQW)
   monolithic, 349
  photonic integrated circuits (PIC), 348
  semiconductor, 350–353
   PIC (illus.), 351
 transmission systems:
  BER, 388–389
   FEC, 388–389
   fiber chromatic dispersion, 388
   improvement of (illus.), 389
   polarization dispersion, 388
  coherent, 359–368, (illus.) 359
   heterodyne receiver, 360
   homodyne receiver, 360–361,
    (illus.) 366, (illus.) 367
   multiplexing bandwidth (illus.),
    362
   optical phase-locked loop (OPLL),
    363
   PLL, 365–368
  equipment measurements, 389–397
   cable, 390–391
   line terminal equipment (LTE),
    (illus.) 392, (illus.) 393
  intensity modulated, 358–359
   chirping, 358
   quantum limit, 358
  junction routes (interexchange
   traffic), 374–378
   5B6B code (illus.), 376

Optical fibers, transmission systems
 (*Cont.*):
  long-haul links, 378–380
   distance limitation due to
    dispersion (illus.), 379
   soliton formation (illus.), 381
   soliton transmission, 380–381
  local area networks and subscriber
   loops, 381–388
   subscriber networks (local access
    network), 382, 384, 385
  multiplexing, 368–372
   frequency division (FDM),
    370–372, (illus.) 371
   wave division (WDM), 368–370,
    (illus.) 369–370
  repeaters, 372–374
   optical, 373–374
   regenerative, 372–373
Optical fiber cables, 310–323
 bending, 304–306
 bending loss (illus.), 305
 microbending attenuation, 305
 splicing/jointing, 311–321
  fusion, 314–318
  mechanical, 318–321
Optical fiber characteristics,
  295–306
 attenuation, 297–300
 dispersion, 300–303
  chromatic, 301
  material, 301–302
  modal, 301–302
  waveguide, 303
 numerical aperture, 296–297
 polarization, 303–304
Optical fiber equipment measurements,
  389–397
 cable measurements, 390–391
 line terminal measurements,
  391–397
Optical fiber light detectors,
  330–333
 avalanche photodiode (APD),
  330–332
 PIN diodes, 332–333
Optical fiber light sources, 323–330
 laser diode, 326–330
  distributed feedback, 326–330
  Fabry-Perot, 326–330
 LED, 325–326
Optical fiber link (illus.), 13
 design, 306–310

Optical fiber link design, 306–310
 power budget, 308–310
Optical fiber modulators, 338–341
 direct, 338–339
 external, 339–341
Optical fiber system configuration 12
Optical filters, 344–347
 frequency tunable, 344–347
 mode coupling tunable, 345–346
 semiconductor laser structures, 346–347
Optical integrated circuits:
 lithium niobate, 349–350
 semiconductor, 350–353
Optical isolators, 343–344
Optical multiplexing, 368–372
 frequency division (FDM), 370–372
 wave division (WDM), 368–370
Optical packet switching, 460–461
Optical repeaters, 373–374
 regenerative, 372–373
Optical solid state circuit integration,
  348–353
 lithium niobate, 349–350
 semiconductor, 350–353
Optical systems design, 374–380
 BER improvement, 388–389
 fiber to the home, 382
 intensity-modulated, 358–359
 junction (interexchange) routes,
  374–378
 local area networks (LAN), 381–388
 long-haul links, 378–380
 soliton transmission, 380–381
 subscriber loops, 381–388
Optical switches, 347–348
Optical time domain reflectometer
 (OTDR), 390–391
Optical transmission systems:
 coherent, 359–368
  balanced PLL, 366–368
  nonlinear PLL, 365–366
Opto-electronic integrated circuit (OEIC),
  348
 semiconductor, 350
Overhead, 67–69

Packet networks, 453–454
Packet switching 16, 451–461
Parallel-to-serial conversion, 244–245
Passive repeaters, 186–191
Payload pointer, 70–72
Peak limiting, 21
Performance objectives, 171–178

Permanent virtual circuit (PVC) service,
  454, 456, 458
Personal communications network (PCN)
  126, 409, 436–439
 spread spectrum techniques in, 126
Phase lock loop (PLL), 54, 93, 106–107,
  109–110
 optical balanced, 366–368
 optical nonlinear, 365–366
Phase shift keyed (PSK) modulation, 90,
  94–104
 2-phase (illus.), 95
 4-phase (illus.), 96
 8-phase (illus.), 96
 double sideband suppressed carrier
  (DSB-SC), 100
 serial-to-parallel converter, 95
Photonic integrated circuit (PIC), 348,
  350–353
 chemical beam epitaxy (CBE), 348
 metal organic vapor phase epitaxy
  (MOVPE), 348, 351
Photonic switches, 347–348
PIN diodes, 248, 332–333
Pixels (PELS), 83
Plesiochronous frequency, 49
Polarization:
 controllers, 333–335
 fiber, 303–304
 microwaves, 143–144
Portable radiotelephones, 436–439
Positive pulse stuffing, 50–51
Postcursor equalization, 216–217
Power budget, optical fiber, 308–310
Precursor equalization, 217–220
Preemphasis/deemphasis, 287
Private branch exchange (PBX), 3
Propagation delay, satellite, 5
Propagation problems, 409–416
 field strength predictions, 410–413
 irregular terrain effects, 410–416
Protocol, 444, 454–459, 468
Protocol analysis, 500–501
Pseudorandom bit sequence (PRBS), 263,
  265
Pseudorandom noise generation,
  127–128
Public switched telephone network
 (PSTN), 427
Pulse amplitude modulation (PAM), 21, 92
Pulse code modulation (PCM), 17–39
 adaptive differential (ADPCM), 28,
  133, 429, 441

Pulse code modulation (PCM) (*Cont.*):
frames, 30–37
alignment of, 32–37
multiframes, 30–39
alignment of, 37–38
Pulse position noise (illus.), 271

Quadrature amplitude modulation
(QAM), 90, 97–105
4-QAM system, 97, 102
16-QAM system, 97–100, 102, 104,
107–108, 220–253
64- or 256-QAM system, 253
constellation diagram, 98
higher-level system (illus.), 99
modem (illus.), 97
signal state-space diagram, 98
trellis coding for, 122–124
Quantization, 20–27
noise, 22–25
nonuniform quantizing, 25–27

Radio frequency (RF), 9
amplifiers, 237–238
power, 255
predistorter, 236
Rain attenuation (illus.), 152
Rainfall, 256–257
Reed-Solomon (RS) codes, 112–113,
118–120, 439
Reflections, 157–159
Refraction, 151–157
Regenerative repeater 12, 372–373
Remote node (RN), 382
Residual BER, 175
Return to zero (RZ), 39–42

Sampling, 18–22
Shannon's sampling criterion, 18
Satellite communications, 4–5
very small aperture terminal (VSAT),
4–5
Satellite propagation delay, 5
Scrambling, 226–231
Semiconductor optical integrated circuits,
350–353
OEICs, 350
PICs, 350–353, (illus.) 351
Serial-to-parallel conversion, 231
Serial-to-parallel converter, 95
Service access point (SAP), 382–383
Service channel methods, 251–253
Single-mode fiber, 299

Single-sideband amplitude modulation
(SSB-AM), 282–283
Solid-state optical circuit integration,
348–353
Soliton transmission, 380–381
Space diversity, 179–181, 201
Splicing, fiber cables, 311–320
Spread spectrum techniques, 124–132,
(illus.) 127
direct sequence (DSSS), 126–127, 129–132
chip rate, 132
frequency-hopping (FHSS), 126–127,
129–130,
chip rate, 129
PCNs in, 126
pseudorandom noise generation,
127
pseudonoise generator (illus.), 127
signal (illus.), 128
Spur route 8
Standards, international, 16, 444–449,
502–505
Stored program controlled (SPC) switch,
347
Stuffing:
control, 51
message word (illus.), 54
Subbaseband (SBB) signal, 108
Subscriber (*see* Customer)
Subscriber networks, 382–385
Surface acoustic wave (SAW), 92
characteristics (illus.), 92
filter, DMRs in, 92
Switching exchange (*see* Central office)
Synchronous digital hierarchy (SDH),
69–70, 490, 492–493
administrative unit (AU), 77–78
concatenation, 72–73
container, 77
fixed location mapping, 70
multiplexing structure (illus.), 77
container, 78
payload pointer, 70–73
summary of, 77–80
tributary unit (TU), 78
tributary unit group (TUG), 78
virtual container (VC), 78
virtual tributary (VT), 78
Synchronous digital multiplexing, 48,
63–81
network node interface (NNI), 65
Synchronous multiplexer overhead
(illus.), 68

Synchronous optical network (SONET), 65, 492–493

Synchronous payload envelope (SPE), 67

Synchronous transport module (STM), 73–76
  frame structure (illus.), 74
  nesting, 74–76

Synchronous transport signal (STS), 66–73, (illus.) 67, (illus.) 69, (illus.) 71

Target token rotation time (TTRT), 466, 471

Telephone density, 1

Television, 80–87
  analog color, 82–83
  chrominance:
    processing, 85
    signal, 82
  community antenna TV (CATV), 84, 87
  digital:
    signals, multiplexing of, 80–87
    transmission system, 84–86
  error correction, 85
  high-definition (HDTV), 81, 83, 86–87
    sub-Nyquist encoding system (MUSE), 87
  luminance:
    processing, 85, (illus.) 86
    signal, 82
  signal digitization, 81–82
  video compression techniques in, 83–84

Terrain effects, 157–166
  diffraction, 163–166
  Fresnel zones, 159–163
  irregular, 410–416
  reflections, 157–159

Time division multiple access (TDMA), 132–133, 421, 426–435

Time division multiplexing (TDM), 9, 18

Token rings, 464–467

Token rotation time (TRT), 466, 471

Token slack time, 475

Transmission media, 3

Transmitter distortion level, 263–264

Transverse electric (TE), 303, 334, 345, 347

Transverse magnetic (TM), 304, 334, 345, 347

Trellis-coded modulation, 110–111, 114–117, 121–124

Tributary, 48–50

Tributary unit (TU), 78–80

Two-mode interference (TMI) coupler, 342

Vector sum excited linear prediction (VSELP) coding, 434–435

Very large-scale integration (VLSI) circuits, 117, 400, 424

Very small-aperture terminal (VSAT), 4–5

Video compression techniques, 83–84
  discrete cosine transform (DCT), 84
  pixels (PELS), 83

Virtual call (VC) service, 454, 456, 458

Virtual channel identifier (VCI), 494

Virtual container (VC), 78

Virtual path identifier (VPI), 494

Virtual tributary (VT), 78–80

Visitor location register (VLR), 472

Viterbi algorithm, 117–119

Voltage controllable oscillator (VCO), 93, 108, 110, 252

Voltage standing-wave ratio (VSWR), 146, 192, 240

Wave division multiplexing (WDM), 368–370

Waveguide run, 240

Wide area networks (WANs), 473–476

X.25 protocol, 454–459

ABOUT THE AUTHOR

Robert G. Winch is a telecommunications professional who has been designing and implementing transmission systems since 1975. He is a project coordinator/digital transmission expert with the International Telecommunication Union, a member organization within the United Nations. Previously, Dr. Winch worked as a transmission engineer for ITT Corporation and as an assistant professor/telecommunications consultant at the University of the Virgin Islands. He received his doctoral degree in microwave electronics from the University of Oxford, United Kingdom.